T0252330

A NETWORKING APPROACH
TO GRID COMPUTING

A NETWORKING APPROACH TO GRID COMPUTING

DANIEL MINOLI
Managing Director
Leading-Edge Networks Incorporated

WILEY-INTERSCIENCE

A JOHN WILEY & SONS, INC., PUBLICATION

For general information on our other products and services please contact our Customer Care Department within the U.S. at 877-762-2974, outside the U.S. at 317-572-3993 or fax 317-572-4002.

Wiley also publishes its books in a variety of electronic formats. Some content that appears in print, however, may not be available in electronic format.

Library of Congress Cataloging-in-Publication Data is available.

ISBN 978-0-471-68756-6

For Anna, Emma, Emile, Gabrielle, Gino, Angela, and Peter

Contents

About the Author

Daniel Minoli has many years of IT, telecom, and networking experience for end users and carriers, including work at AIG, ARPA think tanks, Bell Telephone Laboratories, ITT, Prudential Securities, Bell Communications Research (Bellcore/Telcordia), and AT&T (1975–2001). Recently, he also played a founding role in the launching of two networking companies through the high-tech incubator Leading Edge Networks Inc., which he ran in the early 2000s: Global Wireless Services, a provider of broadband, hotspot mobile Internet and hotspot VoIP (Vo Wi-Fi) services to high-end marinas; and, InfoPort Communications Group, an optical and Gigabit Ethernet metropolitan carrier supporting data center/SAN/channel extension and grid computing network access services (2001–2003). Mr. Minoli's grid computing work goes back to 1987.

An author of a number of textbooks on information technology, telecommunications, and data communications, he has also written columns for *ComputerWorld, NetworkWorld,* and *Network Computing* (1985–1995). He has taught at New York University, Rutgers University, Stevens Institute of Technology, Carnegie Mellon University, and Monmouth University (1984–2003). Also, he was a Technology Analyst At-Large, for Gartner/DataPro (1985–2001); based on extensive hands-on work at financial firms and carriers, he tracked technologies and wrote around fifty distinct CTO/CIO-level technical/architectural scans in the area of telephony and data systems, including topics on security, disaster recovery, IT outsourcing, network management, LANs, WANs (ATM and MPLS), wireless (LAN and public hotspot), VoIP, network design/economics, carrier networks (such as metro Ethernet and CWDM/DWDM), and e-commerce. Over the years, he has advised venture capitalists for investments of $150M in a dozen high-tech companies, and has acted as expert witness in a (won) $11B lawsuit regarding a wireless air-to-ground communication system.

■■■■■ Preface

In February 1974 this author, as a math major at the Polytechnic Institute of Brooklyn, (co)invented a now well-rooted but computationally complex concept of "hyperperfect numbers" and he used an early form of grid computing—also known as utility computing—to study this concept (see pages 83 and 86). His interest in grid computing that grew out of this 1970s work lasted throughout the late 1980s and into the early 2000s.

This is the first book that takes a comprehensive view of grid computing technology from a *networking* perspective. Grid computing seamlessly integrates resources and services across distributed, heterogeneous, dynamic "virtual organizations" that span disparate administrative entities within a single enterprise and/or external entities or service providers. The past decade has seen a significant level of government funding directed at grid-related projects at NASA, national laboratories, supercomputer centers, and academic institutions.

Up to now, grid computing has been largely of interest to researchers at mathematics and computer science departments, national laboratories, informatics institutes, and government-funded research laboratories, but it turns out that this technology can be of value to Fortune 500 Companies looking to reduce their run-the-engine costs. A fair number of such companies are already availing themselves of the clear financial benefits; others may soon follow.

Commoditization of any sort of resource works to the clear advantage of the user and it affords at-large macroeconomics benefits. In recent years, we have seen the aggressive commoditization of all sorts of consumer entertainment electronics, personal computers, and personal communication devices such as cellular telephones and Personal Digital Assistants. This has resulted in a precipitous decrease in prices (and costs) of these products.

During the past ten years or so, a similar commoditization has been experienced in computing hardware platforms that support information technology applications at businesses of all sizes. Some writers have encapsulated this rapidly occurring phenomenon with the phrase "IT doesn't matter." This price-disrupting predicament affords new opportunities for organizations, although it also has conspicuous process- and people-dislocating consequences.

The commoditization thrust of computing hardware has been captured under the auspices of grid computing, which is viewed by proponents as the "next big thing." Grid computing is perceived as having as much potential for changing the way companies do business as the Internet did. Grid computing can be considered as a network of computation. It supports the concept of "utility computing," with which users can get "on-demand" "machine cycles off a grid" without having to own the physical assets or infrastructure. It includes mechanisms and protocols for coordinated resource sharing and problem solving among pooled assets distributed across the globe, such as PCs, servers, mainframes, supercomputers, and data stores.

This author was advocating the concept of grid computing as far back as 1987. In Bellcore Special Report SR-NPL-000790 in a section called "Network for a Computing Utility" we stated in part:

> The proposed service provides the entire apparatus to make the concept of the Computing Utility possible. This includes as follows: (1) the physical network over which the information can travel, and the interface through which a guest PC/workstation can participate in the provision of machine cycles and through which the service requesters submit jobs; (2) a load sharing mechanisms to invoke the necessary servers to complete a job; (3) a reliable security mechanism; (4) an effective accounting mechanism to invoke the billing system; and, (5) a detailed directory of servers. . . . Security is one of the major issues for this service, particularly if the PC is not fully dedicated to this function, but also used for other local activities. Virus threats, infiltration and corruption of data, and other damage must be appropriately addressed and managed by the service; multi-task and robust operating systems are also needed for the servers to assist in this security process . . . protocols and standards will be needed to connect servers and users, as well as for accounting and billing. These protocols will have to be developed before the service can be established. . . .

Evolving grid computing standards can also be used to facilitate "open-source outsourcing services" as a way for corporations to implement portability ("open source") in their outsourced operations.

A lot of the requisite infrastructure and mechanisms have become available in recent years. This book explores practical advantages of grid computing and what is needed by an organization to migrate to this new computing paradigm, but it does so with a degree of emphasis on the communication apparatus. The book is intended for practitioners and decision makers in organizations who want to explore the overall opportunities afforded by this new technology. The text follows in the tradition of the author of exploring the practical utility of the technology, with a minimum of theoretical or research-level exposition. The book is principally targeted to the user community, specifically, information technology departments at Fortune 500 companies that want a systematized summary of the state of the business, without having to digest extensive files of research material.

Acknowledgments

The author would like to thank Dr. Robert B. Cohen and Dr. Edward Feser, for use of a high-value table (Chapter 2) and a high-value graph (Chapter 7).

The author would also like to thank Tim Wu and Andy Walden for material for Chapter 10 on VPNs.

CHAPTER 1

Introduction

1.1 WHAT IS GRID COMPUTING AND WHAT ARE THE KEY ISSUES?

Grid computing (or, more precisely a "grid computing system") is a virtualized distributed computing environment. Such an environment aims at enabling the dynamic "runtime" selection, sharing, and aggregation of (geographically) distributed autonomous resources based on the availability, capability, performance, and cost of these computing resources, and, simultaneously, also based on an organization's specific baseline and/or burst processing requirements. When people think of a grid, the idea of an interconnected system for the distribution of electricity, especially a network of high-tension cables and power stations, comes to mind. In the mid-1990s the grid metaphor was applied to computing, by extending and advancing the 1960s concept of "computer time sharing." The grid metaphor strongly illustrates the relation to, and the dependency on, a highly interconnected networking infrastructure.

This book is a survey of the grid computing field as it applies to corporate environments and it focuses on what are the potential advantages of the technology in these environments. The book is a synthesis of the state of the industry and takes a balanced view of the field, but at the same time serves as a familiarization vehicle for the interested information technology (IT) professional. It should be noted at the outset, however, that no claim is made herewith *that there is a single or unique solution to a given computing problem;* grid computing is *one* of a number of available solutions in support of optimized distributed computing. Corporate IT professionals, for whom this text is intended, will have to perform appropriate functional, economic, business-case, and strategic analyses to determine which computing approach ultimately is best for their respective organizations. Furthermore, it should be noted that grid computing is an evolving field and, so, *there is not* always one canonical, normative, universally accepted, or axiomatically derivable view of "everything grid related," and it follows that we occasionally present multiple views, multiple interpretations, or multiple perspectives on a topic, as they might be perceived by different stakeholders or communities of interest.

This introductory chapter begins the discussion by providing a sample of what a number of stakeholders define grid computing to be, along with an assay of the industry. The precise definition of what exactly this technology encompasses is still evolving and there is not a globally accepted normative definition that is perfectly

A Networking Approach to Grid Computing. By Daniel Minoli
ISBN 0-471-68756-1 © 2005 John Wiley & Sons, Inc.

1

nonoverlapping with other related technologies. Grid computing emphasizes (but does not mandate) geographically distributed, multiorganization, utility-based, outsourcer-provided, networking-reliant computing methods.

In its basic form, the concept of grid computing is straightforward: with grid computing an organization can transparently integrate, streamline, and share dispersed, heterogeneous pools of hosts, servers, storage systems, data, and networks into one synergistic system, in order to deliver agreed-upon service at specified levels of application efficiency and processing performance. Additionally, or, alternatively, with grid computing an organization can simply secure commoditized "machine cycles" or storage capacity from a remote provider, "on-demand," without having to own the "heavy iron" to do the "number crunching." Either way, to an end-user or application this arrangement (ensemble) looks like one large, cohesive, virtual, transparent computing system [91, 102]. A grid mechanism is an enabling technology for online collaboration and for discovery and access to distributed resources. A grid mechanism is basically a middleware; it is a distributed computing technology. Broadband networks play a fundamental enabling role in making grid computing possible and this is the motivation for looking at this technology from the perspective of communication.

According to IBM's definition [118, 128],

[A] grid is a collection of distributed computing resources available over a local or wide area network that appear to an end user or application as one large virtual computing system. The vision is to create virtual dynamic organizations through secure, coordinated resource-sharing among individuals, institutions, and resources. Grid computing is an approach to distributed computing that spans not only locations but also organizations, machine architectures, and software boundaries to provide unlimited power, collaboration, and information access to everyone connected to a grid. . . . [T]he Internet is about getting computers to talk together; grid computing is about getting computers to work together. Grid will help elevate the Internet to a true computing platform, combining the qualities of service of enterprise computing with the ability to share heterogeneous distributed resources—everything from applications, data, storage and servers.

Another definition, this one from The Globus Alliance (a research and development initiative focused on enabling the application of grid concepts to scientific and engineering computing), is as follows [129]:

The grid refers to an infrastructure that enables the integrated, collaborative use of high-end computers, networks, databases, and scientific instruments owned and managed by multiple organizations. Grid applications often involve large amounts of data and/or computing and often require secure resource sharing across organizational boundaries, and are thus not easily handled by today's Internet and Web infrastructures.

Yet another industry-formulated definition of grid computing is as follows [96, 103]:

A computational grid is a hardware and software infrastructure that provides depend-able, consistent, pervasive, and inexpensive access to high-end computational capabil-ities. A grid is concerned with coordinated resource sharing and problem solving in dynamic, multi-institutional virtual organizations. The key concept is the ability to ne-gotiate resource-sharing arrangements among a set of participating parties (providers and consumers) and then to use the resulting resource pool for some purpose. The sharing that we are concerned with is not primarily file exchange but rather direct ac-cess to computers, software, data, and other resources, as is required by a range of col-laborative problem-solving and resource-brokering strategies emerging in industry, science, and engineering. This sharing is, necessarily, highly controlled, with resource providers and consumers defining clearly and carefully just what is shared, who is al-lowed to share, and the conditions under which sharing occurs. A set of individuals and/or institutions defined by such sharing rules form what we call a *virtual organiza-tion* (VO).

Whereas the Internet is a network of communication, grid computing is seen as a network of computation: the field provides tools and protocols for resource sharing of a variety of IT resources. Grid computing approaches are based on coordinated resource sharing and problem solving in dynamic, multiinstitutional VOs. A (short) list of examples of possible VOs include: application service providers, storage ser-vice providers, machine-cycle providers, and members of industry-specific consor-tia. These examples, among others, represent an approach to computing and prob-lem solving based on collaboration in data-rich and computation-rich environments [3, 91]. The enabling factors in the creation of grid computing systems in recent years have been the proliferation of broadband (optical-based) communications, the Internet, and the World Wide Web infrastructure, along with the availability of low-cost, high-performance computers using standardized (open) operating systems [94, 96, 104]. The role of communications as a fundamental enabler will be emphasized throughout the chapters of this textbook.

Prior to the deployment of grid computing, a typical business application had a dedicated platform of servers and an anchored storage device assigned to each indi-vidual server. Applications developed for such platforms were not able to share re-sources, and, from an individual server's perspective, it was not possible, in gener-al, to predict, even statistically, what the processing load would be at different times. Consequently, each instance of an application needed to have its own excess capacity to handle peak usage loads. This predicament typically resulted in higher overall costs than would otherwise need to be the case [130]. To address these lacu-nae, grid computing aims at exploiting the opportunities afforded by the synergies, the economies of scale, and the load smoothing that result from the ability to share and aggregate distributed computational capabilities, and deliver these hardware-based capabilities as a transparent service to the end-user.[1] To reinforce the point, the term "synergistic" implies "working together so that the total effect is greater

---

[1]As implied in the opening paragraphs, a number of solutions in addition to grid computing (e.g., virtual-ization) can be employed to address this and other computational issues; grid computing is just one ap-proach.

than the sum of the individual constituent elements." From a service-provider perspective, grid computing is somewhat akin to an application service provider (ASP) environment, but with a much-higher level of performance and assurance [131]. Specialized ASPs, known as grid service providers (GSPs) are expected to emerge to provide grid-based services, including, possibly, "open-source outsourcing services."

Grid computing started out as the simultaneous application of the resources of many networked computers to a single (scientific) problem [96]. Grid computing has been characterized as the "massive integration of computer systems" [97]. Computational grids have been used for a number of years to solve large-scale problems in science and engineering. The noteworthy fact is that, at this juncture, the approach can already be applied to a mix of mainstream business problems. Specifically, grid computing is now beginning to make inroads into the commercial world, including financial services operations, making the leap forward from such scientific venues as research labs and academic settings [130].

The possibility exists, according to the industry, that with grid computing, companies can save as much as 30% of certain key line items of the operations budget (in an ideal situation), which is typically a large fraction of the total IT budget [98]. Companies spend, on the average, 6%[2] of their top line revenues yearly on IT services; for example, a $10 B/yr Fortune 500 company might spend $600 M/yr on IT. Industry vendors make the assertion that *cluster computing* (aggregating processors in parallel-based configurations), yield reductions in IT costs and costs of operations that are expected to reach 15% by 2005 and 30% by 2007–8 in most early-adopter sectors. Use of *enterprise grids* (middleware-based environments to harvest unused "machine cycles," thereby displacing otherwise-needed growth costs), is expected to result in a 15% savings in IT costs by the year 2007–8, growing to a 30% savings by 2010 to 2012 [111]. This *potential* savings is what this book is all about.

In 1994, this author published the book *Analyzing Outsourcing, Reengineering Information and Communication Systems* (McGraw-Hill), calling attention to the possibility that companies could save 15–20% or more of their IT costs by considering outsourcing, and the trends of the mid-2000s have, indeed, validated this (then) timely assertion [120]. At this juncture, we call early attention to the fact that the possibility exists for companies to save as much as 15–30% of certain key line items of the IT operations (run-the-engine) budget by using grid computing and/or related computing/storage virtualization technologies. In effect, grid computing, particularly the utility computing aspect, can be seen as another form of outsourcing. Perhaps utility computing will be the next phase of outsourcing and be a major trend in the 2010 decade. As we discuss later in the book, the evolving grid computing standards can be used by companies to deploy a next-generation kind of "open source outsourcing" that has the advantage of offering portability, enabling companies to easily move their business among a variety of pseudocommodity providers.

This book explores practical advantages of grid computing and what is needed by an organization to migrate to this new computing paradigm, if it so chooses. The

---

[2]The range is typically 2% to 10%; for example, see, among others [120].

book is intended for practitioners and decision makers in organizations (not necessarily software programmers) who want to explore the overall business opportunities afforded by this new technology. At the same time, the importance of the underlying networking mechanism is emphasized. For any kind of new technology, corporate and business decision makers typically seek answers to questions such as these (which are the theme of this book): (i) "What is this stuff?"; (ii) "How widespread is its present/potential penetration?"; (iii) "Is it ready for prime time?"; (iv) "Are there firm standards?"; (v) "Is it secure?"; (vi) "How do we bill it, since it's new?"; (vii) "Tell me how to deploy it (at a macro level)"; and (viii) "Give me a self-contained reference—make it understandable and as simple as possible." Table 1.1 summarizes these and other questions that decision makers, CIOs, CTOs, and planners may have about grid computing that are addressed in this book. Table 1.2 lists some of the concepts embodied in grid computing and other related technologies.

Grid computing is also known by a number of other names (although some of these terms have slightly different connotations), such as "grid" (the term "the grid" was coined in the mid-1990s to denote a proposed distributed computing infrastructure for advanced science and engineering [40]), "computational grid," "computing-on-demand," "on-demand computing," "just-in-time computing," "platform computing," "network computing," "computing utility" (the term used by this author in the late 1980s [6]), "utility computing," "cluster computing," and "high-performance distributed computing." With regard to nomenclature, in this text, besides the term Grid Computing, we will also interchangeably use the terms *grid* and *computational grid*. In this text, we use the term *grid technology* to describe the entire collection of grid computing elements, middleware, networks, and protocols.

To deploy a grid, a commercial organization needs to assign computing resources to the shared environment and deploy appropriate grid middleware on these resources, enabling them to play various roles that need to be supported in the grid that are covered in this book (scheduler, broker, etc.). Some (minor) application retuning and/or parallelization may, in some instances, be required; data accessibility will also have to be taken into consideration. A security framework will also be required. If the organization subscribes to the service-provider model, then grid deployment would mean establishing adequate access bandwidth to the provider, undertaking some possible application retuning, and establishing security policies (the assumption being that the provider will itself have a reliable security framework).

The concept of providing computing power as a utility-based function is generally attractive to end users requiring fast transactional processing and "scenario modeling" capabilities. The concept may also be attractive to IT planners looking to control costs and reduce data center complexity. The ability to have a cluster, an entire data center, or other resources spread across an area connected by the Internet (and/or, alternatively, connected by an intranet or extranet), operating as a single transparent virtualized system that can be managed as a service, rather than as individual constituent components, likely will, over time, increase business agility, reduce complexity, streamline management processes, and lower operational costs [130]. Grid technology allows organizations to utilize numerous computers to solve problems by sharing computing resources. The problems to be solved might involve data processing, network bandwidth, data storage, or a combination thereof.

Table 1.1   Issues of interest and questions that CIOs/CTOs/planners have about grid computing

What is grid computing and what are the key issues?
  Grid benefits and status of technology
  Motivations for considering computational grids
  Brief history of grid computing
  Is grid computing ready for prime time?
  Early suppliers and vendors
  Challenges
  Future directions

What are the components of grid computing systems/architectures?
  Portal/user interfaces
  User security
  Broker function
  Scheduler function
  Data management function
  Job management and resource management

Are there stable standards supporting grid computing?
  What is OGSA/OGSI?
  Implementations of OGSI
  OGSA services
  Virtual organization creation and management
  Service groups and discovery services
  Choreography, orchestration, and workflow
  Transactions
  Metering service
  Accounting service
  Billing and payment service
  Grid system deployment issues and approaches
  Generic implementations: Globus Toolkit

Security considerations—Can grid computing be trusted?

  What are the grid deployment/management issues?
    Challenges and approaches
    Availability of products by categories
    Business grid types
    Deploying a basic computing grid
    Deploying more complex computing grid
    Grid operation

  What are the economics of grid systems?
    The chargeable grid service
    The grid payment system

  How does one pull it all together? Communication and networking infrastructure
    Communication systems for local grids
    Communication systems for national grids
    Communication systems for global grids

Table 1.2 Definition of some key terms

Grid Computing

- (Virtualized) distributed computing environment that enables the dynamic "runtime" selection, sharing, and aggregation of (geographically) distributed autonomous (autonomic) resources based on the availability, capability, performance, and cost of these computing resources, and, simultaneously, also based on an organization's specific baseline and/or burst processing requirements.
- Enables organizations to transparently integrate, streamline, and share dispersed, heterogeneous pools of hosts, servers, storage systems, data, and networks into one synergistic system, in order to deliver agreed-upon service at specified levels of application efficiency and processing performance.
- An approach to distributed computing that spans multiple locations and/or multiple organizations, machine architectures, and software boundaries to provide power, collaboration, and information access.
- Infrastructure that enables the integrated, collaborative use of computers, supercomputers, networks, databases, and scientific instruments owned and managed by multiple organizations.
- A network of computation: namely, tools and protocols for coordinated resource sharing and problem solving among pooled assets. Allows coordinated resource sharing and problem solving in dynamic, multiinstitutional virtual organizations.
- Simultaneous application of the resources of many networked computers to a single problem. Concerned with coordinated resource sharing and problem solving in dynamic, multiinstitutional virtual organizations.
- Decentralized architecture for resource management, and a layered hierarchical architecture for implementation of various constituent services.
- Combines elements such as distributed computing, high-performance computing, and disposable computing, depending on the application.
- Local, metropolitan, regional, national, or international footprint. Systems may be in the same room, or may be distributed across the globe; they may be running on homogenous or heterogeneous hardware platforms; they may be running on similar or dissimilar operating systems; and they may owned by one or more organizations.
- Types. (i) *Computational grids:* machines with set-aside resources allocated to "number-crunch" data or provide coverage for other intensive workloads. (ii) *Scavenging grids:* commonly used to locate and exploit machine cycles on idle servers and desktop computers for use in resource-intensive tasks. (iii) *Data grids:* a unified interface for all data repositories in an organization, through which data can be queried, managed, and secured.
- Computational grids can be local enterprise grids (also called intragrids), and Internet-based grids (also called intergrids.) Enterprise grids are middleware-based environments used to harvest unused "machine cycles," thereby displacing otherwise needed growth costs.
- Other terms (some with slightly different connotations): "computational grid," "computing-on-demand," "on-demand computing," "just-in-time computing," "platform computing," "network computing," "computing utility," "utility computing," "cluster computing," and "high-performance distributed computing."

Virtualization

- An approach that allows several operating systems to run simultaneously on one (large) computer (e.g., IBM's z/VM operating system lets multiple instances of Linux coexist on the same mainframe computer). (*continued*)

Table 1.2 Continued

Virtualization (*cont.*)

- More generally, it is the practice of making resources from diverse devices accessible to a user as if they were a single, larger, homogenous resource that appears to be locally available.
- Dynamically shifting resources across platforms to match computing demands with available resources: the computing environment can become dynamic, enabling autonomic shifting of applications between servers to match demand.
- The abstraction of server, storage, and network resources in order to make them available dynamically for sharing, both internal to and external to an organization. In combination with other server, storage, and networking capabilities, virtualization offers customers the opportunity to build more efficient IT infrastructures. Virtualization is seen by some as a step on the road to utility computing.

Clusters

- Aggregating of processors in parallel-based configurations, typically in a local environment (within a data center); all nodes work cooperatively as a single unified resource.
- Resource allocation is performed by a centralized resource manager and scheduling system.
- Comprised of multiple interconnected independent nodes that cooperatively work together as a single unified resource; unlike grids, cluster resources are typically owned by a single organization.
- All users of clusters have to go through a centralized system that manages allocation of resources to application jobs. Cluster management systems have centralized control, complete knowledge of system state and user requests, and complete control over individual components.

(Basic) Web Services (WS)

- Web services provide standard infrastructure for data exchange between two different distributed applications (grids provide an infrastructure for aggregation of high-end resources for solving large-scale problems).
- Web services are expected to play a key constituent role in the standardized definition of grid computing since they have emerged as a standards-based approach for accessing network applications.

Peer-to-peer (P2P)

- P2P is concerned with the same general problem as grid computing, namely, the organization of resource sharing within virtual communities.
- Grid communities focus on aggregating distributed high-end machines such as clusters, whereas the P2P community concentrates on sharing low-end systems such as PCs connected to the Internet.
- Like P2P, grid computing allows users to share files (many-to-many sharing). With grid computing, the sharing is not only in reference to files, but also other IT resources.

In a grid environment, the ensemble of resources is able to work together cohesively because of defined protocols that control connectivity, coordination, resource allocation, resource management, security, and chargeback. Generally, the protocols are implemented in the middleware. The systems "glued" together by a computational grid may be in the same room, or may be distributed across the globe; they may be running on homogenous or heterogeneous hardware platforms; they may be running on similar or dissimilar operating systems; and they may owned by one or more organizations. The goal of grid computing is to provide users with a single view and/or single mechanism that can be utilized to support any number of computing tasks: the grid leverages its extensive informatics capabilities to support the "number crunching" needed to complete the task and all the user perceives is, essentially, a large virtual computer undertaking his or her work [90].

In recent years, there have been an increasing roster of published articles, conferences, tutorials, resources, and tools related to this topic [90]. Distributed virtualized grid computing technology, as we define it today, is still fairly new, being only a decade in the making. However, as already implied, a number of the basic concepts of grid computing go back as far as the mid 1960s and early 1970s. Recent advances, such as ubiquitous high-speed networking in both private and public venues (e.g., high-speed intranets and high-speed Internet), make the technology more deployable at the practical level, particularly when looking at corporate environments.

As far back as 1987, this researcher was advocating the concept of grid computing in internal Bell Communications Research White Papers (e.g., in Special Reports SR-NPL-000790, an extensive plan written by the author listing progressive *data* services that could be offered by local telcos and RBOCs, entitled "A Collection of Potential Network-Based Data Services" [6]). In a section called "*Network for a Computing Utility*," it was stated by us:

> The proposed service provides the entire apparatus to make the concept of the Computing Utility possible. This includes as follows: (1) the physical network over which the information can travel, and the interface through which a guest PC/workstation can participate in the provision of machine cycles and through which the service requesters submit jobs; (2) a load sharing mechanism to invoke the necessary servers to complete a job; (3) a reliable security mechanism; (4) an effective accounting mechanism to invoke the billing system; and, (5) a detailed directory of servers. . . . Security is one of the major issues for this service, particularly if the PC is not fully dedicated to this function, but also used for other local activities. Virus threats, infiltration and corruption of data, and other damage must be appropriately addressed and managed by the service; multi-task and robust operating systems are also needed for the servers to assist in this security process. . . . The Computing Utility service is beginning to be approached by the Client/Server paradigm now available within a Local Area Network (LAN) environment. . . . This service involves capabilities that span multiple 7-layer stacks. For example, one stack may handle administrative tasks, another may invoke the service (e.g., Remote Operations), still another may return the results (possibly a file), and so on. . . . Currently no such service exists in the public domain. Three existing analogues exist, as follows: (1) timesharing service with a centralized computer; (2) highly parallel computer systems with hundreds or thousands of nodes (what people now call cluster computing), and (3) gateways or other processors connected as servers on a LAN. The distinction between these and the proposed service is the security and accounting arenas, which are much more complex in the distributed, public (grid) environment. . . . This service is basically feasible once a transport and switching network with strong security and accounting (chargeback) capabilities is deployed, as shown in Figure. . . . A high degree of intelligence in the network is required . . . a physical net-

work is required . . . security and accounting software is needed . . . protocols and standards will be needed to connect servers and users, as well as for accounting and billing. These protocols will have to be developed before the service can be established. . . .

1.2 POTENTIAL APPLICATIONS AND FINANCIAL BENEFITS OF GRID COMPUTING

Grid proponents take the position that grid computing represents a "next step" in the world of computing, and that grid computing promises to move the Internet evolution to the next logical level. According to some ([92, 171, 172] among others), *"utility computing is a positive, fundamental shift in computing architecture,"* and many businesses will be completely transformed over the next decade by using grid-enabled services as they integrate not only applications across the Internet but also raw computer power and storage. Furthermore, proponents prognosticate that infrastructure will appear that will be able to connect multiple regional and national computational grids, creating a universal source of pervasive and dependable computing power that will support new classes of applications [93].

Most researchers, however, see grid computing as an evolution, not a revolution. In fact, grid computing can be seen as the latest and most complete evolution of more familiar developments such as distributed computing, the Web, peer-to-peer (P2P) computing, and virtualization technologies [43]. Some applications of grid computing, particularly in the scientific and engineering arenas, include, but are not limited to, the following [105]:

- Distributed supercomputing/computational science
- High-capacity/throughput computing: large-scale simulation/chip design and parameter studies
- Content sharing, for example, sharing digital content among peers (e.g., Napster)
- Remote software access/renting services: ASPs and web services
- Data-intensive computing: drug design, particle physics, stock prediction
- On-demand, real-time computing: medical instrumentation and mission-critical initiatives
- Collaborative computing (e-science, e-engineering): collaborative design, data exploration, education, e-learning
- Utility computing/service-oriented computing: new computing paradigm, new applications, new industries, and new business

The benefits gained from grid computing can translate into competitive advantages in the marketplace. For example, the potential exists for grids to [43, 94]

- Enable resource sharing
- Provide transparent access to remote resources

- Make effective use of computing resources, including platforms and data sets
- Reduce significantly the number of servers needed by (25–75%)
- Allow on-demand aggregation of resources at multiple sites
- Reduce execution time for large-scale data processing applications
- Provide access to remote databases and software
- Provide load smoothing across a set of platforms
- Provide fault tolerance
- Take advantage of time zone and random diversity (in peak hours, users can access resources in off-peak zones)
- Provide the flexibility to meet unforeseen emergency demands by renting external resources for a required period instead of owning them
- Enable the realization of a virtual data center

(Naturally there also are challenges associated with a grid deployment—this field being new and evolving.) As implied by the last bulleted point, there is a discernable IT trend afoot toward virtualization and on-demand services. Virtualization[3] (and supporting technology) is an approach that allows several operating systems to run simultaneously on one (large) computer. For example, IBM's z/VM operating system lets multiple instances of Linux coexist on the same mainframe computer. More generally, virtualization is the practice of making resources from diverse devices accessible to a user as if they were a single, larger, homogenous, resource that appears to be locally available. Virtualization supports the concept of dynamically shifting resources across platforms to match computing demands with available resources: the computing environment can become dynamic, enabling autonomic shifting of applications between servers to match demand [170]. There are well-known advantages in sharing resources, as a routine assessment of the behavior of the M/M/1 queue (memoryless/memoryless/1 server queue) versus the M/M/m queue (memoryless/memoryless/n servers queue) demonstrates: a single more powerful queue is more efficient than a group of discrete queues of comparable aggregate power. Grid computing represents a development in virtualization: as we have stated, it enables the abstraction of distributed computing and data resources such as processing, network bandwidth, and data storage to create a single system image; this grants users and applications seamless access (when properly implemented) to a large pool of IT capabilities. Just as an Internet user views a unified instance of content via the Web, a grid computing user essentially sees a single, large virtual computer [43]. "Virtualization"—the driving force behind grid computing—has been a key factor since the earliest days of electronic business computing.

Studies have shown that when problems can be parallelized, such as in the cases of data mining, records analysis, and billing (as may be the case in a bank, securities company, financial services company, credit card company, insurance company, etc.), then significant savings are achievable. Specifically, whereas a classical mod-

---

[3]Virtualization can be achieved without grid computing; but many view virtualization as a step toward the goal of deploying grid computing infrastructures.

el may require, say, $100 K to process 100 K records, a grid-enabled environment may take as little as $20 K to process the same number of records. Hence, the bottom line is that Fortune 500 companies have the potential to save 30% or more in run-the-engine costs on the appropriate line item of their IT budget.

Grid computing can also be seen as part of a larger rehosting initiative and underlying IT trend at many companies (where alternatives such as Linux® or possibly Windows Operating Systems could, in the future, be the preferred choice over the highly reliable, but fairly costly UNIX solutions). Although each organization is different and the results vary, the directional cost trend is believable. Vendors engaged in this space include (but are not limited to) IBM, Hewlett-Packard, Sun Microsystems, and Oracle. IBM uses "*on-demand*" to describe its initiative, HP has its *Utility Data Center* (UDC) products, Sun Microsystems has its *N1 Data-center Architecture,* and Oracle has the *10g family* of "grid-aware" products. Several software vendors also have a stake in grid computing, including but not limited to Microsoft, Computer Associates, Veritas Software, and Platform Computing [100].

Commercial interest in grid computing is on the rise, according to market research published by Gartner. The research firm estimated that 15% of corporations adopted a utility (grid) computing arrangement in 2003, and the market for utility services in North America would increase from US$8.6 billion in 2003 to more than US$25 billion in 2006. By 2006, 30% of companies would have some sort of utility computing arrangement, according to Gartner [100]. Based on published statements, IBM expects the sector to move into "hypergrowth" in 2004, with "technologies . . . moving 'from rocket science to business service'", and the company had a target of doubling its grid revenue during that year [169]. According to economists, proliferation of high-performance cluster and grid computing and web services (WSs) applications will yield substantial productivity gains in the United States and worldwide over the next decade [111].

A recent report from research firm IDC concluded that 23% of IT services will be delivered from offshore centers by 2007 [112]. Grid computing may be a mechanism to enable companies to reduce costs, yet keep jobs, intellectual capital, and data from migrating abroad. Whereas distributed computing does enable the idea of "remoting" functions, with grid computing this "remoting" can be done to some in-country regional rather than third world location (just like electric and water grids have regional or near-countries scope, rather than having far-flung third world remote scope.) The migration of IT jobs abroad has, in the opinion of this researcher, national security/homeland security risk implications in the long term, particularly if terabytes of data about U.S. citizens and our government become the resident ownership of politically unstable third world countries.

Although there are advantages to grids (e.g., potential reduction in the number of servers from 25 to 50% and related run-the-engine costs), some companies have reservations about immediately implementing the technology. Some of this hesitation relates to the fact that the technology is new, and in fact may be overhyped by the potential provider of services. Other issues may be related to "protection of turf": eliminating vast arrays of servers implies reduction in data center space, reduction in management span of control, reduction in operational staff, reduction in budget, etc. This is the same issue that was faced in the 1990s regarding outsourcing (e.g., see [120]). Other reservations may relate to the fact that infrastructure changes are needed, and this may have a short-term financial disbursement implication. Finally, not all situations, environments, and applications are amenable to a grid paradigm.

1.3   GRID TYPES, TOPOLOGIES, COMPONENTS, AND LAYERS—A PRELIMINARY VIEW

Grid computing embodies a combination of a decentralized architecture for resource management and a layered hierarchical architecture for implementation of various constituent services [101]. A grid computing system can have local, metropolitan, regional, national, or international footprints. In turn, the autonomous resources in the constituent ensemble can span a single organization, multiple organizations, or a service provider space. Grids can be focused on the pooled assets of one organization or span virtual organizations that use a common suite of protocols to enable grid users and applications to run services in a secure, controlled manner [91]. Furthermore, resources can be logically aggregated for a long period of time (say, months or years), or for a temporary period of time (say, minutes, days, or weeks).

Grid computing often combines elements such as distributed computing, high-performance computing, and disposable computing, depending on the application of the technology and the scale of the operation. Grids can, in practical terms, create a virtual supercomputer out of existing servers, workstations, and even PCs, to deliver processing power not only to a company's own stakeholders and employees, but also to its partners and customers. This metacomputing environment is achieved by treating such IT resources as processing power, memory, storage, and network bandwidth as pure commodities. Like an electricity or water network, computational power can be delivered to any department or any application where it is needed most at any given time, based on specified business goals and priorities. Furthermore, grid computing allows charge-back on a per-usage basis rather than for a fixed infrastructure cost [130].

Present-day grids encompass the following types [90]:

- Computational grids, in which machines with set-aside resources allocated to "number crunch" data or provide coverage for other intensive workloads
- Scavenging grids, commonly used to find and harvest machine cycles from idle servers and desktop computers for use in resource-intensive tasks (scavenging is usually implemented in a way that is unobtrusive to the owner/user of the processor)
- Data grids, which provide a unified interface for all data repositories in an organization, and through which data can be queried, managed, and secured

As already noted, no claim is made herein *that there is a single solution to a given problem;* grid computing is one of the available solutions. For example, although some of the machine-cycle inefficiencies can be addressed by virtual servers/rehosting (e.g., VMWare, MS VirtualPC, VirtualServer, LPARs from IBM, and partitions from Sun and HP, which do not require a grid infrastructure), one of the possible approaches to this inefficiency issue is, indeed, grid computing. Grid computing does have an emphasis on geographically distributed, multiorganization, utility-based (outsourced), networking-reliant methods, whereas clustering and rehosting have a more (but not exclusively) datacenter-focused, single-organization-oriented approach. Organizations will need to perform appropriate functional, economic, and strategic assessments to determine which approach is, in final analysis, best for their specific environment. (This text is on grid computing; hence, our emphasis is on this approach, rather than other possible approaches, such as virtualization.)

Figures 1.1, 1.2, and 1.3 provide a pictorial view of some grid computing environments. Figure 1.1 depicts the traditional computing environment in which a multitude of often-underutilized servers support a disjoint set of applications and data sets. As implied by this figure, the typical IT environment prior to grid computing operated as follows. A business-critical application runs on a designated server. Although the average utilization may be relatively low, during peak cycles the server in question can get overtaxed. As a consequence of this instantaneous overtaxation, the application can slow down, experience a halt, or even stall. In this traditional in-

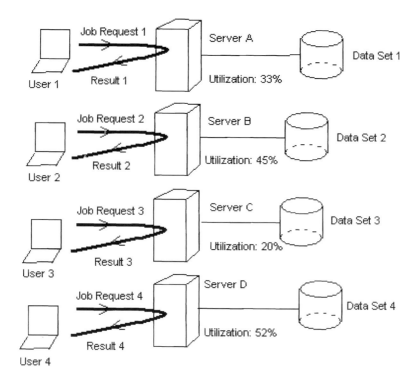

Figure 1.1 Standard computing environment.

stance, the large data set that this application is analyzing exists only in a single data store (note that although multiple copies of the data could exist, it would not be easy with the traditional model to synchronize the databases if two programs independently operated aggressively on the data at the same time.) Server capacity and access to the data store place limitations on how quickly desired results can be returned. Machine cycles on other servers are unable to be constructively utilized, and available disk capacity remains unused [43].

Figure 1.2 depicts an organization-owned computational grid. Here, a middleware application running on a grid-computing broker manages a small(er) set of processors and an integrated data store. A computational grid is a hardware and software infrastructure that provides dependable, consistent, pervasive, and inexpensive access to high-end computational capabilities. In a grid environment, workload can be broken up and sent in manageable pieces to idle server cycles. Not *all* applications are necessarily instantly migratable to a grid environment without at least some redesign. Although legacy business applications may, a priori, fit such a class of applications, a number of Fortune 500 companies are indeed looking into *how* such legacy applications can be modified and/or retooled such that they can be made to run on grid-based infrastructures.

A scheduler sets rules and priorities for routing jobs on a grid-based infrastructure. When servers and storage are enabled for grid computing, copies of the data can be stored in formerly unused space and can easily be made available [43]. A

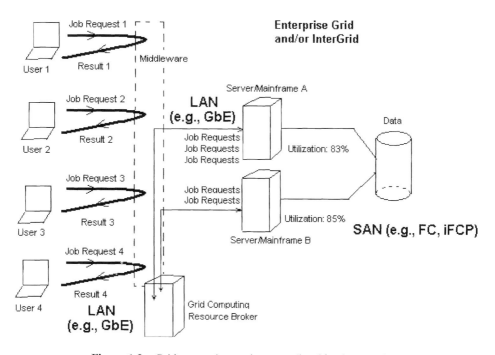

Figure 1.2   Grid computing environment (local implementation).

grid also provides mechanisms for managing the distributed data in a seamless way [96, 106]. A grid middleware provides facilities to allow use of the grid for applications and users. Middleware such as Globus [107], Legion [108], and UNICORE (UNiform Interface to Computer Resources) [109] provide software infrastructure to handle various challenges of computational and data grids [106].

Figure 1.3 depicts the utility-oriented implementation of a computational grid. This concept is analogous to an electric power network (grid) in which power generators are distributed but the users are able to access electric power without concerning themselves about the source of energy and its pedestrian operational management [110].

As suggested by Figures 1.1 to 1.3, grid computing aims to provide seamless and scalable access to distributed resources. Computational grids enable the sharing, selection, and aggregation of a wide variety of geographically distributed computational resources (such as supercomputers, computing clusters, storage systems, data sources, instruments, and developers) and presents them as a single, unified resource for solving large-scale computing- and data-intensive applications (e.g., molecular modeling for drug design, brain activity analysis, and high-energy physics). An initial grid deployment at a company can be scaled over time to bring in additional applications and new data. This allows gains in speed and accuracy without significant cost increases. Several years ago, this author coined the phrase "*the corporation is the network*" [95]. Grid computing supports this concept very well: with grid computing, all a corporation needs to run its IT apparatus is a reli-

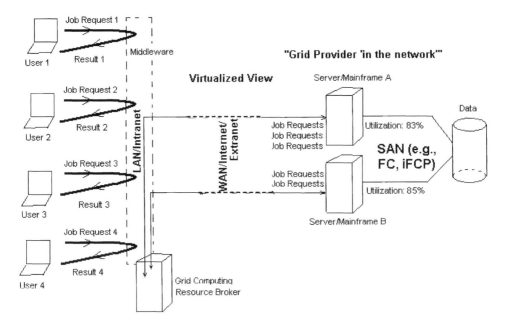

Figure 1.3 Grid computing environment (remote implementation).

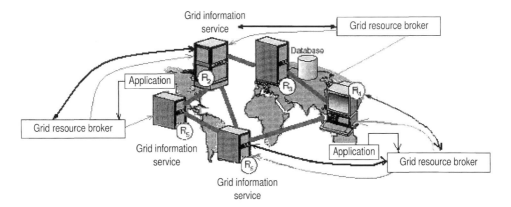

Figure 1.4   Pictorial view of World Wide InterGrid.

able high-speed network to connect it to the distributed set of virtualized computational resources not necessarily owned by the corporation itself.

Grid computing started out as mechanism for sharing computational resources
distributed all over the world for basic science applications, as illustrated pictorially
by Figure 1.4 [96, 106]. But other types of resources, such as licenses or specialized
equipment, can now also be virtualized in a grid computing environment. For example, if an organization's software license agreement limits the number of users
that can be utilizing the license simultaneously, license management tools operating
in grid mode could be employed to keep track of how many concurrent copies of
the software are active. This will prevent the number from exceeding the allowed
number, as well as schedule jobs according to priorities defined by the automated
business policies. Specialized equipment that is remotely deployed on the network
could also be managed in a similar way, thereby reducing the need for the organization to purchase multiple devices, in much the same way today that people in the
same office share Internet access or a printing resources across a LAN [130].

Some grids focus on data federation and availability; other grids focus on computing power and speed. Many grids involve a combination of the two. For end
users, all infrastructure complexity stays hidden [43]. Data (data base) federation
makes disparate corporate databases look like the constituent data is all in the same
database. Significant gains can be secured if one can work on all the different data
bases, including selects, inserts, updates, and deletes, as if all the tables existed in a
single data base.[4] Almost every organization has significant unused computing ca-

---

[4]The "federator" system operates on the tables in the remote systems called the "federatees." The remote
tables appear as virtual tables in the "federator" data base. Client application programs can perform operations on the virtual tables in the "federator" data base, but the real persistent storage is in the remote
data base. Each "federatee" views the "federator" as just another data base client connection. The "federatee" is simply servicing client requests for data base operations. The "federator" needs client software
to access each remote data base. Client software for IBM Informix®, Sybase, Oracle, and so on would
need to be installed to access each type of federatee [126].

pacity, widely distributed among a tribal arrangement of PCs, midrange platforms, mainframes, and supercomputers. For example, if a company has 10,000 PCs, at an average computing power of 333 MIPS, this equates to an aggregate 3 tera (10^{12}) floating-point operations per second (TFLOPS) of potential computing power. As another example, in the United States there are an estimated 300 million computers. At an average computing power of 333 MIPS, this equates to a raw computing power of 100,000 TFLOPS. Mainframes are generally idle 40% of the time; Unix servers are actually "serving" something less than 10% of the time; most PCs do nothing for 95% of a typical day [43]. This is an inefficient situation for customers. TFLOPS speeds that are possible with grid computing enable scientists to address some of the most computationally intensive scientific tasks, from problems in protein analysis that will form the basis for new drug designs, to climate modeling, to deducing the content and behavior of the cosmos from astronomical data [97].

Many scientific applications are also data intensive, in addition to being computationally intensive. By 2006, several physics projects will produce multiple petabytes (10^{15} bytes) of data per year. This has been called "peta-scale" data. PCs now ship with up to 100 gigabytes (GB) of storage (as much as an entire 1990 supercomputer center) [113]: one petabyte would equate to 10,000 of these PCs, or to the PC base of a "smaller" Fortune 500 company. Data grids also have some immediate commercial applications. Grid-oriented solutions are the way to manage this sort of storage requirement, particularly from a data access perspective (more than just from a physical storage perspective.) As time evolved, the management of "peta-scale" data became burdensome. The confluence and combination of large data set size, geographic distribution of users and resources, and computationally intensive scientific analyses, prompted the development of data grids, as noted earlier [106, 115]. Here, a *data* middleware (usually part of general-purpose grid middleware), provides facilities for data management. Various research communities have developed successful data middleware such as Storage Resource Broker (SRB) [116], Grid Data Farm [117], and European Data Grid Middleware. These middleware tools have been effective in providing a framework for managing high volumes of data but they are often incompatible.

The key components of grid computing include the following [90]:

- Resource management: the grid must be aware of what resources are available for different tasks
- Security management: the grid needs to take care that only authorized users can access and use the available resources
- Data management: data must be transported, cleansed, parceled, and processed
- Services management: users and applications must be able to query the grid in an effective and efficient manner

More specifically, grid computing can be viewed as being comprised of a number of logical hierarchical layers. Figure 1.5 depicts a first view of the layered architec-

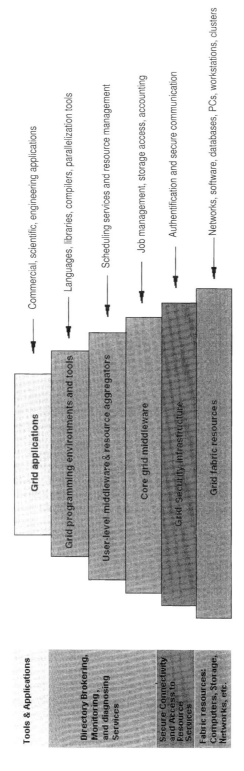

Figure 1.5 One view of grid computing layers.

Commercial, scientific, engineering applications

Languages, libraries, compilers, parallelization tools

Scheduling services and resource management

Job management, storage access, accounting

Authentification and secure communication

Networks, software, databases, PCs, workstations, clusters

Grid applications

Grid programming environments and tools

User-level middleware & resource aggregators

Core grid middleware

Grid Security infrastructure

Grid fabric resources

Tools & Applications

Directory Brokering, Monitoring, and diagnosing Services

Secure Connectivity and Access to Resource Services

Fabric resources: Computers, Storage, Networks, etc.

ture of a grid environment. At the base of the grid stack, one finds the *grid fabric,* namely, the distributed resources that are managed by a local resource manager with a local policy; these resources are interconnected via local-, metropolitan-, or wide-area networks. The grid fabric includes and incorporates networks; computers such as PCs and processors using operating systems such as Unix, Linux, or Windows; clusters using various operating systems; resource management systems; storage devices; and data bases. The *security infrastructure* layer provides secure and authorized access to grid resources. Above this layer, the *core grid middleware* provides uniform access to the resources in the fabric—the middleware designed to hide complexities of partitioning, distributing, and load balancing. The next layer, the *user-level middleware* layer, consists of resource brokers or schedulers responsible for aggregating resources. The *grid programming environments and tools* layer includes the compilers, libraries, development tools, and so on, that are required to run the applications (resource brokers manage execution of applications on distributed resources using appropriate scheduling strategies and grid development tools to grid-enable applications). The top layer consists of *grid applications* themselves [94].

Building on this intuitive idea of layering, it would be advantageous if industry consensus were reached on the series of layers. Architecture standards are now under development by the Global Grid Forum (GGF).[5] The GGF is an industry advocacy group; it supports community-driven processes for developing and documenting new standards for grid computing. The GGF is a forum for exchanging information and defining standards relating to distributed computing and grid technologies. GGF is fashioned after the Grid Forum, the eGrid European Grid Forum, and the Grid Community in the Asia-Pacific region. GGF is focusing on open grid architecture standards [119]. Technical specifications are being developed for architecture elements, for example, security, data, resource management, and information. Grid architectures are being built based on Internet protocols and services (e.g., communication, routing, name resolution, etc.) The layering approach is used to the extent possible because it is advantageous for higher-level functions to use common lower-level functions. The GGF's approach has been to propose a set of core services as basic infrastructure, as shown in Figure 1.6. These core services are used to construct high-level, domain-specific solutions. The design principles are: keep participation cost low, enable local control, support adaptation, and use the "IP hourglass" model of many applications using a few core services to support many fabric elements (e.g., operating systems). In the meantime, Globus Toolkit™ has emerged as the de facto standard for several important connectivity, resource, and collective protocols. The toolkit, having a "middleware plus" capability, addresses issues of security, information discovery, resource management, data management, communication, fault detection, and portability (see Chapter 6) [33].

---

[5]GGF members include Cisco Systems, Hewlett-Packard, IBM, Microsoft, Qwest Communications, Silicon Graphics, Sun Microsystems, Oracle, Level(3), and BellSouth, among 46 participants at press time.

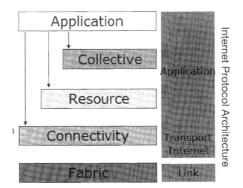

"Coordinating multiple resources":
ubiquitous infrastructure services,
app-specific distributed services

"Sharing single resources":
negotiating access, controlling use

"Talking to things": communication
(Internet protocols) & security

"Controlling things locally": Access
to, & control of, resources

Figure 1.6   Layered grid architecture, Global Grid Forum. This material is licensed for use under the terms of the Globus Toolkit Public License. See http://www.globus.org/toolkit/download/license.html for the full text of this license.

1.4   COMPARISON WITH OTHER APPROACHES

It is important to note that certain IT computing constructs are not grids, as we discuss next. In some instances, these technologies are the optimal solution for an organization's problem; in other cases, grid computing is the best solution, particularly if in the long term one is especially interested in supplier-provided utility computing.

The distinction between *clusters* and grids relates to the way resources are managed. In the case of clusters (aggregations of processors in parallel-based configurations), the resource allocation is performed by a centralized resource manager and scheduling system. Also, nodes cooperatively work together as a single unified resource. In the case of grids, each node has its own resource manager and does not aim at providing a single system view [131]. A cluster is comprised of multiple interconnected independent nodes that cooperatively work together as a single unified resource. This means all users of clusters have to go through a centralized system that manages the allocation of resources to application jobs. Unlike grids, cluster resources are almost always owned by a single organization. Actually, many grids are constructed by using clusters or traditional parallel systems as their nodes, although this is not a requirement. An example of a grid that contains clusters as its nodes is the NSF TeraGrid [101]; another example is the World Wide Grid, which has many nodes that are clusters located in organizations such as NRC Canada, AIST-Japan, N*Grid Korea, and the University of Melbourne. Although cluster management systems such as Platform's Load Sharing Facility, Veridian's Portable Batch System, or Sun's Sun Grid Engine can deliver enhanced distributed computing services, they are not grids themselves; these cluster management systems have centralized control, complete knowledge of system state and user requests, and complete control over individual components (such features tend not to be characteristic of a grid proper) [103].

Grid computing also differs from basic *Web services,* although it now makes use of these services. Web services have become an important component of distributed computing applications over the Internet [173]. The World Wide Web is not yet in itself a grid, its open, general-purpose protocols support access to distributed resources but not the coordinated use of those resources to deliver negotiated qualities of service [103]. So, whereas the Web is mainly focused on communication, grid computing enables resource sharing and collaborative resource interplay toward common business goals. Web services provide standard infrastructure for data exchange between two different distributed applications, whereas grids provide an infrastructure for aggregation of high-end resources for solving large-scale problems in science, engineering, and commerce. While most Web services involve static processing and moveable data, many grid computing mechanisms involve static data (on large databases) and moveable processing. However, there are similarities as well as dependencies. First, similar to the case of the World Wide Web, grid computing keeps complexity hidden—multiple users experience a single, unified experience. Second, Web services are utilized to support grid computing mechanisms. These Web services will play a key constituent role in the standardized definition of grid computing, since Web services have emerged in the past few years as a standards-based approach for accessing network applications. The recent trend is to implement grid solutions using web services technologies, for example, the Globus Toolkit 3.0 middleware. In this context, low-level grid services are instances of Web services (a grid service is a Web service that conforms to a set of conventions that provide for controlled, fault-resilient, and secure management of services) [31, 101].

Both peer-to-peer computing and grid computing are concerned with the same general problem, namely, the organization of resource sharing within VOs. As is the case with P2P environments, grid computing allows users to share files; but unlike P2P, grid computing allows many-to-many sharing. Furthermore, with grid computing the sharing is not only in reference to files but other resources as well. The grid community generally focuses on aggregating distributed high-end machines such as clusters, whereas the P2P community concentrates on sharing low-end systems such as PCs connected to the Internet [94]. Both disciplines take the same general approach to solving this problem, namely, the creation of overlay structures that coexist with, but need not correspond in structure to underlying organizational structures. Each discipline has made technical advances in recent years, but each also has, in current instantiations, a number of limitations: there are complementary aspects regarding the strengths and weaknesses of the two approaches that suggests that the interests of the two communities are likely to grow closer over time [121]. P2P networks can amass computing power, as does the SETI@home project, or share content, as Napster and Gnutella have done in the recent past. Given the number of grid and P2P projects and forums that began worldwide at the turn of the decade, it is clear that interest in the research, development, and commercial deployment of these technologies is burgeoning [94]. (This topic is revisited at the end of Chapter 2.)

Grid computing also differs from *virtualization.* Resource virtualization is the *abstraction* of server, storage, and network resources in order to make them available *dynamically* for sharing, both inside and outside an organization. *Virtualization is a step along the way on the road to utility computing* (grid computing) and, in combination with other server, storage, and networking capabilities, offers customers the opportunity to build, according to advocates, an IT infrastructure "without" hard boundaries or fixed constraints [162]. Virtualization has somewhat more of an emphasis on local resources, whereas grid computing has more of an emphasis on geographically distributed interorganizational resources. The universal problem that virtualization is solving in a data center is that of dedicated resources. While this approach does address performance, this method lacks fine granularity. Typically, IT managers take an educated guess as to how many dedicated servers they will need to handle peaks, purchase extra servers, and then later find out that a significant number of these servers are significantly underutilized. A typical data center has a large amount of idle infrastructure, bought and set up online to handle peak traffic for different applications. Virtualization offers a way of moving resources from one application to another dynamically. However, specifics of the desired virtualizing effect depend on the specific application deployed [161]. Three representative products are HP's Utility Data Center, EMC's VMware, and Platform Computing's Platform LFS. With virtualization, the logical functions of the server, storage, and network elements are separated from their physical functions (e.g., processor, memory, I/O, controllers, disks, switches). In other words, all servers, storage, and network devices can be aggregated into independent pools of resources. Some elements may even be further subdivided (server partitions, storage LUNs) to provide an even more granular level of control. Elements from these pools can then be allocated, provisioned, and managed, manually or automatically, to meet the changing needs and priorities of one's business. Virtualization can span the following domains [162]:

1. Server virtualization for horizontally and vertically scaled server environments. Server virtualization enables optimized utilization, improved service levels, and reduced management overhead.

2. Network virtualization, enabled by intelligent routers, switches, and other networking elements supporting virtual LANs. Virtualized networks are more secure and more able to support unforeseen spikes in customer and user demand.

3. Storage virtualization (server, network, and array-based). Storage virtualization technologies improve the utilization of current storage subsystems, reduce administrative costs, and protect vital data in a secure and automated fashion.

4. Application virtualization enables programs and services to be executed on multiple systems simultaneously. This computing approach is related to horizontal scaling, clusters, and grid computing, in which a single application is able to cooperatively execute on a number of servers concurrently.

5. Data center virtualization, whereby groups of servers, storage, and network resources can be provisioned or reallocated on the fly to meet the needs of a new IT service or to handle dynamically changing workloads [162].

Grid computing deployment, although potentially related to a rehosting initiative, is not just rehosting. As Figure 1.7 depicts, rehosting implies the reduction of typically a large number of servers (possibly using some older and/or proprietary OS) to a smaller set of more powerful and more modern servers (possibly running on open-source OSs). This is certainly advantageous from the operations, physical maintenance, and power and space perspectives. There are savings associated with rehosting. However, applications are still assigned specific servers. Grid computing, on the other hand, permits the true virtualization of the computing function, as seen in Figure 1.7. Here, applications are not preassigned a server, but the "run-time" assignment is made based on real-time considerations. (Note: in the bottom diagram, the hosts could be colocated or spread all over the world. When *local* hosts are aggregated in tightly coupled configurations, they tend to generally be of the cluster parallel-based computing type; such processors, however, can also be nonparallel-computing-based grids, for example, by running the Globus Toolkit. When *geographically dispersed* hosts are aggregated in distributed computing configurations, they tend to generally be of the grid computing type and not run in a clustered arrangement. Figure 1.7 does not show geography and the reader should conclude that the hosts are arranged in a grid computing arrangement.)

In summary, like clusters and distributed computing, grids bring computing resources together. Unlike clusters and distributed computing, which need physical proximity and operational homogeneity, grids can be geographically distributed and heterogeneous. Like virtualization technologies, grid computing enables the virtualization of IT resources. Unlike virtualization technologies, which virtualize a single system, grid computing enables the virtualization of broad-scale and disparate IT resources [43]. Scientific-community deployments, such as the distributed data processing system being deployed internationally by "data grid" projects (e.g., GriPhyN, PPDG, EU DataGrid), NASA's Information Power Grid, the Distributed ASCI Supercomputer system that links clusters at several Dutch universities, the DOE Science Grid and DISCOM Grid that link systems at DOE laboratories, and the TeraGrid mentioned above being constructed to link major U.S. academic sites, are all bona-fide examples of grid computing. A multisite scheduler such as Platform's MultiCluster can reasonably be called (first-generation) grids. Other examples of grid computing include the distributed computing systems provided by Condor, Entropia, and United Devices, which make use of idle desktops; peer-to-peer systems such as Gnutella, which support file sharing among participating peers; and a federated deployment of the Storage Resource Broker, which supports distributed access to data resources [103].

1.5   A FIRST LOOK AT GRID COMPUTING STANDARDS

One of the challenges of any computing technology is getting the various components to communicate with each other. Nowhere is this more critical than when trying to get different platforms and environments to interoperate. It should, therefore, be immediately evident that the grid computing paradigm requires standard, open,

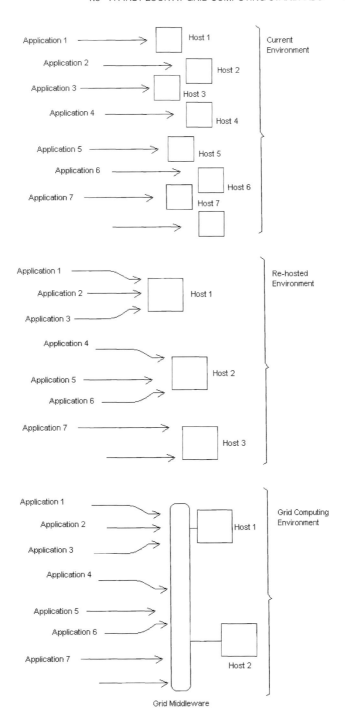

Figure 1.7   A Comparison with rehosting.

general-purpose protocols and interfaces. Standards for grid computing are now being defined and are beginning to be implemented by the vendors [90, 122]. To make the most effective use of the computing resources available, these environments need to utilize common protocols [123]. Standards are the "holy grail" of grid computing (see Chapters 4 and 5).

Regarding this issue, proponents make the case that we are now indeed entering a new phase of grid computing in which standards will define grids in a consistent way by enabling grid systems to become easily built "off-the-shelf" systems. Standard-based grid systems have been called by some "third-generation grids" or 3G grids. First-generation or "1G grids" involved local "metacomputers" with basic services such as distributed file systems and site-wide single sign-on, upon which early-adopters developers created distributed applications with custom communications protocols. Test beds extended 1G grids across distances, and attempts to create "metacenters" explored issues of interorganizational integration. 1G grids were totally custom-made proofs of concept [122]. 2G grid systems began with projects such Condor, I-WAY (the origin of Globus), and Legion (origin of Avaki), in which underlying software services and communications protocols could be used as a basis for developing distributed applications and services. 2G grids offered basic building blocks, but deployment involved significant customization and filling in many gaps. Independent deployments of 2G grid technology today involve enough customized extensions that interoperability is problematic, and interoperability among 2G grid systems is rather difficult. This is why the industry needs 3G grids [122].

By introducing standard technical specifications, 3G grid technology will have the potential to allow both competition and interoperability not only among applications and toolkits, but among implementations of key services. The goal is to mix and match components, but this potential will only be realized if the grid community continues to work at defining standards [122]. The Global Grid Forum community is applying lessons learned from 1G and 2G grids and from Web services technologies and concepts to create 3G architectures [122].

GGF has driven initiatives such as the Open Grid Services Architecture (OGSA). OGSA is a set of specifications and standards that integrate and leverage the worlds of Web services and grid computing (Web services are viewed by some as the "biggest technology trend" in the last five years [75]). With this architecture, a set of common interface specifications supports the interoperability of discrete, independently developed services. OGSA brings together Web standards such as XML (eXtensible Markup Language), WSDL (Web Service Definition Language), UDDI (Universal Description, Discovery, and Integration), and SOAP (Simple Object Access Protocol), with the standards for grid computing developed by the Globus Project [102]. The Globus Project is a joint effort on the part of researchers and developers from around the world that is focused on grid research, software tools, testbeds, and applications. As TCP/IP (Transmission Control Protocol/Internet Protocol) forms the backbone for the Internet, the OGSA is the backbone for grid computing. The recently-released Open Grid Services Infrastructure (OGSI) service specification is the keystone in this architecture [122].

In addition to making progress on the standards front, grid computing as a service needs to address various issues and challenges. Besides standardization, some of these issues and challenges include security, true scalability, autonomy, heterogeneity of resource access interfaces, policies, capabilities, pricing, data locality, dynamic variation in availability of resources, and complexity in creation of applications [101].

1.6 A PRAGMATIC COURSE OF INVESTIGATION

In order to identify possible benefits to their organizations, planners should understand grid computing concepts and the underlying networking mechanisms. Practitioners interested in grid computing are asking basic questions [90]: What do we do with all of this stuff? Where do we start? How do the pieces fit together? What comes next? As implied by the introductory but rather encompassing discussion above, grid computing is applicable to enterprise users at two levels:

1. Obtaining computing services over a network from a remote computing service provider
2. Aggregating an organization's dispersed set of uncoordinated systems into one holistic computing apparatus

As noted, with grid computing organizations can optimize computing and data resources, pool such resources to support large-capacity workloads, share the resources across networks, and enable collaboration [43]. Grid technology allows the IT organization to consolidate and numerically reduce the number of platforms that need to be kept operating.

This book focuses on networking as not only the best way to understand what grid computing is, but, more importantly, the best way to understand why and how grid computing is important to IT practitioners (rather than "just another hot technology" that gets researchers excited but never has much effect on what network professionals see in the real world). The goals of this book are three-fold:

1. Describe the basic concepts and technical components of grid computing
2. Describe possible benefits of grid computing
3. Describe the networking apparatus required to support grid computing in an efficient manner

This introductory chapter looked at grid computing technology from a generalized perspective and offers some motivational and advocacy information. The rest of the book looks at specific technical issues, focusing on standards and networking issues. A number of books that focus strictly on the IT and/or software side are available to the interested reader; the present text has networking as a major subtheme. In the paragraphs that follow, we highlight some additional subthemes of this course of investigation and survey the field.

One can build and deploy a grid in a variety of sizes and types: for large or small firms, for a single department or the entire enterprise, and for enterprise business applications or scientific endeavors. Like many other recent technologies, however, grid computing runs the risk of being overhyped. CIOs need to be careful not to be to oversold on grid computing. A sound, reliable, conservative economic analysis is, therefore, required that encompasses the true total cost of ownership (TCO) and assesses the true risks associated with this approach. This is a subtheme of this text.

Like the Internet, grid computing has its roots in the scientific and research communities. After about a decade of research, open systems are poised to enter the market. Coupled with a rapid drop in the cost for communication bandwidth, commercial-grade opportunities are emerging for Fortune 500 IT shops searching for new ways to save money. All of this has to be properly weighted against the commoditization of machine cycles (just buy more processors and retain the status quo), and against reduced support cost by way of subcontinent outsourcing of said IT support and related application development (just move operations abroad, but otherwise retain the status quo). During the past ten years or so, a tour-de-force commoditization has been experienced in computing hardware platforms that support IT applications at businesses of all sizes. Some have encapsulated this rapidly occurring phenomenon with the phrase "IT doesn't matter." The price-disrupting predicament brought about by commoditization affords new opportunities for organizations, although it also has conspicuous process- and people-dislocating consequences. Grid computing is but one way to capitalize on this Moore's Law driven commoditization. This is yet another subtheme of this text.

We have noted already that at its core grid computing is, and must be, based on an open set of standards and protocols that enable communication across heterogeneous, geographically dispersed environments. Therefore, planners should track the work that the standards groups are doing. In this book, we are going to look at the significance of standards in grid computing, how they affect capabilities and facilities, what standards exist, and how they can be applied to the problems of distributed computation [123]. This is yet another subtheme of this text.

There is effort involved with resource management and scheduling in a grid environment. When enterprises need to aggregate resources distributed within their organization and prioritize allocation of resources to different users, projects, and applications based on their quality of service (QoS) requirements (call these service-level agreements), they need to be concerned about resource management and scheduling. The user QoS-based allocation strategies enhance the value delivered by the utility. The need for QoS-based resource management becomes significant whenever more than one competing application or user needs to utilize shared resources [101]. This is yet another subtheme of this text.

Regional economies may benefit significantly from grid computing technologies, as suggested earlier in the chapter, assuming that two activities occur [111]. First, the broadband infrastructure needed to support grid computing and Web services must be developed in a timely fashion (this being the underlying theme of this text). Second, states and regions must attract (or grow from within) a sufficient pool of skilled computer and communications professionals to fully deploy and utilize the new technologies and applications. This is another motivation to become familiar with the technology and to read the literature on it.

The approach we are taking in this text is: What has been learned in the past 10 years related to grids that can be taken forward? What valid approaches will we make normative going forward? What is required to make it good enough for the commercial market? What is economically viable for the commercial market? What is sufficient to convey the value of grid technology to a prospective end user? What are the performance levels for businesses: tera (FLOPS, bps, B) or mega (FLOPS, bps, B)? Tera or giga? Peta or exa? What are the requisite networking mechanisms without which grid computing cannot happen?

CHAPTER 2

Grid Benefits and Status of Technology

This chapter examines the potential business benefits for considering grid technology in a corporate environment, while at the same time looking at some of the challenges to be faced and addressed. A quick assessment of the maturity of the technology ("is it ready for prime-time?") is provided, along with a high-level view of some of the key players. The chapter also briefly looks at the history of grid computing.[1]

2.1 MOTIVATIONS FOR CONSIDERING COMPUTATIONAL GRIDS

This subsection identifies some possible reasons and opportunities for commercial organizations to look into the technology and consider planning its eventual deployment. The main points of this subsection are that the potential for savings exists, the grid computing industry is developing rapidly, and people are "doing it." As we have seen in Chapter 1, grids are persistent environments that enable software applications to integrate instruments, displays, and computational and information resources that are managed by diverse organizations in widespread locations [129]. Grid computing allows organizations to share computing power, databases, and other tools securely across corporate, institutional, and geographic boundaries without sacrificing local autonomy [124]. Grid computing enables people from different organizations and locations to work together to solve a specific problem, such as design collaboration. Grid computing software platforms allow resource discovery, resource sharing, and collaboration in a distributed environment [119]. Although grid computing has been used within the academic and scientific community for about a decade, grid standards, grid toolkits, grid products, and enabling technologies (such as ubiquitous broadband networking), are now becoming available that allow businesses to increasingly use and reap the advantages of this form of "outsourced" computing [125].

The benefits of grid computing, according to the industry, will be greater productivity gained through greater flexibility and speed of deployment, access to massive computing power, collaboration, and cost savings [102]. The overarching busi-

---

[1]Although some of this information is time-dependent, the information serves the purpose of validating that the technology is already "ready for prime time," and can only be more so in the future.

A Networking Approach to Grid Computing. By Daniel Minoli
ISBN 0-471-68756-1 © 2005 John Wiley & Sons, Inc.

ness goal is to save IT money because IT budgets are always under pressure due to increasing demand for new services and technologies. Organizations spend as much as 10% of their top-line revenue on IT-related projects, with 6% being typical. Service-oriented businesses tend to be at the higher end of this scale. These costs have remained somewhat fixed over the years, as shown in [127]. A new opportunity now exists to address these cost components. Some of these savings can be achieved by replacing a technology with another. Other savings can be achieved by technology simplification (such as virtualization). We have already cited anecdotal information in Chapter 1 pointing to nontrivial potential savings with grid computing. One researcher stated that "Service providers that are not planning to invest in a portfolio of IT utility services will be under increasing pressure to show they can still provide flexible, cost-effective IT infrastructure services supporting ever-changing business applications and processes" [34]. And another wrote that "Utility Computing will drive major change in system architecture, system management, IT product/service packaging, and pricing. The investments, however, will be relatively modest in 2004. Leading vendors will sharpen the message—from a marketing cloud to implementation road-maps for business problems. Also, the perceived leadership in the market place could change" [27].

According to proponents [163], virtualization (a step along the way to full grid computing) brings

. . . [A]cross-the-board increases in infrastructure efficiency and flexibility, companies . . . stand to see dramatic reductions in overall operational costs—50% or more. . . . Areas where . . . customers may achieve potential savings [are as follows]:

- Deployment costs reduced 30% to 80%
- Capacity planning costs reduced 5% to 40%
- Self-adapting technologies reduce management costs 80% to 100%
- Security costs reduced 20% to 30%
- Usage metering costs reduced from 4% to 30%
- Upgrading and migration costs reduced 20% to 40%

Clearly, these numbers are very optimistic and may not be achievable in all cases, or even most cases. But they do represent a "stake-in-the-ground" regarding economics by vendors (such as HP in this case).

Another user reported on these advantages:

Grid computing has enabled Digex to reduce the time taken to resolve a system performance problem from one hour to just 15 minutes, cutting customer response time by 75%. Grid computing's ease of use and built-in monitoring features enable junior data center staff to undertake tasks for which database administrators (DBAs) were previously required. This allows us to increase the databases managed by each DBA from 40 to 68, an increase of 70%, and grow our business without increasing our resources. [86]

To highlight the potential of grid computing, one only needs to look at today's typical enterprise computing environment. Despite conspicuous decreases in the cost of processing power, storage capacity, and network bandwidth, the bulk of a firm's IT budget is still tied up in operations and maintenance costs. Administrators are managing in the range of 30–50 server systems each, and these systems may be typically utilizing only a small fraction (e.g., 5–25%) of their resources [130]. Some companies have thousands of discrete production-level servers deployed. Virtualization clearly addresses these issues directly. Grid computing is emerging as a viable technology that businesses can use to get more profits and productivity out of IT resources. Grid computing is a promising technology for three reasons [96]:

1. Its ability to make more cost-effective use of a given amount of computer resources
2. As a way to solve problems that cannot be approached without a significant amount of computing power
3. The fact that the resources of many computers can be cooperatively and perhaps synergistically harnessed and managed in collaboration toward a common objective

The argument is made that what these companies need is not more hardware, but more efficient use of existing hardware. Companies need a way to tie all of these underutilized machines together into a number-crunching pool, manage those resources, and provide secure and reliable access to the thus virtualized resources [91]. Without large-scale end-user retraining or a huge investment in new technology infrastructure, promoters of grid computing envision a more productive workforce [130].

According to published reports, firms that have implemented grid architectures have reported measurable improvements. For illustrative purposes, some report that processor utilization rates have grown by 80%, while costs have dropped in some cases by as much as 90%. As an illustrative example, the National Institute for Environmental Health Services (NIEHS) has implemented a grid that helps them achieve a 95% reduction in total elapsed execution time on key research projects [130]. Firms have found that Intel-based Linux servers, often used in grid deployments, can be between 1 to 10% of the total cost of "heavy iron" machines like mainframes or high-end UNIX servers [130]. It is also cheaper to simply add the smaller Linux servers (or computing blades) as needed to incrementally grow the amount of processing power as opposed to adding much more expensive high-end machines.[2] With organizations trying to reduce their IT budgets and hardware being the second-highest expenditure after compensation for operations and maintenance, the strategy can make sense, at face value [130]. IT planners should evaluate the ways they can grid-enable the organization's applications.

A pragmatic approach is to start small. For example, an organization can build and deploy a grid at the departmental level. After analyzing the results and out-

---

[2]An additional advantage to such commodity hardware is that they are considered disposable by some. "Low-end" US$2,500 Linux devices are more likely to be replaced than repaired.

comes of such limited (controlled) deployment, the corporate planners can enlarge the scope of the initiative. Toolkits such as the Globus Toolkit can be useful in this context [102]. A computer language such as Java is not sufficient by itself to run effective grids. Although Java provides useful technology for portable, object-oriented application development, it does not address many of the problems that arise when one tries to achieve high-performance execution in heterogeneous distributed environments (for example, Java does not help one run programs on different types of supercomputers, discover the policy elements that apply at a particular site, achieve single sign-on authentication, perform high-speed transfer across wide-area networks, etc.) A grid middleware (such as the Globus Toolkit) addresses some of these concerns and uses Java to provide portable clients [129].

According to published reports, grid computing technology has already been used in areas such as finance, defense research, medicine discovery, decision making, and collaborative design. There is industry-wide interest on grid computing from research and industry, including IBM, Platform, Avaki, Entropia, Sun Microsystems, and, HP, among others. Table 2.1 provides a short list of grid-related resources. A typical advocacy bulletin from a provider reads as follows:

[Does] Grid Computing unleash the power of existing systems? What if you could analyze huge data sets instantaneously? Run scenarios not hundreds but thousands of times? Bring teams together to radically reduce time-to-market? All while increasing productivity? You might think you owned a supercomputer. And in today's on demand world, it can feel like you need one, just to get the detailed results—always faster, always more accurately—that make you competitive in the marketplace. The good news: IT components you may already own—mainframes, Unix servers, Intel Servers, databases, storage systems, even desktop computers and workstations—harbor enormous, untapped processing and storage power. Power you can easily start pulling together and dedicating to your most pressing business problems. [43]

Table 2.1 Grid consortiums and open forums (partial list)

Asia Pacific Grid
Australian Grid Forum
Content Alliance: About Content Peering
Distributed.net
eGrid: European Grid Computing Initiative
EuroTools SIG on Metacomputing
Global Grid Forum
Global Grid Forum (GGF)
Grid Computing Info Centre
GridForum Korea
IEEE Task Force on Cluster Computing
New Productivity Initiative (NPI)
Peer-to-Peer (P2P) Working Group
SETI@home
The Distributed Coalition

Even if only a fraction of these benefits are actually realizable in "real life", it is still worth the effort to investigate and pursue grid computing technologies.

Many application domains in which large processing problems can easily be divided into subproblems and solved independently are already taking advantage of grid computing (the Internet has facilitated access to untapped computing power). In the scientific world, these include, among others, Monte Carlo simulations, drug design, operations research, and e-Science projects like the Genome project and ecological modeling [156, 157]. Some examples of the new scientific and engineering applications enabled by grid environments include [129]:

- "Smart instruments." These are advanced scientific instruments (e.g., electron microscopes, particle accelerators, and wind tunnels), adjoined with remote supercomputers, databases, and users, all with the goal to enable interactive use (rather than batch approaches), online scenario comparisons, and collaborative data analysis.
- "Teraflop desktops." These support applications such as chemical modeling, symbolic algebra, and other tasks that transfer computationally intensive operations to capability rich remote resources.
- "Collaborative engineering" (also known as, teleimmersion). These applications entail high-bandwidth access to shared virtual spaces that support interactive manipulation of shared datasets and management of complex simulations, to support collaborative design of high-end systems.
- "Distributed supercomputing." These are large virtual supercomputers logically assembled to solve problems too large to fit on any single computer.
- "Parameter studies." These are rapid, large-scale parametric studies in which a single program is run many times in order to explore a multidimensional parameter space.

A number of university consortia, corporations, professional groups, and other stakeholders have developed and/or are developing frameworks and software for managing grid computing projects. The European Community, for example, is sponsoring a project for grid-based high-energy physics, earth observation, and biology applications. In the United States, the National Center for Supercomputing Applications at the University of Illinois at Urbana–Champaign has demonstrated a prototype of a national computational grid called the National Technology Grid; this grid enables the science and engineering community to reap the benefits of high-performance computing and communications technologies and makes these developments available to broad sectors of society [96]. Table 2.2 identifies some of the grid/P2P initiatives.

As already noted, grids are not just addressing unused or underutilized processing power across existing computers. Another type of grids is a data grid, which can be used to aggregate underused or unused storage into a larger virtual data store; these storage configurations lead to improved performance and reliability over that of any single machine [130]. Table 2.2 also identifies some of the data grid initiatives.

Table 2.2 Key (scientific) grid/P2P initiatives of recent years (partial list)

Grid Applications
  Access Grid
  APEC Cooperation for Earthquake Simulation
  Australian Computational Earth Systems Simulator
  Australian Virtual Observatory
  Cellular Microphysiology
  DataGRID—WP9: Earth Observation Science Application
  Distributed Proofreaders
  DREAM Project: Evolutionary Computing and Agents Applications
  EarthSystemGrid
  Fusion Collaboratory
  Geodise: Aerospace Design Optimisation
  Globus Applications
  GRid seArch & Categorization Engine (GRACE)
  HEPGrid: High Energy Physics and the Grid Network
  Italian Grid (GRID.IT) Applications
  Japanese BioGrid
  Knowledge Grid
  Molecular Modelling for Drug Design
  NC BioGrid
  NEESgrid: Earthquake Engineering Virtual Collaboratory
  Neuro Science—Brain Activity Analysis
  NLANR Distributed Applications
  OpenMolGrid
  Particle Physics Data Grid
  The International Grid (iGrid)
  UK Grid Apps Working Group
  US Virtual Observatory

P2P Integrated Systems and Applications
  Bayanihan Computing Group
  Cetacean acoustic communication study
  DALiWorld
  Distributed Particle Accelerator Design
  Distributed.net
  Evolutionary@Home
  FightAIDS@Home
  Folderol—Bringing the Human Genome Project to the Desktop
  Folding@home
  Genome@home
  Great Internet Mersenne Prime Search (GIMPS)
  Life Mapper
  Moneybee: Stock forecasts
  SaferMarkets.org—Understanding and Predicting Market Volatility
  Server-less Video on Demand
  SETI@home: Search for Extraterrestrial Intelligence at Home
  XPulsar@home

Table 2.2   Continued

Data Grid Initiatives
Datacentric Grid
DIDC Data Grid Work
EU DataGrid
GridPP
GriPhyN (Grid Physics Network)
HEPGrid (High Energy Physics and Grid Networks)
Particle Physics Data Grid (PPDG)
Virtual Laboratory: Tools for Data Intensive Science on Grid

Our emphasis going forward is on commercial, IT, and data center applications of grid computing. As implied by the discussion above, in the past it was science that drove interest in grid computing. For example, in the United Kingdom, interest in grid computing eventually led to the government's creation of an e-science program focusing on the use of computer technology to share resources and collaborate on approaches. In due course, however, it became clear that industry leaders would need to play a role in realizing the vision of utility computing; for example, it was realized that there were business-centric deficiencies in the grid model, specifically related to database interoperability [7]. Vendors now are actively engaged in developing business-centric grid solutions, products brought forth by commercial software and database suppliers to facilitate mainstream deployment of grid computing services.

Early adopter commercial customers are reportedly turning to grid technologies to help them improve the utilization and responsiveness and reduce the cost of their IT assets. For example, with the advances in grid computing, many data-mining, pattern-detection, and scenario-modeling processes of interest to banks, credit card companies, and financial institutions can be implemented more readily. The necessity of such advanced information-based approaches is driven by the increase in financial transaction flow, the need to better understand customer profitability, and the pressure to more effectively manage risks. Sophisticated risk modeling done in real time, such as Bayesian knowledge-based analysis, fuzzy logic, and Monte Carlo simulations are now commonplace in financial firms [130].

Financial services companies have traditionally built their own distributed-processing environments and looked at parallel computing as a way to avoid the purchase of supercomputers. However, up to now these companies have been limited to solutions tailored to specific products or specific business lines; they have, in general, not been able to take advantage of the full potential of a virtualized solution. The potential promise of grid computing includes higher resource utilization, multifold increase in processing power and throughput, lower costs, and faster time to deployment of products and services. Reportedly, grid products from providers such as IBM, Sun Microsystems, Oracle, EMC/VMWare, and Platform Computing, among others, have been tested and deployed by a number of Fortune 500 companies. Firms that were once taking a low-tech approach to parallel computing now can avail themselves of grid middleware. With these evolving grid computing tools, companies are no longer required to manually subdivide algorithms (needed to solve some business problems) via some preconditioning process, to run these algorithms on separate machines, and then to manually merge and integrate the results [130].

Some illustrative examples of actual commercial grid deployments are provided in Table 2.3. The examples, and others like them, provide just some of the motivations and commercial precedents for exploring the opportunities afforded to Fortune 500 companies by grid computing. As the grid matures, standard technologies are emerging for basic grid operations. In particular, the community-based, open-source Globus Toolkit is being utilized by most major grid projects [43, 113]. It is expected that in the next few years the grid computing industry will continue to introduce higher levels of standardization, virtualization, and automation, all of which will not only increase utilization of existing physical resources but also, and more importantly, simplify the management of a firm's IT infrastructure [130].

2.2   BRIEF HISTORY OF COMPUTING, COMMUNICATIONS, AND GRID COMPUTING

The implication was already given that grid computing is not fundamentally a new concept. It has been in use for a number of years at scientific research and development organizations for the most computer-intensive applications, such as, but not limited to, aerospace simulation, circuit design, and human DNA sequencing. Server farms and parallel processing are early precursors to today's modern grid technology; however, grid computing capabilities are available today for commercial use based on more powerful enabling technologies such as network bandwidth, faster processors, and advances in grid middleware software, all at reasonable costs to the end users [130].

As we stated in Chapter 1, grid computing could properly be called "network computing." As this term would imply, the discipline deals with communications and with computing. Figure 2.1 depicts some key phases in recent developments in networking and computing. Figure 2.2 calls out some discrete milestones leading to the emergence of peer-to-peer networks and computational grids. Compared to the history of the electrical power grid, which spans more than two centuries, the computational grid, or rather, the entire computer communication infrastructure, the Internet—has a history of less than half a century [94]. Our emphasis in this book being on the commercial side, we forgo here a full bibliographic study of the academic research over the past decade that has led to developments in grid computing, but refer the interested reader to other references for this information (the resources of Table 2.1 and Table 2.2 are a place to start, among other resources).

A review of recent history shows that standards provide major impetus for widespread acceptance and deployment of a technology. The state of grid computing today may remind one of the early days of the Web, or even of the emergence of XML (eXtensible Markup Language) and Web services; things began slowly, but

Table 2.3   Commercial grid product deployments (partial list)

Grid Middleware	Condor
	Cosm P2P Toolkit
	Globus Toolkit
	Grid Datafarm
	Gridbus
	GridSim: Toolkit for Grid Resource Modeling and Scheduling Simulation
	Jxta Peer to Peer Network
	Legion: A Worldwide Virtual Computer
	PUNCH
	Simgrid
Grid Systems	Amica
	Bayanihan
	Compute Power Market
	CrossGrid
	DAMIEN
	DIET
	Echelon: Agent Based Grid Computing
	Global Operating Systems
	GridLab
	Harness Parallel Virtual Machine Project
	JAVELIN: Java-Based Global Computing
	Management System for Heterogeneous Networks
	MetaNEOS
	MILAN: Metacomputing In Large Asynchronous Networks
	MOBIDICK
	MultiCluster
	NeuroGrid
	Poland Metacomputing
	PUNCH—Network Computing Hub
	XtremWeb
Grid Schedulers	AppLeS
	Computing Centre Software (CCS)
	Condor/G
	DISCWorld
	NetSolve
	Nimrod/G Grid Resource Broker
	SILVER Metascheduler
	ST-ORM
Grid Portals	ActiveSheets
	Enginframe
	G-Monitor
	Grid Enabled Desktop Environments
	Gridscape
	Interactive Control and Debugging of Distribution- IC2D
	Lecce GRB Portal

(continued)

Table 2.3 *Continued*

Grid Portals (*cont.*)	NLANR Grid Portal Development Kit SDSC GridPort Toolkit UNICORE—Uniform Interface to Computing Resources
Grid Programming Environments	Albatross: Wide Area Cluster Computing Cactus Code GAF3J—Grid Application Framework for Java GrADS: Grid Application Development Software Project Jave-based CoG Kit MetaMPI—Flexible Coupling of Heterogenous MPI Systems Nimrod—A tool for distributed parametric modeling Ninf ProActive PDC REDISE—Remote and Distributed Software Engineering Virtual Distributed Computing Environment
Grid Performance Monitoring and Forecasting	NetLogger Network Weather Service Remos
Grid Testbeds and Developments	Alliance Grid Technologies Asia Pacific Bioinformatics Network EuroGrid GrangeNet G-WAAT I-Grid Internet Movie Project Irish Computational Grid (ICG) Kerala Education Grid LHC Grid Micro Grid N*Grid Korea NASA Information Power Grid (IPG) Nordic Grid NPACI: Metasystems OurGrid Polder Metacomputer TeraGrid ThaiGrid The Alliance Virtual Machine Room The Distributed ASCI Supercomputer (DAS) World Wide Grid (WWG)
Grid and P2P Commercial Companies	Avaki CapCal Centrata DataSynapse Distributed Science Elepar EMC/VMWare

Table 2.3 *Continued*

Grid and P2P (*cont.*) Commercial Companies	Entropia.com
	Grid Frastructure
	GridSystems
	Groove Networks
	HP
	IBM
	Intel
	Jivalti
	Mind Electric
	Mithral
	Mojo Nation
	NewsToYou.com
	NICE, Italy
	Noemix, Inc.
	Oracle
	Parabon
	Platform Computing
	Popular Power
	Powerllel
	ProcessTree
	Sharman Networks Kazza
	Sun Gridware
	Sysnet Solutions
	Ubero
	United Devices
	Veritas
	Xcomp

once standards and tools appeared and coalesced, growth quickly ensued [90]. Until recently, all grid computing systems have been situation specific. If one installed the distributed.net client, one could process or access work from the SETI@Home grid. One could not deploy a United Devices client solution without also using their distribution and management system [123]. Fortunately, grid standards, frameworks, open implementations, and off-the-shelf applications are now evolving rapidly. Recently, grid computing has started to leverage Web services to define

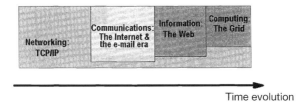

Time evolution

Figure 2.1 Major phases in computing and communications.

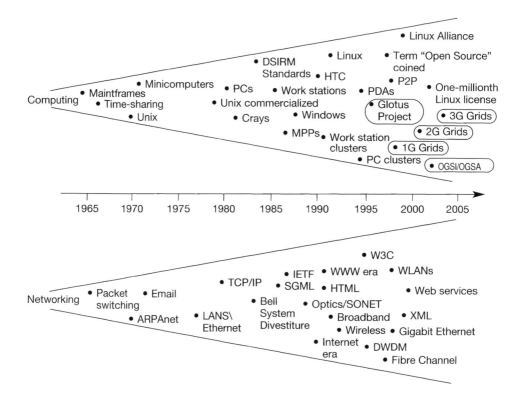

Figure 2.2 Recent history of computing and communications.

standard interfaces for business services like business process outsourcing. (Web services have emerged in the past few years as a standards-based approach for accessing network applications.)

Mainframes appeared in the 1960s. They were (and still are) massive computational devices that were (and continue to be) the purview of large corporations, government agencies, and university labs. The origins of grid computing are grounded in these early days of computing, when using the "spare" machine cycles was seen as an efficient and cost-effective way of getting the most out of what was then very expensive hardware. SInce mainframes cost hundreds of thousands of dollars (millions, in today's dollars), every second had to be accounted for, and those otherwise "wasted" cycles could be used to get the most out of the cost [123]. Table 2.4 provides, for perspective purposes, an excerpt from a quite visionary document of 1970.

The idea of network-based computing has advanced over the years, from the 1970s to the present day. For example, in 1987 this researcher was advocating the concept of grid computing, stating:

The proposed service provides the entire apparatus to make the concept of the Computing Utility possible. This includes as follows: (1) the physical network over which

Table 2.4   Grid computing had an illustrious start in the networking space

Dave Walden, *A Note on Interprocess Communication in a Resource Sharing Computer Network,* Request for Comments: 61 Bolt Beranek and Newman, Network Working Group, July 17, 1970

The attached note is a draft of a study I am still working on. It may be of general interest to network participants.

Resource Sharing Computer Network

"A resource sharing computer network is defined to be a set of autonomous, independent computer systems, interconnected so as to permit each computer system to utilize all of the resources of each other computer system. That is, a program running in one computer system should be able to call on the resources of the other computer systems much as it would normally call a subroutine." This definition of a network and the desirability of such a network is expounded upon by Roberts and Wessler.[1]

The actual act of resource sharing can be performed in two ways: in a pairwise ad hoc manner between all pairs of computer systems in the network or according to a systematic network wide standard. This paper develops one possible network wide system for resource sharing.

I believe it is natural to think of resources as being associated with processes[2] and therefore view the fundamental problem of resource sharing to be the problem of interprocess communication. I also share with Carr, Crocker, and Cerf the view that interprocess communication over a network is a subcase of general interprocess communication in a multiprogrammed environment.

These views pervade this study and have led to a two part study. First, a model for a timesharing system having capabilities particularly suitable for enabling interprocess communication is constructed. Next, it is shown that these capabilities can be easily used in a generalized manner which permits interprocess communication between processes distributed over a computer network.

This note contains ideas based on many sources. Particularly influential were 1) an early sketch of a Host protocol for the ARPA Network[1,2,3] by W. Crowther of Bolt Beranek and Newman Inc. (BBN) and S. Crocker of UCLA; 2) Ackerman and Plummer's paper on the MIT PDP-1 time sharing system[5]; and 3) discussion with R. Kahn of BBN about Host protocol, message control, and routing for the ARPA Network. Hopefully, there are also some original ideas in this note. . . .

[1]L. Roberts and B. Wessler, Computer Network Development to achieve Resource Sharing, Proceedings 1970 SJCC.
[2]V. Vyssotsky, F. F. Corbato, and R. Graham, Structure of the MULTICS Supervisor, Proceedings 1965 FJCC.
[3]C. Carr, S. Crocker, and V. Cerf, Host/Host Communication Protocol in the ARPA Network, Proceedings 1970 SJCC.
[4]F. Heart, et al, The Interface Message Processor for the ARPA Computer Network, Proceedings 1970 SJCC.
[5]W. Ackerman and W. Plummer, An Implementation of Multi-processing Computer System, Proceedings Gatlinburg Symposium on Operating System Principles.

the information can travel, and the interface through which a guest PC/workstation can participate in the provision of machine cycles and through which the service requesters submit jobs; (2) a load sharing mechanisms to invoke the necessary servers to complete a job; (3) a reliable security mechanism; (4) an effective accounting mechanism to invoke the billing system; and, (5) a detailed directory of servers. Currently no such service exists in the public domain ... security and accounting ... are much more complex in the distributed, public (grid) environment. ... This service is basically feasible once a transport and switching network with strong security and accounting (chargeback) capabilities is deployed. A high degree of intelligence in the network is required ... a physical network is required ... security and accounting software is needed ... protocols and standards will be needed to connect servers and users, as well as for accounting and billing. These protocols will have to be developed before the service can be established. ...

The 1970s and 1980s saw the emergence of minicomputers, microcomputers, and desktop machines that gave computing power to an expanding community of people at work and in their homes. This was followed by the client–server computing model in the early 1990s that placed some amount of the functionality (e.g., presentation services) at the distributed endpoints; specifically, at the clients. Networking technologies and protocols to interconnect all these machines together and allow them to communicate saw major deployments throughout the 1980s and early 1990s. The 1990s saw the rise of commercialized Internet; the Internet expanded our ability to communicate and share files and data with any networked machine, regardless of physical location. Now we are turning the corner on the next thing: grid computing. Advocates claim it has as much potential for changing the way we do business as the Internet did [143].

Communication

The de-facto computational grid's (scientific grid's) communication infrastructure is the Internet. The Internet began as a research network supported by the U.S. Department of Defense's Advanced Research Projects Agency (DARPA). (For corporate-oriented grids, high-capacity high-quality extranets and/or private networks can be utilized, particularly, for more secure applications and real-time, high-performance applications.) In 1969, the ARPAnet consisted of four nodes: the University of California, Los Angeles; Stanford Research Institute; University of California, Santa Barbara; and the University of Utah. By the mid-1970s, ARPAnet's reach encompassed over 30 universities, military sites, and government contractors, and its user base had expanded to include the computer science research community at large. By the way, in 1974 the Transmission Control Protocol was introduced, which later (1978) was split into TCP/IP [94]. In 1983, the ARPAnet consisted of several hundred computers.

In the 1980s, the National Science Foundation created the NSFnet. NSFnet was a communications network intended to give scientific researchers easy access to the NSF's supercomputer centers. In 1985, NSF arranged with DARPA to support a collaboration of supercomputing centers and computer science researchers across ARPAnet. Very quickly, one network after another linked in, and the result was the Internet as we now know it [97]. In 1986, the Internet Engineering Task Force (IETF) formed as a loosely self-organized group of people who contributed to the engineering and evolution of Internet technologies. In 1989, responsibility for and management of ARPAnet officially passed from military interests to the academically oriented NSF.

The World Wide Web—developed in the late 1980s by Tim Berners-Lee and his team as a way to share information—energized a major revolution in computing. HyperText Markup Language (HTML) (based on the Standard Generalized Markup Language, developed in the mid 1980s), provided a standard-based means of creating and organizing documents; HyperText Transfer Protocol (HTTP), browsers, and servers provide a mechanism to link these documents and access them online transparently, regardless of their location. The World Wide Web Consortium, formed in 1994, develops new standards for information interchange. For example, initiatives related to XML aim to provide a framework for developing software that can be delivered as a utility service via the Internet [94].

SONET/DWDM (Synchronous Optical Network/Dense Wavelength Division Multiplexing) optical technology deployed in many industrialized nations in the late 1990s and early 2000s now provides broadband connectivity and services at a reasonable price. Some corporate Wide-Area Networks (WANs) now already operate at 155 Mbps, three orders of magnitude faster than the state-of-the-art 56 kbps that connected U.S. supercomputer centers in the mid 1980s. OC-48 rings (2.4 Gbps) have been deployed by some Fortune 50 companies since the mid 1990s. This bandwidth availability is now a key driver (enabler) of grid computing. But to work with labs across the world on petabyte data sets, scientists now require even more—in the range of tens of gigabits per second (Gbps) [113].

During the past few years, the theoretical performance of wide-area networks has doubled every 12 months or so, supported by innovations in optoelectronic technologies (this equates with a potential to increase by two orders of magnitude every five years; however, the user-affordable, commercially available bandwidth has grown at a much slower rate). The NSFnet network, which, as noted, connects the National Science Foundation supercomputer centers in the United States, exemplifies this trend. In the mid-1980s, NSFnet's backbone operated at a DS0 rate (56 Kbps); now, the centers are connected by the 40 Gbps TeraGrid network [113]. Communication speed will likely continue to increase over time and costs will continue to decrease. But planners need to realize that quality high-speed bandwidth will never be free. At some point in the future, there may be an attractive fixed price like $99.99 per month for all you can use up to 10 Mbps, or $199.99 per month for all you can use up to 100 Mbps, or $399.99 per month up to 1 Gbps. This is theoretical, because, although, one can go from 32 lambdas to 64 lambdas to 128 lambdas in a Dense Wavelength Division Multiplexer product, this bandwidth is not generally available to the typical user. An economically viable model needs to be developed by the carriers to dispense bandwidth. (This author did not conceive in the early 1970s that bottled water would ever become the business that it did in the United

States, and that a liter of water would cost up to $2 retail. In order to maintain a supply, a business needs to be profitable.)

Computation

The concept of sharing distributed resources is not new. Since the late 1960s, much work has been devoted to developing distributed systems, but with mixed success. In 1965, MIT's Fernando Corbató and the other designers of the Multics operating system envisioned a computer facility operating "like a power company or water company"; and in the 1968 article "The Computer as a Communications Device," J. C. R. Licklider and Robert W. Taylor foresaw grid-like scenarios [56, 58, 113]. Hence, the idea of harnessing unused machine cycles emerged in the late 1960s and early 1970s, when computers were first linked by data communication networks. ARPAnet supported early experiments with distributed computing.

In 1973, the Xerox Palo Alto Research Center developed a worm program that roamed among about 100 Ethernet-connected computers, replicating itself in each machine's memory. Each worm used idle resources to perform a computation and could reproduce and transmit clones to other nodes of the network. With the worms, developers distributed graphic images and shared computations for rendering realistic computer graphics [94].

Since the1990s, distributed computing has reached a new, global level, as briefly described in the previous sections. The availability of powerful PCs and workstations and high-speed networks (such as Gigabit Ethernet) as commodity components has led to the emergence of clusters for high-performance computing. The availability of such clusters within many organizations has fostered a growing interest in aggregating distributed resources to solve large-scale problems of multiinstitutional interest. Computational grids and peer-to-peer computing are the results of these initiatives.

In 2002, the National Science Foundation installed hardware for the TeraGrid, a transcontinental supercomputer system that is expected to do for computing power what the Internet did for documents. To start with, clusters of high-end microcomputers were set up at four sites: the National Center for Supercomputing Applications at the University of Illinois at Urbana–Champaign; the U.S. Department of Energy's Argonne National Laboratory near Chicago; Caltech in Pasadena, CA; and the San Diego Supercomputer Center at the University of California, San Diego. These four clusters are to be networked together so tightly that they will behave as a single entity. This virtual computer will work through problems at up to 13.6 TFLOPS, eight times faster than the most powerful academic supercomputer available at the time of this writing [97].

Today, a combination of technology trends, technical advances, and standardization makes it feasible to start to realize the grid vision, for both commercial as well as scientific applications. In scientific circles, researchers hope to put in place a new international scientific infrastructure with tools that, in aggregate, can meet the challenging demands of 21st-century science. Numerous government-funded R&D projects are variously developing core technologies, deploying production grids, and applying grid technologies to challenging applications [113].

Grid Technology

The "modern" history of grid computing goes back to the 1996–1999 time period. In this phase, one saw extensive application experimentation and the development of some core grid protocols. Globus Toolkit 1.0 represented the "state of the art" in grid computing at that time. Data grids appeared starting in 1999 with Globus Toolkit 2.0+. This phase afforded medium-scale data management and analysis. The next phase came with the Open Grid Services Architecture, starting in 2001. This phase is represented by the Globus Toolkit 3.0 product; it saw the integration with Web services and resource virtualization (Web services support a standards-based approach for accessing network applications.) This phase also brought forth a number of higher-level services. The problem with grids that had emerged in recent years related to a lack of systemization; namely, lack of a common vocabulary, lack of common interfaces or APIs, lack of common intercommunication protocols, and lack of a common infrastructure formulation. Newly emerging open grid services establish a common vocabulary and a systemization of concepts. At the same time it was realized that there are natural similarities with Web services at the "lower layers," and, hence, standardization efforts sought to avoid reinvention at these layers. The latest phase (from 2003 forward) is characterized by more extensive standardization, ubiquitous computing (including wireless transport and sensors). This last phase is the one that will define the true commercialization of grid computing. However, some conservative players do not expect full ubiquitous grid technology deployment on a broad scale in corporate America until the decade of the 2010s. Figure 2.3 shows an evolution path for the technology.

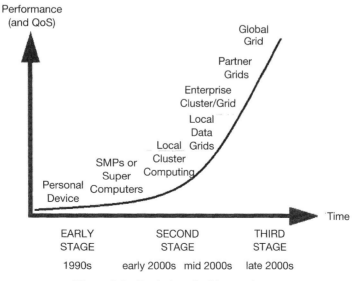

Figure 2.3 Evolution of grids over time.

2.3 IS GRID COMPUTING READY FOR PRIME TIME?

In the previous sections, we highlighted the status of grid computing as a discipline. This section looks at the readiness of the technology for mainstream corporate applications, by citing some examples of currently available products and/or initiatives.

Some market research firms indicate that there is substantial data pointing to acceleration in the grid computing market, with commercialization heading toward an inflection point [1]. The period 2005–2006 is seen by industry stakeholders as a critical period of market development for grid computing technologies. Over this period, the commercial viability of the technology is expected to mature and early adopter customers are expected to give way to broader adaptation of grids for enterprise applications both at single-site and multisite installations. The nature of competition is expected to mature as vendors integrate grid computing technologies into existing offerings and strategies ranging from utility computing to Web services [1]. In a 2003, according to a Gartner European survey, 50% of respondents were aware of the IT utility model for outsourcing that promotes IT infrastructure services from service providers as a commodity. Once market maturity is reached, the IT utility model is likely to support organizations looking to respond with agility to emerging market demands. The utility model has been in existence for quite a few years now. The big companies have utilized this model in conjunction with the offshore service providers; this has resulted in documented reduced costs of operations for these companies. In the meantime, though, customers remain unsure about issues such as cost, security, and integration with existing IT systems [34].

The industry has generally expressed views such as these: "With Grid Computing, businesses can optimize computing and data resources, pool them for large capacity workloads, share them across networks, and enable collaboration" [43]. And, "It's called grid, utility, or on-demand, but it's all about the same thing: Creating computing infrastructures that can dynamically change tasks as processing needs ebb and flow. It's a grand vision, but getting there won't be easy" [100]. The opportunity exists to virtualizing one's computing environment, automate it, and integrate the business processes and information, so that one will have an on-demand operating environment that can transform one's business [90]. However, the observation at the conclusion of the previous section should also be kept in mind as a reference point.

As a counterbalance to the (hopefully not Pollyannaish) arguments made by the proponents, the following observation is worth noting, particularly the concluding punchline:

> IBM . . . is taking an interesting tack in the battle to provide customers the tools to move to more nimble business practices. . . . IBM's holistic approach is refreshing. It gives customers a number of options in terms of where to start in creating what it calls an "on demand" environment. While it does not dictate specific technology, it focuses on positioning technology as the tool and business process as the enabler. . . . However, even more important is the need for technology (the tool) to not be overly complex, proprietary, or costly. Put simply, it cannot be in the way. Therefore, the heterogeneous platforms, multiple form factors, diverse applications, and different management tools already installed in customer sites cannot be barriers. They must be brought

together efficiently. Customers are clear on a few points. They need to become more efficient, and they need to maintain innovation. . . . Although the messaging around initiatives is improving, the problems are as complex and varied as ever. Looking forward, this will be the yardstick for all of these sorts of utility-like initiatives: How cost effectively and efficiently were my business problems addressed? Everything else is just technology. [87].

Leading customers in the financial services, banking, telecommunications and education industries are deploying grid computing to address their mission-critical business challenges [43]. Table 2.5 identifies recent business applications of grid technology [111]. Figure 2.4, synthesized from a number of sources, including but limited to [111], depicts possible penetration trajectories.

Corporate IT planners can build an enterprise grid infrastructure at this time. Planners can use both open-source and vendors' proprietary tools and products. Over time, as the grid standards solidify, one can expect vendors to enable their tools to comply with the new standards, making it easier for the planner to combine components that will work together [90]. The first step in creating a grid is to transform individual computers, network elements, and storage systems into an aggregated and virtual pool of resources that can be allocated and monitored automatically, and whose usage can be metered accordingly. Provisioning of defined business services running on the grid would then take place according to specified goals and priorities. Business requirements are further used to develop and automate policies and service-level objectives that will manage the applications and the resources they need across the network [130]. Grid tools now available can be classified into these general categories [90]:

- Infrastructure. Infrastructure components include file systems, schedulers, resource managers, messaging systems, security applications, certificate authorities, and file transfer mechanisms (e.g., GridFTP) (see Chapter 6).
- Middleware. Software plug-ins that facilitate the use of grid technology. For example, the open source Globus Toolkit 3.0, a mature set of tools useful for building a grid, is the first full-scale implementation of the OGSI standard. It has the basic facilities for implementing a simple yet world-spanning, grid [147]. The tool's strength is a good security model with a mechanism for hierarchically collecting data about the grid. The toolkit was developed by the Globus Project, a research and development project focused on enabling the application of grid concepts to scientific and engineering computing. The toolkit is a set of services and software libraries to support grids and grid applications. The Globus Toolkit 3.0 includes software for security, information infrastructure, resource management, data management, communication, fault detection, and portability [90].
- Directory services. Applications and systems on a grid system must be capable of discovering what services are available to them; specifically, in order to share and collaborate, grid systems must be able to define (and monitor) the grid's topology. Grid director'y services implementations are generally based on the Lightweight Directory Access Protocol (LDAP) and Domain Name Server (DNS).

Table 2.5 Examples of recent applications of grid technology

Type of Grid	Purpose	Business Impacts	Broadband Networking Investment Impacts	Examples
Data Grid	"Transport link"—connects databases at different locations in a single company. Can be built behind a Web services portal (access point).	Significant savings in finding information. Efficiency gains due to shortening the time R&D or design staff need to find information.	Large investment in broadband links to connect data centers that are in different locations.	• Eli Lilly—Data Grid between Sphinx Labs in NC and other R&D labs. • AstraZeneca—connects R&D centers in Sweden, UK and US. • Bank of America—built links between data storage centers to support its free checking over the Internet.
Cluster Computing/ Computational Grid	"Processing power"—harnesses power of computer to achieve high computing speed.	Big saving in processing time. Adds to efficiency by providing for greater output. Savings on R&D and design costs.	No initial impact on broadband until cluster computing evolves to an enterprise grid.	• Pratt & Whitney saved 50% in engineering time on engine projects (Platform Computing) • Novartis speeded drug lead identification by a factor of 10. • Oxford University—anthrax screen of billions of possible drug compounds in 24 days.
Enterprise Grid	"Processing power + transport" within a single firm—links R&D centers at different geographical locations.	Efficiency due to processing power plus access to data. Savings on R&D time and time to market. Upside in terms of greater output/sales.	Investment in broadband links—can require very high speeds due to large amount of data transmitted.	• AstraZeneca—Sweden, UK and US. • GM, Daimler Chrysler, Ford—links to engineering groups for design, mostly in Europe.
Partner Grid	"Processing power + transport" for more than one firm.	Savings in design time and R&D time, plus time to market. Permits more efficient collaboration between partners, often in supply chain relationship.	Significant investment in secure, high-performance, broadband links between two or more firms.	• GM, Daimler Chrysler, Ford—links to engineering groups at partner firms for design, mostly in Europe.
Web Services	Provide secure Internet access to new services for consumers and businesses. Seem to develop closely with cluster computing and data grids.	Big gains in productivity. Big savings in the cost of offering services and time to bring new services to market. Requires building a "data grid-like" structure to provide rapid updating of information.	Large amount of spending on broadband to link data centers. Significant spending on software and integration services.	• Bank of America's banking over the Internet that relies on Web services has resulted in linking a significant number of data centers.

Source: Grid Computing—Projected Impact on North Carolina's Economy and Broadband Use Through 2010 [111].

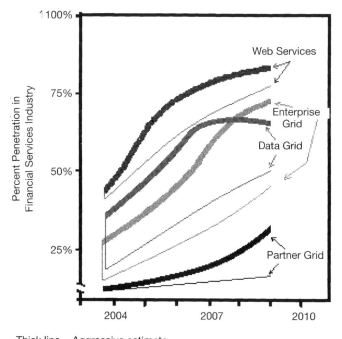

Thick line = Aggressive estimate
Thin line = Conservative estimate

Figure 2.4 Possible penetration of grids in the financial services industry.

- Schedulers and load balancers. One of the main benefits of a grid is maximizing efficiency. Schedulers and load balancers provide this function along with other functions. Schedulers ensure that jobs are completed in some desired order (priority, deadline, urgency, for instance) and load balancers distribute tasks and data management across systems to decrease the chance of bottlenecks.
- Developer tools. Tools for developers of grid-enabled applications focus on different aspects (file transfer, communications, environment control), and range from utilities to application programming interfaces (APIs.)
- Security. Security covers authentication and authorization, intended to control who/what can access a grid's resources. Additionally, security includes message integrity and message confidentiality; these capabilities are crucial to industry segments such as financial and healthcare.

2.4 EARLY SUPPLIERS AND VENDORS

The following discussion is intended to illustrate the state of affairs at press time; this discussion does not sanction one vendor and/or approach. Table 2.6 provides a synopsis of major players and products/strategies. (Although this information is a press-time snapshot, one can safely assume that the companies listed will continue to pursue grid computing initiatives in the future; as time goes by, an augmented set of capabilities and products is expected to emerge.) Figure 2.5 depicts a published taxonomy of grid computing that can be used to organize some of the commercial grid applications that are emerging ([171, 172]).

IBM, a major force in the market, announced a "go-to-market" strategy built around "focus areas" that address the needs of the aerospace, automotive, financial markets, government, life science, agricultural chemical, electronics, higher education, and petroleum industries [43]. These grid offerings are designed to operate in a heterogeneous environment and will incorporate OGSA. IBM Global Services plans to support all elements of a grid implementation with both IBM and non-IBM hardware and software.

By press time, Oracle had introduced grid-enabled enterprise products such as the Oracle Database 10g and the Oracle Application Server 10g [48]. Oracle Database 10g includes management and clustering capabilities that enable the data base to be used in an enterprise grid computing environment [8, 48]. According to the vendor, the Oracle Application Server 10g provides a complete middleware layer

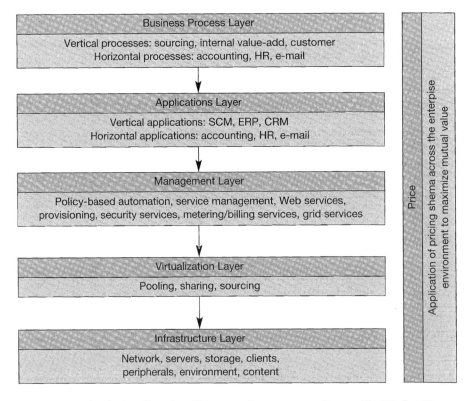

Figure 2.5 The Yankee Group's utility computing taxonomy. Source: The Yankee Group, 2004 [171, 172].

"that transforms a middle-tier infrastructure into a low-cost, efficient, easy-to-manage computing grid." Incorporated into the application server are several new services that allow an administrator to "virtualize" all middle-tier services and resources, managing them from a single console as if they were one.

Sun Microsystems offers Grid Engine software. Described as a distributed resource management tool, Grid Engine allows engineers to pool the computer cycles on up to 100 workstations at a time (some, however, view this more like a cluster and less like a grid [103]). Hewlett-Packard reportedly plans on including grid-enabled software such as the Globus Toolkit and new SGA standard version 3.0 into its servers and storage devices, of course, but also consumer products such as handhelds, PCs, and printers [141]. In Chapter 1, we also (briefly) discussed HP's UDC approach to virtualization.

A more inclusive press-time list of grid products include (but are not limited to) those shown in Table 2.6.

2.5 POSSIBLE ECONOMIC VALUE

Recently, a very valuable study was undertaken by Cohen and Feser to evaluate the possible macroeconomic value of the introduction of grid computing [111]. We highlight a few results here to reinforce the opportunities that may exist and make a rough extrapolation to the national macroeconomic potential. The study is focused on North Carolina, but the results are applicable nationwide.

2.5.1 Possible Economic Value: One State's Positioning

The study [111] estimates that, *given adequate access to broadband infrastructure and sufficient IT workforce,* the deployment of high-performance grid computing and Web services applications would contribute the following gains to North Carolina's economy over 2010 baseline growth forecasts:[3]

- An additional $10.1 billion in output
- An additional 1.5% in aggregate labor productivity
- An additional $7.2 billion in personal income
- An additional 24,000 jobs, the net result of 55,700 new jobs created from increased industrial growth and 31,700 jobs lost due to the adoption of new grid and Web services technologies and downsizing
- An additional $1.2 billion in expenditures for communications services, with 80 to 90% of the new spending devoted to the purchase of broadband access

Several key structural changes in North Carolina's economy can be expected as a result of the adoption of cluster and grid computing and Web services technolo-

---

[3]The remainder of this section is quoted directly from [111].

Table 2.6 A press-time snapshot of products and activities of key suppliers

HP	Utility Data Center (UDC). Focuses on resource management across its server.
IBM	On Demand. By press time, IBM and its partners were offering 19 grid solutions in nine vertical industries (automotive, financial markets, government, and others). The strategy is to grid-enable all of its products. The company states that it will continue to incorporate virtualization technologies into its server software products and plan to incorporate autonomic capabilities into DB2 and associated content-management products. They state they are emphasizing an OGSA compliance approach based on partnering via On Demand Innovation Centers (centers allow users and developers to test their products on IBM's existing On Demand products). The company states that it will continue to incorporate virtualization technologies into its server software products and plans to incorporate autonomic capabilities into DB2 and associated content-management products [43].
Microsoft	Virtual Server is a virtual machine solution for application migration and server consolidation. With Virtual Server, a Windows Server 2003-based server can run multiple operating systems concurrently. The goal is to make it easier to migrate legacy applications. Virtual Server aims at reducing capital expenditures through the use of fewer servers. Virtual Server does not require custom drivers and it does not use any proprietary protocols.
Oracle	10g family of "grid-aware" products focuses on databases. Architects see this as Oracle's own brand of grid computing: a database system that comprises multiple nodes and lets IT planners shift database resources between them. Oracle 10g grid-computing functionality is essentially the latest version of the company's database clustering technology [170].
Platform Computing	Products provide support for Linux on the zSeries mainframe. These include Platform LSF, which is designed to provide on-demand access to an organization's global computing resources and balances workloads across the entire organization; Platform JobScheduler, a software solution that accelerates batch processing by integrating, automating, and grid-enabling silos of applications, jobs, and process flows across distributed computing clusters; and Platform MultiCluster, which allows enterprises to create a single, cohesive computing environment with easy-to-manage resource-sharing policies across geographies.
Sun Microsystems	N1 Data-center Architecture. Approach based on clusters (Sun Grid Engine). N1 is Sun Microsystems' architecture, products, and services for supporting network computing. The marketing angle of the company is that N1 allows "managing *n* computers as 1." The N1 grid system provides the services for managing heterogeneous environments and eliminating the underlying IT complexity through technical means. Ultimately, N1 Grid will encompass multiple organizations [99]. At press time, the company was reportedly shifting focus from technical or high-performance computing markets to the commercial markets; the company was working on technology that better virtualizes desktops by pushing computing power and management capabilities out of the data center and down to blade servers that can better mange desktop computing cycles [166].
EMC's/ VMWare	VMWare. Virtualization technology that aims at lowering the cost of Intel server farms. Converts the workloads of all of a specified set of servers to run as a single hardware pool without dropping any application

gies. First, the impact of the adoption of cluster and grid computing and Web services will ripple through multiple sectors, contributing to shifts in the state's industry mix. Significant growth generated by investment in clusters, grids and Web services by five early adopter industries will contribute to gains in a number of other sectors, with the largest impacts accruing to industries that directly support the deployment of computer applications. Spending will increase for software by $1.13 billion, for computers by $681 million, for professional services by $575 million, and for communications equipment by $432 million over baseline forecasts. That spending, in turn, will lead to the expansion of other industries, such as office equipment, which would post revenues 19% higher than a baseline forecast predicts. At the same time, some industries will be negatively impacted. The estimated output impact of $10.1 billion by 2010 represents $10.9 billion in gains and $800 million in losses across sectors.

Second, increased demand for grids and Web services will be a catalyst for innovation among telecommunications service providers. Firms using grids and Web services will call on telecommunications service providers to develop new services and capabilities, including higher levels of performance and security, and possibly advanced services for customer billing and electronic commerce support. Appropriate innovative responses by service providers will give both the firms they service and the state as a whole a competitive advantage in gaining new business and in attracting the sorts of firms that require those new services and capabilities. That will strengthen North Carolina's position as an early leader in development of leading edge information and communication technologies.

Third, increases in labor productivity derived from the deployment of grid computing and Web services, particularly in early adopter sectors, will reduce firms' business costs and improve efficiency, freeing up capital for spending on new plant and equipment, employee training, and research and development. That will improve the long-run competitiveness of North Carolina industry, in addition to setting off a dynamic wave of expansion and business growth.

Fourth, prices for goods and services are likely to rise much more slowly in an economy where many firms are using grid computing and Web services to improve productivity, making some industries more competitive than might otherwise be expected. Price impacts could be particularly important for the apparel, textiles, and furniture sectors. Industry executives in those industries believe that nearly all firms will have some form of electronic transactions system within four years, involving order entry via an electronic catalog that minimizes color, fabric, and size selection errors, reducing production and retailing costs by 11 and 15%, respectively, by 2005. The cost savings will increase as firms link more of their operations to networks that permit the tracking of every phase in the process of getting a product to the consumer. Although few industries have such networks in place today, some firms in the U.S. and overseas are beginning to build the outline of such systems. One European retailer has used an electronic network to cut the time from design to delivery to 15 days.

Fifth, business practices will change significantly in certain industries, particularly in early adopter sectors. When the new technologies are initially adopted, they are largely time and resource saving. Nevertheless, as firms begin to use them, they begin to consider how the capabilities offered by highly innovative computer and software technologies might provide other benefits. A number of firms have begun to transform business processes in order to develop products more efficiently. For instance, auto assemblers and drug companies have begun to test new products with more sophisticated modeling techniques and very detailed data bases. Some firms are looking at ways they can use the huge increases in computing power that cluster and grid computing and Web services provide to offer far more effective access to corporate data bases. Thus, in drug discovery, pharmaceutical firms are becoming hypercomputerized, using computers and programmed chips to analyze how specific proteins will react with the human body, thus minimizing the need for wet labs. Further down the road, firms using grids expect to begin outsourcing more work to their suppliers, binding those partners even more intimately to their own operations.

Sixth, rural areas will be even more challenged by inadequate access to broadband services. Beyond the need to enhance competitiveness, the very survival of many textile, apparel, and furniture firms is threatened by their location in rural areas of North Carolina where broadband services may be inadequate. In addition, many other rural firms act as suppliers to the large retailers and manufacturers (e.g., Wal-Mart, The Gap, and Ford Motor Company) that are moving to implement grid computing and Web services. Some rural businesses are dealers and distributors for those companies. As the larger firms adopt more sophisticated computer-based supply chains, large retailers and manufacturers are demanding that all businesses they interact with upgrade their Internet access speed and acquire the skills to support new technologies. That will be particularly challenging for many rural companies.

2.5.2   Possible Economic Value: Extrapolation

In Table 2.7, based on modeling, shows that grid computing could add around $382 billion of economic output to the U.S. economy by 2010 and create about 900,000 net new jobs. State economic data based on Bureau of Economic Analysis data were used to make a rough model by normalizing all numbers to the Cohen/Feser study for North Carolina [111]. Straight linearity with regard to IT sizing and ensuing grid services opportunities was assumed. Although this is definitely a crude, first pass at calculating potential national-level numbers, it is first start. Generally, worldwide numbers for IT and telecom are around twice the U.S. numbers (including the U.S.). In Table 2.8, we perform the extrapolation based not on ratios of economic output, but by ratio of high-tech jobs. The results are effectively identical.

2.6   CHALLENGES

Despite the potential advantages of grid computing, it should be noted that the utilization of this technology in the general corporate environment is just nascent. To begin with, like other technologies, a certain amount of hype is inevitable (for example, proponents talk about computer grids, data grids, science grids, access grids,

Table 2.7 First-pass estimate of economic value of grid computing*

	2,001 Economic data (latest available)	Percentage	Additional output ($B) in 2010	Additional Jobs in 2010	Additional telecom expenditures ($B) in 2010
New England	549,472	5.89%	$22.52	53,508	$2.68
Connecticut	152,985	1.64%	$6.27	14,898	$0.74
Maine	34,020	0.36%	$1.39	3,313	$0.17
Massachusetts	265,722	2.85%	$10.89	25,876	$1.29
New Hampshire	45,270	0.48%	$1.86	4,408	$0.22
Rhode Island	33,451	0.36%	$1.37	3,257	$0.16
Vermont	18,048	0.19%	$0.74	1,758	$0.09
Mideast	1,741,057	18.65%	$71.35	169,546	$8.48
Delaware	35,745	0.38%	$1.46	3,481	$0.17
District of Columbia	56,077	0.60%	$2.30	5,461	$0.27
Maryland	175,256	1.88%	$7.18	17,067	$0.85
New Jersey	332,897	3.57%	$13.64	32,418	$1.62
New York	766,526	8.21%	$31.41	74,645	$3.73
Pennsylvania	374,500	4.01%	$15.35	36,469	$1.82
Great Lakes	1,434,052	15.36%	$58.77	139,649	$6.98
Illinois	441,797	4.73%	$18.11	43,023	$2.15
Indiana	178,184	1.91%	$7.30	17,352	$0.87
Michigan	297,475	3.19%	$12.19	28,968	$1.45
Ohio	349,331	3.74%	$14.32	34,018	$1.70
Wisconsin	167,299	1.79%	$6.86	16,292	$0.81
Plains	604,905	6.48%	$24.79	58,906	$2.95
Iowa	86,968	0.93%	$3.56	8,469	$0.42
Kansas	80,680	0.86%	$3.31	7,857	$0.39
Minnesota	175,371	1.88%	$7.19	17,078	$0.85
Missouri	167,370	1.79%	$6.86	16,299	$0.81
Nebraska	53,563	0.57%	$2.20	5,216	$0.26
North Dakota	17,757	0.19%	$0.73	1,729	$0.09
South Dakota	23,165	0.25%	$0.95	2,256	$0.11
Southeast	1,994,577	21.37%	$81.74	194,234	$9.71
Alabama	112,026	1.20%	$4.59	10,909	$0.55
Arkansas	63,701	0.68%	$2.61	6,203	$0.31
Florida	446,482	4.78%	$18.30	43,479	$2.17
Georgia	273,876	2.93%	$11.22	26,670	$1.33
Kentucky	110,074	1.18%	$4.51	10,719	$0.54
Louisiana	125,295	1.34%	$5.13	12,201	$0.61
Mississippi	61,527	0.66%	$2.52	5,992	$0.30
North Carolina	246,291	2.64%	$10.09	23,984	$1.20
South Carolina	106,485	1.14%	$4.36	10,370	$0.52
Tennessee	168,412	1.80%	$6.90	16,400	$0.82
Virginia	241,539	2.59%	$9.90	23,521	$1.18
West Virginia	39,012	0.42%	$1.60	3,799	$0.19

(*continued*)

Table 2.7 *Continued*

	2,001 Economic data (latest available)	Percentage	Additional output ($B) in 2010	Additional Jobs in 2010	Additional telecom expenditures ($B) in 2010
Southwest	992,959	10.64%	$40.69	96,695	$4.83
Arizona	153,684	1.65%	$6.30	14,966	$0.75
New Mexico	54,930	0.59%	$2.25	5,349	$0.27
Oklahoma	85,948	0.92%	$3.52	8,370	$0.42
Texas	698,547	7.48%	$28.63	68,025	$3.40
Rocky Mountain	299,089	3.20%	$12.26	29,126	$1.46
Colorado	159,308	1.71%	$6.53	15,514	$0.78
Idaho	36,832	0.39%	$1.51	3,587	$0.18
Montana	20,708	0.22%	$0.85	2,017	$0.10
Utah	63,933	0.68%	$2.62	6,226	$0.31
Wyoming	18,254	0.20%	$0.75	1,778	$0.09
Far West	1,719,594	18.42%	$70.47	167,456	$8.37
Alaska	24,490	0.26%	$1.00	2,385	$0.12
California	1,260,041	13.50%	$51.64	122,704	$6.14
Hawaii	38,839	0.42%	$1.59	3,782	$0.19
Nevada	69,538	0.74%	$2.85	6,772	$0.34
Oregon	124,847	1.34%	$5.12	12,158	$0.61
Washington	202,470	2.17%	$8.30	19,717	$0.99
Total	9,335,705	100%	$382.59	909,121	$45.46

Estimated GDP 2010 at 3.33% CAGR: $12,536,751

*State raw economic data based on Bureau of Economic Analysis (*http://www.bea.doc.gov/bea/news-rel/gdp303p.pdf*). All numbers are normalized to the Cohen/Feser study for North Carolina [111]. Assumption made of straight linearity with regards to IT sizing and ensuing grid services opportunities.

knowledge grids, bio grids, sensor grids, cluster grids, campus grids, tera grids, and commodity grids [103].) A degree of healthy skepticism is warranted. Although vendors make all sorts of claims, many of these claims are just marketing angles. Just deploying a scheduler LAN does not create a "cluster grid," anymore that a workstation with a processor, memory, disk, and network card, is a grid.

At the technical level, although many firms have a keen interest in "virtualization," there are obstacles to full realization of grid computing benefits at this time. Some of these include [44, 130]:

- Applications requiring computing power that rises to the level supported by grid computing generally do not exist in medium-size companies.
- Cost and time needed to rewrite and test applications.
- Enabling resource sharing across distinct institutions. One wants to facilitate coordinated use of diverse resources, including infrastructure resources (certificate authorities, information services). These are expensive to run. Ad-

Table 2.8 U.S. Department of Labor, Bureau of Labor Statistics: High Tech Jobs (code 15-0000 Computer and Mathematical Science Occupations) Occupational Employment and Wages, 2002

State	Occupation code	Total employees	Percentage
Alabama	15-0000	29,120	
Alaska	15-0000	3,210	
Arizona	15-0000	45,630	
Arkansas	15-0000	13,530	
California	15-0000	366,250	
Colorado	15-0000	75,940	
Connecticut	15-0000	42,030	
Delaware	15-0000	10,010	
District of Columbia	15-0000	26,130	
Florida	15-0000	127,820	
Georgia	15-0000	92,630	
Guam	15-0000	380	
Hawaii	15-0000	6,610	
Idaho	15-0000	9,650	
Illinois	15-0000	122,930	
Indiana	15-0000	37,810	
Iowa	15-0000	20,790	
Kansas	15-0000	27,180	
Kentucky	15-0000	23,770	
Louisiana	15-0000	16,480	
Maine	15-0000	7,420	
Maryland	15-0000	90,000	
Massachusetts	15-0000	103,340	
Michigan	15-0000	77,560	
Minnesota	15-0000	69,140	
Mississippi	15-0000	8,450	
Missouri	15-0000	54,050	
Montana	15-0000	4,700	
Nebraska	15-0000	20,280	
Nevada	15-0000	10,150	
New Hampshire	15-0000	11,800	
New Jersey	15-0000	115,990	
New Mexico	15-0000	12,290	
New York	15-0000	169,750	
North Carolina	15-0000	75,140	2.70%
North Dakota	15-0000	4,030	
Ohio	15-0000	94,370	
Oklahoma	15-0000	22,110	
Oregon	15-0000	32,630	
Pennsylvania	15-0000	103,960	
Puerto Rico	15-0000	6,950	
Rhode Island	15-0000	9,300	
South Carolina	15-0000	19,970	

(*continued*)

Table 2.8 *Continued*

State	Occupation code	Total employees	Percentage
South Dakota	15-0000	5,110	
Tennessee	15-0000	36,000	
Texas	15-0000	209,500	
Utah	15-0000	25,210	
Vermont	15-0000	5,220	
Virgin Islands	15-0000	230	
Virginia	15-0000	139,860	
Washington	15-0000	83,870	
West Virginia	15-0000	7,280	
Wisconsin	15-0000	44,760	
Wyoming	15-0000	1,860	
Total		2,780,180	

Note: This major group comprises the following occupations: Computer and Information Scientists, Research; Computer Programmers; Computer Software Engineers, Applications; Computer Software Engineers, Systems Software; Computer Support Specialists; Computer Systems Analysts; Database Administrators; Network and Computer Systems Administrators; Network Systems and Data Communications Analysts; Actuaries; Mathematicians; Operations Research Analysts; Statisticians; Mathematical Technicians; and residual, "All Other," occupations in this major group.

Key observation. Note that the percentage of high-tech jobs in North Carolina is 2.70% on a national basis. Interestingly, this is practically the same ratio (likely, by pure chance) of the total economic output of North Carolina in reference to national GDP (see Table 2.7). This means that an extrapolation based on state economic output to country economic output or based on in-state high-tech jobs to country high-tech jobs leads to the same outcome in terms of the total extrapolated economic value. However, the state-by-state values could be different than reported in the previous table, although we do not undertake that state-by-state extrapolation in this table (which could easily be done).

dressing security and policy concerns of resource owners and users. Resource discovery, access, reservation, allocation; authentication, authorization, policy; communication; fault detection and notification; and so on. Addressing security and policy concerns of resource owners and users.

- Existing systems are performing adequately and no business case (positive net present value) has been developed (which itself could take time and effort).
- Grid operations management challenges. Users, resources, and owners are geographically distributed. Resources, users, and applications are heterogeneous. Resource availability and capabilities vary with time. Policies and strategies are heterogeneous and decentralized. Quality of service (service-level agreements) are heterogeneous. Costs and prices vary based on resources, users, time, and demand [105].
- Grid systems have input/output bandwidth to storage devices dependencies.
- Grid systems have network bandwidth dependencies.
- Grid systems have network latency dependencies.
- Grid systems may suffer from synchronization protocol inefficiencies.

- It is unlikely that any commercially sustainable grid infrastructure will be provided by any nonresearch (nongovernment funded) organization without financial compensation for the use of resources by external users. For grid services to be provided on demand (i.e., to deliver the utility infrastructure that is the ultimate goal of grid computing) "donor" organizations will want to be paid for providing the resources.

- Lack of desire to bring in outside help as code is proprietary (especially true at hedge funds and buy-side firms).

- Lack of in-house technical expertise, especially at smaller to medium-size companies.

- Many firms find that they need to optimize the performance of applications so that they can be properly allocated to jobs running on different machines. This requires a different approach to programming and, therefore, it is not necessarily true that no software changes will be required in order to leverage grid computing, and the existing setup and investment cannot be maintained.

- National security. Perhaps number crunching is fine, but to give away sensitive database information (data grids) to a third-world country could be problematic. A local utility is fine, just like with electric power utilities, but one would not want their electric power to be generated by a third-world country. Some countries (e.g., European countries) have strict privacy laws about transborder data flow.

- Need to find ways to operate efficiently when dealing with large amounts of data and computation. Need for shared infrastructure services to avoid repeated development and installation (e.g., one port/service/protocol for remote access to computing, not one per tool/application).

- The so-called programming problem: how does the planner develop robust, secure, long-lived, well-performing applications for dynamic, heterogeneous grids?

- There are a number of new approaches to problem solving in *addition* to grid computing (in competition with), including cluster computing (especially for supercomputer applications), distributed computing, peer-to-peer, data grids, and collaboration (scavenging) grids.

- There are challenges in structuring and writing programs: parallelizing of applications code in an effective manner can be difficult, even though it is advantageous (for certain type of problems) to do so if/when possible.

- There is a need for interoperability when different groups want to share resources (diverse components, policies, mechanisms; e.g., standard notions of identity, means of communication, resource descriptions).

- There is a need to facilitate the development of sophisticated applications, including code sharing. To address this programming problem, one needs stable programming environments (such as APIs, SDKs).

- Use of third-party applications makes parallelizing code difficult.

In particular, challenges often exist to perfect scalability. In a perfectly scalable environment, if a job running on a 1 GFLOPS machine required 10 seconds to complete satisfactorily, then it would need 1 second if it ran in a parallelized version on ten 1 GFLOPS machines (and/or 1 second if it ran on one 10 GFLOPS machine). One challenge relates to the algorithms used for splitting the application to run among many processors: if the application processes can only be split into a limited number of independently running parts, then this predicament forms a scalability barrier. Another challenge arises if the resulting microjobs are not completely independent (for example, if all of the microjobs need to read and write from one common file or database, the access limits of that file or database will become the limiting factor in the application's scalability. Other types of interjob contention in a parallel grid application include message communications latencies among the jobs, network communication capacities, synchronization protocols, input–output bandwidth to devices and storage devices, and latencies interfering with real-time requirements) [47]. Not all applications can be made to run in parallel on a grid and achieve scalability and there are no all-encompassing utilities for transforming generic applications to a morphed version that can exploit the parallel capabilities of a grid. A limited number of tools exist that skilled designers can use to write a parallel grid application but tools for automatic parallelization of generic applications are not yet available.

Another challenge is that up to now grid computing has been mostly the purview of a handful of researchers at mathematics and computer science departments, national laboratories, informatics institutes, and government-funded research. It turns out, as we have indicated, that this technology can be of value to Fortune 500 companies looking to reduce their run-the-engine costs. One of the challenges of grid computing (which this book aims at addressing) is to graduate the technology beyond the purely academic orbit it has preponderantly held in the recent past and near present.

CHAPTER 3

Components of Grid Computing Systems and Architectures

Chapters 1 and 2 provided an overview of a number of areas that relate to grid computing. This chapter continues the basic discussion that was started in these chapters. We revisit some of the key grid technology issues and then we drill down on the major architectural components of a grid. It should be noted that there is no universal consensus, as of yet, on what the canonical components of a grid should be (given that there are several types of grids from a functional perspective). However, there is general agreement of what some of the high-level fundamental building blocks are. These fundamental building blocks are discussed in this chapter. The chapter provides three views of the components: a functional view, a physical view, and a service view. A planner wishing to use grid principles and wishing to deploy a grid mechanism in his/her Fortune 500 company, will have to deploy and support a number of these components, perhaps all, depending on the application and situation. As standards solidify, a more canonical view of the constituent elements will emerge.

3.1 OVERVIEW

In Chapter 1, it was noted that grid computing embodies a combination of a decentralized architecture for resource management, and a layered hierarchical architecture for implementation of various constituent services [101]. Sharing issues, as they apply to distributed processing, are not adequately addressed by existing technologies because there are complicated user requirements such as "run program X at site Y subject to community policy P, providing access to data at Z according to policy Q." Many "peta-scale" problems or science problems have high performance requirements with unique demands for advanced and high-performance systems. This drives towards the development of grid technology [45]. A grid goes beyond client–server linkage in that it provides distributed data analysis, computation, and collaboration.

A grid allows flexible, secure, coordinated resource sharing among dynamic collections of individuals, institutions, and resources. A grid enables communities ("virtual organizations," which are community overlays on classic organization

A Networking Approach to Grid Computing. By Daniel Minoli
ISBN 0-471-68756-1 © 2005 John Wiley & Sons, Inc.

structures) to share geographically distributed resources as they pursue common goals, postulating the absence of a central location, and/or central control, and/or omniscience, and/or existing trust relationships [103]. Resource sharing in a grid context applies to networks, computers, storage, and sensors. Sharing, however, is always conditional; conditionality issues relate to trust, policy, negotiation, and payments (particularly, unit cost payments). Grids support coordinated problem solving. Grids support dynamic, multiinstitutional virtual organizations; these communities can be large or small, static or dynamic. The grid can be defined at three levels [119]: enterprise (enterprise grid), partner (partner grid), and service (service/utility grid).

Table 3.1 encapsulates the definition of a grid [103]. The three primary types of grids that were introduced in Chapter 1 are summarized in Table 3.2 [125]; it should be noted, however, that there are no restrictive boundaries between these grid types since grids may often be a combination of these basic types. According to Phillip Gill, vendors position grid computing as follows:

A typical middle tier in today's enterprise is a jumble of expensive hardware and devices running Java 2 Platform, Enterprise Edition (J2EE) application servers, HTTP servers, Web caches, portals, and so on. These resources are usually configured for maximum performance but rarely used to maximum efficiency; they often come from different vendors and with different operating systems, making them costly to install, configure, manage, and maintain. Perhaps most importantly, the middle-tier servers lack the flexibility to adapt rapidly to changing business needs. grid computing brings order to this chaos in the middle tier. Originally used for large-scale scientific and re-

Table 3.1 Definition of grid, according to Ian Foster [103]

1. Coordinates resources that are not subject to centralized control . . .	A grid integrates and coordinates resources and users that exist within different control domains (for example, the user's desktop versus central computing; different administrative units of the same company versus different companies) and also addresses the issues of security, policy, payment, and membership, that arise in these settings. (Otherwise, one would be dealing with a local management system.)
2. . . . using standard, open, general-purpose protocols and interfaces . . .	A grid is built from multipurpose protocols and interfaces that address such issues as authentication, authorization, resource discovery, and resource access. It is crucial that these protocols and interfaces be standard and open. (Otherwise, one would be dealing with an application-specific system.)
3. . . . to deliver nontrivial qualities of service.	A grid allows its constituent resources to be used in a coordinated fashion to deliver various qualities of service (relating, for example, to response time, throughput, availability, security, and/or co-allocation of multiple resource types to meet complex user demands), so that the utility of the combined system is significantly greater than the sum of its parts.

Table 3.2 Grid types

Computational grid	This grid is used to allocate resources specifically for computing power. In this situation, most of the processors are high-performance servers. (Note that processors are sometimes called nodes, resources, members, donors, clients, hosts, engines, or machines.)
Scavenging (computational) grid	This grid is used to "locate processors–cycles": grid nodes are exploited for available machine cycles and other resources. Nodes typically equate to desktop computers; a large numbers of processors are generally involved. Owners of the desktop processors are usually given control over when their resources are available to participate in the grid.
Data grid	This grid is used for housing and providing access to data across multiple organizations. Users are not focused on where this data is located as long as they have access to the data.

search computing, the grid computing model increases reliability, scalability, and manageability in the middle tier, all while reducing costs. A grid computing infrastructure turns IT resources—computers, storage, and applications—into a single virtual system that, like a utility—power, water, gas, phone—can be tapped at will, whenever needed. [86]

Grids can be built ranging from just a few processors to large groups of processors organized as a hierarchy that spans a continent or the globe. The simplest grid consists of just a few processors, all of which have the same hardware architecture and utilize the same operating system. These processors are connected in a data center on a LAN or storage area network (SAN); see Figure 3.1, top. (It should be noted, however, that some people would call this arrangement a "cluster" implementation rather than a bonafide "grid.") Because this type of grid utilizes homogeneous systems, generally there are relatively few considerations beyond properly deploying the grid-support software. Also, given the fact that the processors have the same architecture and operating system, selecting application software for these processors is usually a straightforward task [147]. Finally, for this "entry-level" grid, the processors are usually in one department of a given organization; because of this framework, the access to the grid may not require (in general) any special security procedures or usage policies.

A next level of complexity is reached when one includes heterogeneous processors in the grid ensemble. These grids are also referred to as an "intragrids" and/or "enterprise grids." As the term implies, processors participating in the enterprise grid may include devices owned and maintained by multiple departments, but still within one firm; see Figure 3.1, middle. The grid may span a number of geographic locations, where computing facilities (e.g., servers) may be located. These larger grids may have a hierarchical topology, although this is not a strict requirement. For example, processors locally connected in a data center or server room form a "cluster" of processors; in turn, the overall enterprise grid may be organized in a hierar-

Cluster/Local Grid

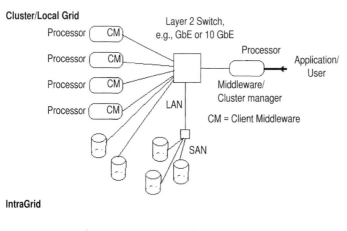

IntraGrid

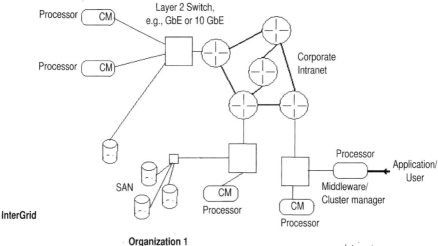

InterGrid

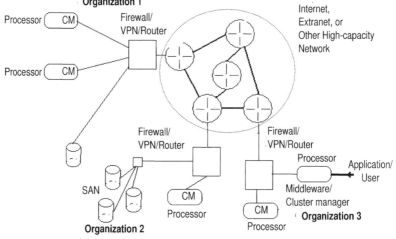

Figure 3.1   Grid types, arranged by complexity.

chy consisting of clusters of clusters. Intranet transmission links (or other high-quality, high-throughput, high-security communication services), are used to interconnect these nodal computing resources and the grid. We noted in Chapter 1 that when local hosts are aggregated in tightly coupled configurations, they tend to generally be of the cluster parallel-based computing type; such processors, however, can also be nonparallel-computing-based grids, e.g., by running the Globus Toolkit. When *geographically dispersed* hosts are aggregated in distributed computing configurations, they tend to generally be of the grid computing type and not running in a clustered arrangement.

Enterprise grids are supported via the organization's intranet. The intranet typically consists of a collection of dedicated frame relay, Asynchronous Transfer Mode (ATM), MultiProtocol Label Switching (MPLS), or IP-based facilities or services. A virtual private network (VPN) service over the Internet may also be used to connect the remote sites of the organization, especially for international applications. In this configuration, additional types of resources are available to the application and/or user besides just basic company-owned nodes; in particular, this grid system will likely include scheduling components. File sharing may still be accomplished using networked file systems. In these enterprise grid environments, dedicated grid processors may also be added by the organization to increase the service levels achieved by the grid, rather than depending entirely on scavenged resources [147]. When the grid expands to encompass discrete departments, operational policies are generally required operational procedures related to how the grid should be used (e.g., what kinds of work is allowed on the grid and at what times). For example, there may be a prioritization by department, or by kinds of applications that should have access to grid resources. Furthermore, security typically becomes important when multiple departments are involved, because sensitive information belonging to one department may need to be protected from access by, and/or intrusion from jobs running for other departments.

Figure 3.1, bottom, depicts what some researchers call a pure grid, or an "intergrid." Such a grid, by definition, crosses organization boundaries. Generally, an intergrid may be used to collaborate on "large" projects of common scientific interest. The most stringent levels of security are usually required in this environment. The intergrid offers the opportunity for sharing, trading, or brokering resources over widespread pools; computational "processor-cycle" resources may also be obtained, as needed, from a utility for a specified fee. Figure 3.2 identifies a number of well-known intergrids and Figures 3.3 and 3.4 depict some specific intergrids.

Table 3.3 summarizes the motivation for and/or purpose of various grids discussed above (partially based on [147].) Table 3.4 identifies some challenges associated with grid computing (also see Chapter 2, Section 2.6).

Some see grid solutions at two levels: *physical* and *logical* [71]. A physical grid refers to computer power and other hardware resources that can be shared over a distributed network. A logical grid refers to software and application sharing, as well as higher-level business-process sharing. Both kinds of grids can coexist in a Grid Computing environment. A physical grid can be used as a component utilized by multiple logical grids; a logical grid can be constructed by utilizing multiple physical grids.

Name	URL & Sponsors	Focus
Access Grid	www.mcs.anl.gov/FL/ accessgrid; DOE, NSF	Create & deploy group collaboration systems using commodity technologies
BlueGrid	IBM	Grid testbed linking IBM laboratories
DISCOM	www.cs.sandia.gov/ discom DOE Defense Programs	Create operational Grid providing access to resources at three U.S. DOE weapons laboratories
DOE Science Grid	sciencegrid.org DOE Office of Science	Create operational Grid providing access to resources & applications at U.S. DOE science laboratories & partner universities
Earth System Grid (ESG)	earthsystemgrid.org DOE Office of Science	Delivery and analysis of large climate model datasets for the climate research community
European Union (EU) DataGrid	eu-datagrid.org European Union	Create & apply an operational grid for applications in high energy physics, environmental science, bioinformatics
EuroGrid, Grid Interoperability (GRIP)	eurogrid.org European Union	Create tech for remote access to supercomp resources & simulation codes; in GRIP, integrate with Globus Toolkit™
Fusion Collaboratory	fusiongrid.org DOE Off. Science	Create a national computational collaboratory for fusion research
Globus Project™	globus.org DARPA, DOE, NSF, NASA, Msoft	Research on Grid technologies; development and support of Globus Toolkit™; application and deployment
GridLab	gridlab.org European Union	Grid technologies and applications
GridPP	gridpp.ac.uk U.K. eScience	Create & apply an operational grid within the U.K. for particle physics research
Grid Research Integration Dev. & Support Center	grids-center.org NSF	Integration, deployment, support of the NSF Middleware Infrastructure for research & education
Grid Application Dev. Software	hipersoft.rice.edu/ grads; NSF	Research into program development technologies for Grid applications
Grid Physics Network	griphyn.org NSF	Technology R&D for data analysis in physics expts: ATLAS, CMS, LIGO, SDSS
Information Power Grid	ipg.nasa.gov NASA	Create and apply a production Grid for aerosciences and other NASA missions
International Virtual Data Grid Laboratory	ivdgl.org NSF	Create international Data Grid to enable large-scale experimentation on Grid technologies & applications
Network for Earthquake Eng. Simulation Grid	neesgrid.org NSF	Create and apply a production Grid for earthquake engineering
Particle Physics Data Grid	ppdg.net DOE Science	Create and apply production Grids for data analysis in high energy and nuclear physics experiments
TeraGrid	teragrid.org NSF	U.S. science infrastructure linking four major resource sites at 40 Gb/s
UK Grid Support Center	grid-support.ac.uk U.K. eScience	Support center for Grid projects within the U.K.
Unicore	BMBFT	Technologies for remote access to supercomputers

Figure 3.2 Well-known intergrids. Copyright © 2002 University of Chicago and The University of Southern California.

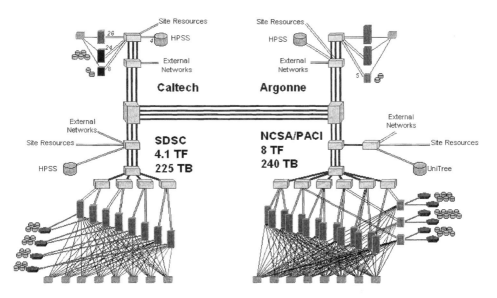

Figure 3.3 The TeraGrid. Copyright © 2002 University of Chicago and The University of Southern California. All rights reserved.

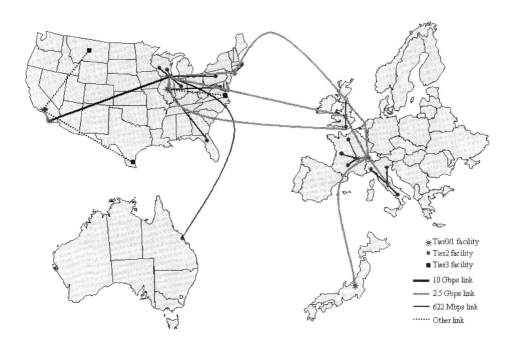

Figure 3.4 The iVDGL. Copyright © 2002 University of Chicago and The University of Southern California.

Table 3.3 Possible grid benefits and applications

Access to a plethora of IT resources	A grid federates a large number of resources contributed by individual resources into a functionally larger, logically virtual resource. Compared to traditional closed or company-specific computing environments, a grid can provide access to a large(r) universe of resources and possibly to special equipment, software, and other services. Scalability can be supported in terms of additional quantity of such resources and/or additional capacity for such resources.
Better utilization of underused resources	Processors, storage, and other resources on a grid are almost invariably better utilized than would otherwise be the case. Organizations typically experience peaks of activity in their monthly, weekly, or daily IT workflow. When the applications are grid-enabled, during usage peaks, these applications can be "moved" (assigned) to underutilized resources (e.g., processors). Furthermore, administrators can utilize grid-support tools to assess usage demand; this, in turn, facilitates improved planning when upgrading systems, increasing system capacity, or retiring end-of-life computing resources.
	Enterprise systems may also have unused storage capacity; data grids can be employed to aggregate such unused storage into a larger virtual data store (larger capacities than available on any single system). This achieves improved performance and reliability compared to a single-threaded processor. Files and databases can seamlessly span many systems; spanning has the potential to improve data transfer rates through the use of striping techniques. Also, data resiliency can be achieved by duplicating it at various points throughout the grid; this data replication serves as a backup and can be hosted on or near the processors most likely to need the data.
Improved availability of computing	An enterprise grid is inherently more resilient and enjoys higher availability than a traditional clustering arrangement. A grid framework (such as Oracle's Application Server 10g) is able to include a tighter integration between the database and the application server from a clustering/failover standpoint. For example, although the database cluster may quickly fail over from one node to another, the middle tier does not typically become aware of the change until a Transmission Control Protocol/Internet Protocol (TCP/IP) timeout triggers a reconnection to the cluster; this timeout could take several minutes. In the grid environment, the database tier notifies the middle tier of the failover, resulting in an immediate reconnect and reducing the total downtime from minutes to seconds [86].
Increased reliability of computing	Compared to traditional environments, a grid provides increased reliability. This is because the constituent nodes in a grid can be relatively inexpensive and can be dispersed geographically. High reliability often is achieved with a relatively high price investment when high-end conventional computing systems are employed. Typically, this hardware is constructed using chips with redundant circuits (these circuits "vote" on results), and the hardware contains additional logic circuitry to achieve graceful recovery from a number of hardware failures.

Table 3.3 *Continued*

Increased reliability of computing (*cont.*)	The systems also use duplicate processors that have hot-pluggability capabilities; when a processor fails it can be replaced without turning the other off. Grid computing, which in effect employs a RAIC-like (redundant array of inexpensive computers) mechanics, provides an alternate approach to reliability that relies more on software technology than expensive hardware.
Parallelization of processing	Many algorithms-based applications can be partitioned into independently running "microjobs." A grid application can be thought of as an aggregate of many smaller "microjobs," each executing on a different processor. For example, a perfectly scalable application will complete 20 times faster if it uses 20 comparable processors.
Resource balancing	For applications that are grid-enabled, the grid infrastructure can offer a resource-balancing capability. This is accomplished by scheduling grid jobs on processors with low utilization. An instantaneous/unplanned demand peak can be handled by routing work requests to relatively idle processors in the grid. If the resources in the grid are already fully utilized, the lowest-priority work being performed on the grid can be temporarily suspended (or even cancelled) to make way for the higher-priority work. (Without a grid infrastructure, such balancing decisions are difficult to prioritize and execute.)
Simplified management of IT resources	Compared to traditional environments, it may be easier to manage a larger dispersed and logically virtualized infrastructure than a plethora of "indigenous" systems. Grid middleware provides a uniform method to handle heterogeneous systems.
Virtual resources and virtual organizations for collaboration	Grid middleware and (open) standards allow heterogeneous systems to work collaboratively to deliver the appearance of a large virtual computing environment, while offering a variety of virtual resources. The users of the grid can be organized dynamically into a number of VOs, each with possibly different policy requirements. VOs can share their resources collectively as a larger grid. Resources are "virtualized" to give them a more uniform interoperability among heterogeneous grid participants.

3.2 BASIC CONSTITUENT ELEMENTS—A FUNCTIONAL VIEW

In this section, we identify, at a high level, the major components of a grid computing system from a functional perspective. Not all of the components discussed herein are needed all the time; depending on the grid design and its expected use, some of these components may not be required, and in some instances these components may be combined. A resource is an entity that is to be shared; this includes computers, storage, data, and software. A resource does *not* have to be a physical entity. A resource is defined in terms of interfaces, not devices; for example, schedulers such

Table 3.4 Partial list of current "drawbacks" of grid computing

Business case not always clear	Grid proponents need to provide a compelling business case [164].
Processes are to be defined	Vendors need to show whether the grid-based process can be effectively managed, including the chargeback model.
Security is to be supported	Security is a concern, particularly for intergrids.
Message/articulation must be crispier	Confusion still exists in industry between cluster computing, virtualization, enterprise grid, intergrids, and P2P. Sharper message needed from vendors.
Proprietary approaches should be eliminated	Leading vendors (e.g., Hewlett-Packard, IBM, Microsoft, Platform Computing, Sun Microsystems, Oracle, VMWare/EMC*) all still approach issue differently and incompatibly. Existing grid computing solutions tend to be limited to an individual vendor's products (IBM's grid computing platform was operating with the largest number of open standards compliance compared with other suppliers at press time) [165].
Parochial focus should be eliminated	For example, IBM's grid computing platform is focused primarily on virtualizing IBM hardware and data bases. Hence, if a firm has a fairly homogenous data center (e.g., running mainly IBM eServers or IBM DB2 data bases) it could benefit from the grid computing solution (likely an enterprise/local grid); otherwise it may not.
Performance to be proven/monitored	Grid computing systems need proper "partitioning" ("zoning") mechanisms to ensure that an application competing for computer resources will not degrade the other applications also looking for resources (particularly in the case of server virtualization).

*IBM, Sun Microsystems, and Hewlett-Packard were known as the "Big Three" in this space at presstime [166].

as Platform's LSF and PBS define a compute resource. Open/close/read/write define access to a distributed file system, for example, NFS, AFS, DFS [3, 45].

In this section, we look at following grid components:

- Grid portal
- Security (grid security infrastructure)
- Broker (along with directory)
- Scheduler
- Data management
- Job and resource management
- Resources

Figures 3.5 and 3.6 provide a pictorial depiction of the concepts discussed in the subsections that follow. Table 3.5 provides a more detailed listing of functions and

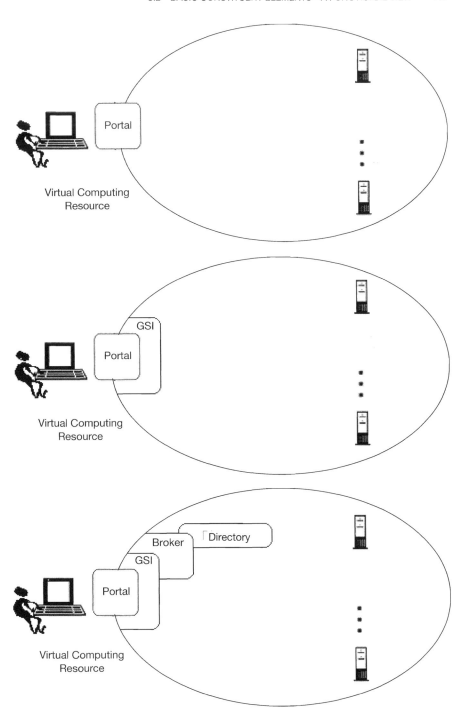

Figure 3.5 Basic grid elements—a user's view.

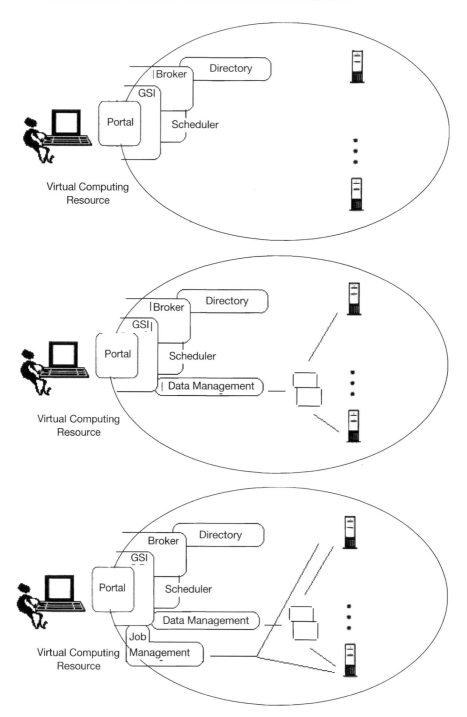

Figure 3.6   Additional grid elements—a user's view.

Table 3.5 Grid functionality to be supported

(Co-)reservation, workflow	Monitoring
Accounting and payment	Performance guarantees
Adaptation	Remote data access
Authorization and policy	Resource allocation
Distributed algorithms	Resource characterization
Fault management	Resource discovery
High-speed data transfer	Resource management
Identity and authentication	System evolution
Intrusion detection	

capabilities to be supported [45]. As a supplement, interested readers may also wish to consult the document, *"Anatomy of the Grid"* by Ian Foster, Carl Kesselman, and Steven Tuecke; the paper contains a description of a grid's constituent parts and what they do, with a focus is on grid architecture [3].

In the discussion below, we use the term "functional block." This is a generic architectural construct (applicable to any architecture). A functional block is a logical aggregation of functions and capabilities that have an affinity, similarity, close relationship, or related purpose.

Portal/User Interface Function/Functional Block

A portal/user interface functional block usually exists in the grid environment. The user interaction mechanism (specifically, the interface) can take a number of forms. The interaction mechanism typically is application specific. In the simplest grid environment, the user access may be via a portal (see Figure 3.5, top). Such a portal provides the user with an interface to launch applications. The applications make transparent the use of resources and/or services provided by the grid. With this arrangement, the user perceives the grid as a virtual computing resource.

The Grid Security Infrastructure: User Security Function/Functional Block

A user security functional block usually exists in the grid environment and, as noted above, a key requirement for grid computing is security. In a grid environment, there is a need for mechanisms to provide authentication, authorization, data confidentiality, data integrity, and availability, particularly from a user's point of view; see Figure 3.5, center. When a user's job executes, typically it requires confidential message-passing services. There may be on-the-fly relationships. But also, the user of the grid infrastructure software (such as a specialized scheduler) may need to set up a long-lived service; administrators may require that only certain users are allowed to access the service. In each of these cases, the application must anticipate and be designed to provide this required security functionality. The invoker of these applications must have an understanding of how to check if these security services are available and how they can be invoked [72].

In grids (particularly intergrids), there is a requirement to support security across organizational boundaries. This makes a centrally managed security system impractical; administrators want to support "single sign-on" for users of the grid, including delegation of credentials for computations that involve multiple resources and/or sites. The grid security infrastructure provides a single-sign-on, run-anywhere authentication service, with support for local control over access rights and mapping from global to local user identities [167]. The grid security infrastructure supports uniform authentication, authorization, and message-protection mechanisms in multiinstitutional settings. Specifically, the grid security infrastructure provides, among other services, single sign-on, delegation, and identity mapping using public key technology (X.509 certificates) (this functionality is revisited in Chapter 6 as part of the Globus Toolkit discussion).

Node Security Function/Functional Block

A node security functional block usually exists in the grid environment. Authentication and authorization is a "two-way street"; not only does the user need to be authenticated, but also the computing resource. There is the need for secure (authenticated and, in most instances, also confidential) communication between internal elements of a computational grid. This is because a grid is comprised of a collection of hardware and software resources whose origins may not be obvious to a grid user. When a user wants to run on a particular processor, the user needs assurances that the processor has not been compromised, making his or her proprietary application, or data, subject to undesired exposure [72].

If a processor enrolls in a dynamic-rather than preadministered manner, then an identification and authentication validation must be performed before the processor can actually participate in the grid's work, as we discussed earlier. A certificate authority (CA) can be utilized to establish the identity of the "donor" processor, as well as the users and the grid itself. Some grid systems provide their own log-in to the grid, whereas other grid systems depend on the native operating systems for user authentication.

Broker Function/Functional Block and Directory

A broker functional block usually exists in the grid environment. After the user is authenticated by the user security functional block, the user is allowed to launch an application. At this juncture, the grid system needs to identify appropriate and available resources that can/should be used within the grid, based on the application and application-related parameters provided by the user of the application. This task is carried out by a broker function. The broker functionality provides information about the available resources on the grid and the working status of these resources. Specifically, grid systems have a capability to define (and monitor) a grid's topology in order to share resources and support collaboration; this is typically accomplished via a directory mechanism (e.g., LDAP and/or DNS); see Figure 3.5, bottom.

Scheduler Function/Functional Block

A scheduler functional block usually exists in the grid environment. If a set of stand-alone jobs without any interdependencies needs to execute, then a scheduler is not necessarily required. In the situation where the user wishes to reserve a specific resource or to ensure that different jobs within the application run concurrently, then a scheduler is needed to coordinate the execution of the jobs.

In a "trivial" environment, the user may select a processor suitable for running the job and then execute a grid instruction that routes the job to the selected processor. In "nontrivial" environments, a grid-based system is responsible for routing a job to a properly selected processor so that the job can execute. Here, the scheduling software identifies a processor on which to run a specific grid job that has been submitted by a user; see Figure 3.6, top. After available resources have been identified, the follow-on step is to schedule the individual jobs to run on these resources. Schedulers are designed to dynamically react to grid load. They accomplish this by utilizing measurement information relating to the current utilization of processors to determine which ones are available before submitting a job.

In an entry-level case, the scheduler could assign jobs in a round-robin fashion to the "next" processor matching the resource requirements. More commonly, the scheduler automatically finds the most appropriate processor on which to run a given job. Some schedulers implement a job priority (queue) mechanism. In this environment, as grid processors become available to execute jobs, the jobs are selected from the highest-priority queues first. Policies of various kinds can be implemented via the scheduler; for example, there could be a policy that restricts grid jobs from executing at certain time windows.

In some more complex environments, there could be different levels of schedulers organized in a hierarchy. For example, a metascheduler may submit a job to a cluster scheduler or other lower-level scheduler rather than to an individual target processor. As another example, a cluster could be represented as a single resource; here, the cluster could have its own scheduler to manage the internal cluster nodes, while a higher-level scheduler could be employed to schedule work to be supported by the cluster in question as an ensemble.

Advanced schedulers monitor the progress of active jobs, managing the overall workflow. If the jobs were to become lost due to system or network outages, a high-end scheduler would automatically resubmit the job elsewhere. However, if a job appears to be in an infinite loop and reaches a maximum timeout, then such jobs will not be rescheduled. Typically, jobs have different kinds of completion codes. This code determines if the job is suitable for resubmission or not [147]).

Some grids also have a "reservation system." These systems allow one to reserve resources on the grid. These are a calendar-based mechanism for reserving resources for specific time periods, and preventing others from reserving the same resource at the same time.

In a "scavenging" grid environment, any processor that becomes idle reports its idle status to the grid management node. The management node in turn assigns to this processor the next job that is satisfied by the processor's resources. If the processor becomes busy with local nongrid work, the grid job is usually suspended or delayed. This situation creates somewhat unpredictable completion times for grid jobs, although it is not disruptive to the processors donating resources to the grid. To create more predictable behavior, grid processors are often "dedicated" to the grid and are not preempted by external work; this enables schedulers to compute the approximate completion time for a set of jobs when their running characteristics are known [147].

Data Management Function/Functional Block

A data management functional block usually exists in a grid environment. There typically needs to be a reliable (and secure) method for moving files and data to various nodes within the grid. This functionality is supported by the data management functional block. Figure 3.6, middle, depicts a data management function needed to support this data management function.

Job Management and Resource Management Function/Functional Block

A job management and resource management functional block usually exists in a grid environment. This functionality is also known as the grid resource allocation manager (GRAM). The job management and resource management function (see Figure 3.6, bottom) provides the services to actually launch a job on a particular resource, to check the job's status, and to retrieve the results when the job is complete. Typically, the management component keeps track of the resources available to the grid and which users are members of the grid. This information is used by the scheduler to decide where grid jobs should be assigned. Also, typically, there are measurement mechanisms that determine both the capacities of the nodes on the grid and their current utilization levels at any given point in time; this information is used to schedule jobs in the grid, to monitor the health of the grid (e.g., outages, congestion, overbooking/overcommitment), and to support administrative tasks (e.g., determine overall usage patterns and statistics, log and account for usage of grid resources, etc.) Furthermore, advanced grid management software can automatically manage recovery from a number of grid failures and/or outages (e.g., specifically identify alternatives processors or setups to get the workload processed) [147].

With grid computing, administrators can "virtualize," or pool, IT resources (computers, storage, and applications) into a single virtual system whose resources can be managed from a single administration console and can be allocated dynamically, based on demand. The job management and resource management functional block supports this simplified view of the enterprise-wide resources.

The work involved in managing the grid may be distributed hierarchically, in order to increase the scalability of the grid. For example, a central job scheduler may not schedule a submitted job directly, but instead, the job request is sent to a secondary scheduler that handles a specified set of processors (e.g., a cluster); the secondary scheduler handles the assignment to the specific processor. Hence, in this instance, the grid operation, the resource data, and the job scheduling are distributed to match the topology of the grid.

On the resource management side of this function, mechanisms usually exist to handle observation, management, measurement, and correlation. It was noted above that schedulers need to react to instantaneous loads on the grid. The donor software typically includes "load sensors" that measure the instantaneous load and activity on resources or processors. Such measurement information is useful not only for instantaneous scheduling of tasks and work, but also for assessing (administratively) overall grid usage patterns [147]. Observation, management, and measurement data can be used, in aggregate, to support capacity planning and initiate deployment of additional hardware. Furthermore, measurement information about specific jobs can be collected and used to forecast the resource requirements of that job the next time it executes. Some grid systems provide the means for implementing custom load sensors for more than just processor or storage resources.

User/Application Submission Function/Functional Block

A user/application submission functional block usually exists. Typically, any member of a grid can submit jobs to the grid and perform grid queries, but in some grid systems, this function is implemented as a separate component installed on "submission nodes or clients" [147].

Resources

A grid would be of no value if it did not contribute resources to the ultimate user and/or application. As noted, resources include processors, data storage, scientific equipment, etc. Besides "physical presence" on the grid (by way of an interconnecting network), there has to be "logical presence." "Logical presence" is achieved by installing grid-support software on the participating processors. After loading and activating the software that manages the grid's use of its affiliated resources, each processor contributing itself or contributing ancillary resources to the grid needs to properly enroll as a member of the grid.

As discussed in the previous subsections, a user accessing the grid to accomplish a task submits a job for execution on the grid. The grid management software communicates with the grid donor software of the resource(s) logically present to forward the job to an appropriate processor. The grid-support software on the processor accepts an executable job from the grid management system and executes it. The grid software on the "donor" processor must be able to receive the executable file (in some cases the executable copies preinstalled on the processor.) The software is run and the output is sent back to the requester. More advanced implementations can dynamically adjust the priority of a running job, suspend a job and resume it later, or checkpoint a job with the possibility of resuming its execution on a different processor [147]. The grid system sends information about any available resources on that processor to the resource management functional block described in the previous subsection. The participating donor processor typically has a self-monitoring capability that determines or measures how busy the processor is. This information is "distributed" to the management software of the grid and it is utilized to schedule the appropriate use of the resources. For example, in a scavenging system, this utilization information informs the grid management software when the processor is idle and available to accept work.

Protocols

After identifying the functional blocks, a generic architecture description proceeds by defining the protocols to be employed between (specifically, on the active interfaces of the) functional blocks. To interconnect these functional blocks, we need protocols, especially standardized protocols. Protocols are formal descriptions of message formats and a set of rules for message exchange. The rules may define sequence of message exchanges. Protocols are generally layered. Figure 3.7 depicts two examples of protocol stacks and network-enabled services.

The grid dénouement (call it vision) requires protocols that are not only open and general purpose but also are vendor-independent and widely adopted *standards.* Standards allows the grid to establish resource-sharing arrangements dynamically with *any* interested party and thus to create something more than a plethora of balkanized, incompatible, noninteroperable distributed systems; standards are also

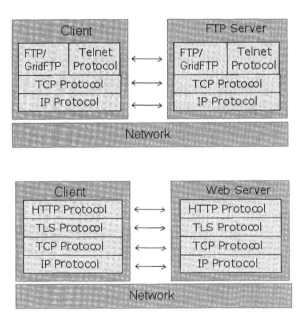

Figure 3.7 Example of protocol stacks and network-enabled services.

important as a means of enabling general-purpose services and tools [103]. Both open source and commercial products can, then, interoperate effectively in this heterogeneous, multivendor grid world, thus providing the pervasive infrastructure that will enable successful grid applications.

At this juncture, the Global Grid Forum is in the process of developing consensus standards for grid environments. On the commercial side, nearly a decade of experience and refinement have resulted in a widely used de-facto standard in the form of the open source Globus Toolkit. The Global Grid Forum has a major effort underway to define the Open Grid Services Architecture (OGSA), which modernizes and extends Globus Toolkit protocols to address new requirements, while also embracing Web services [103]. These topics are discussed in Chapters 4 and 5.

3.3 BASIC CONSTITUENT ELEMENTS—A PHYSICAL VIEW

This section looks at grid resources from a physical viewpoint. A grid is a collection of networks, processors, storage, and other resources.

Networks

The networking mechanism is the most fundamental resource for the grid and also is the theme of this book. In fact, without networking grid computing would not be possible.

The recent growth in communication capacity makes grid computing practical, compared to the limited bandwidth available when distributed computing was first emerging. Transmission of content and job supervision within the grid are important for sending jobs and the required data to points within the grid (some jobs require a large amount of data to be processed and it may not always reside on the processor running the job.) Figure 3.8 depicts one example of an intergrid and the kind of connectivity the particular grid has.

The bandwidth available for the subtending communications links can often be a critical resource that can limit utilization of the grid. LAN connectivity now is in the 1–10 Gbps range, and practically affordable WAN/intranet connectivity for companies is in the 45–155 Mbps range. (Speeds in the 2.4–10 Gbps range are commonplace within the inner workings of carriers, but these kinds of speeds are not generally affordable for Fortune 500 companies for ubiquitous deployment at this time. A single fiber can now carry in the range of 1 Tbps using high-density DWDM, but these speeds begin to be out of reach for all but the largest carriers, at least as of 2004). LANs can be utilized to support clusters (some clusters are implemented with SANs or other channel technology) and local grids. High-capacity, high-quality intranets support intragrids, and long-haul global connectivity (including internet-provided capacity) make intergrids possible. Processors on the intragrid may also have connections to the Internet in addition to the connectivity among the grid processors.

We have already noted that without adequate networking grid computing would not be possible. As discussed in the previous section, a grid typically includes soft-

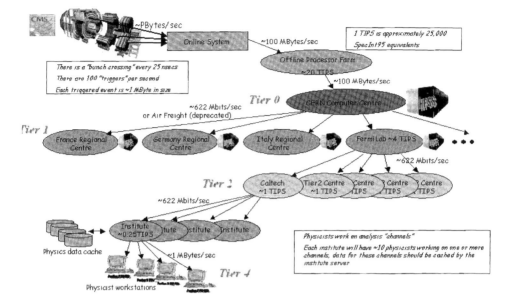

Figure 3.8 Example of high-speed networking in an intergrid. Copyright © 2002 University of Chicago and The University of Southern California. All Rights Reserved.

ware to enable jobs to communicate with each other. For example, an application may split itself into a large number of microjobs, each becoming a separate job in the grid. An application is often comprised of algorithms that require that the microjobs communicate some information among them; for example, the microjobs need to be able to locate other specific microjobs, establish a communications connection with them, and send the appropriate data. At the protocol level, a message passing interface (MPI) is often included as part of the grid system in support of communications and networking, although it should be noted that SOAP (Simple Object Access Protocol) could conceivably become the pervasive messaging protocol in the future (SOAP is discussed in a section that follows). Redundant communication paths are generally needed, but a well-designed intranet will already accommodate this. A grid management system can monitor the topology of the grid and pinpoint possible communication bottlenecks [147].

MPI may be a good protocol solution for long, massively parallel applications; for short serial tasks, a system based on web services (SOAP) can be acceptable. Any given functionality can appear at multiple levels. Conceivably, there could be a back-end parallel computer running an MPI-based job; this could be conceivably front-ended as a service by a middle-tier component running on a completely different computer. One is able "interact" with this service at either level: a high-performance I/O transfer at the parallel computing level, or with a "slower" middle-tier protocol such as SOAP at the service level.

MPI is a de-facto open standard for message passing developed in the 1990s by a committee of vendors (including IBM, Intel, Cray, and nCUBE), implementers, and users. MPI is widely available, with both free available and vendor-supplied implementations. The goal of MPI is to provide a standard for writing message-passing programs. As such, the interface attempts to establish a practical, portable, efficient, and flexible standard for message passing. MPI has been developed under the auspices of the Message Passing Interface Forum (MPIF), with participation from over 40 organizations. MPIF worked in the early 1990s to define a set of accepted library interface standards (the MPIF is not sanctioned or supported by any official standards organization). Version 1.0 of the standard was released in May 1994; MPI-2 was adopted in April 1997.

Much as been said of late about how the theoretical maximum communication speed has been exceeding the rules of Moore's law,[1] and how that can be a catalyst for grid deployment (e.g., see Figure 3.9). Although the availability of broadband does drive the potential for grid deployment, two important points need to be kept in mind.

Point 1. In general, the maximum available speeds are not affordable by anyone, with the possible exception of either government-funded labs or supercomputer centers. Whereas OC-3 (155 Mbps) is probably the typical speed for Fortune 100 applications, OC-12 (622 Mbps) is probably the typical current upper limit even for high-end labs. Thus, OC-768 (about 40 Gbps) and multiples of OC-192 delivered over DWDM (e.g., 256 beams, for a total of 0.3 Tbps) are technologically feasible but not affordable by individual organizations. Figure 3.10 reinforces this point: it shows the actual capacity and traffic of the Internet's backbone over the years, based on industry sources. Although the capacity has increased at a compound annual growth rate of 150% per year over the 1997–2002 timeframe, the backbone capacity is about 0.5–1.0 Gbps; that is a relatively modest amount compared to an OC-48, OC-192, or OC-768 optical-link system.

Point 2. before generalizing, one needs to determine if the problem at hand is processor bound or I/O (storage/data) bound. For example this author cointroduced the concept of hyperperfect numbers in the early 1970s (see Figure 3.11), a "nice" generalization of the concept of a perfect number that is now part of number theory (hyperperfect numbers are expected to have applications in cryptology and signal-processing transforms). A grid computing apparatus could be used to number crunch away and find all such numbers, say up to 10^{15} (currently these have been identified only up to 10^{11}). In this situation, there is very little I/O needed, just number crunching (in fact, the hyperperfect numbers up to 10^7 were found by running an application on a 1960s-vintage, time-sharing, "grid-in-spirit" computing mainframe environment). In other situations, for example, genome research, there in fact may be a need to move a certain large amounts of data. Even then, though, the DNA information could be copied and replicated on optical disks and sent offline (by reg-

---

[1]Gordon Moore made his well-known observation (now known as "Moore's Law") in 1965, just a few years after the first ICs were developed. In his original paper [175], Moore observed an exponential growth in the number of transistors per integrated circuit and predicted that this trend would continue. Through technology advances, Moore's Law, the doubling of transistors every couple of years, has been maintained, and still holds true today. Industry stakeholders (such as Intel) expect that it will continue at least through the end of this decade [168].

- Network versus computer performance
  - —Computer speed doubles every 18 months
  - —Network speed doubles every 9 months
  - —Difference = order of magnitude per 5 years
- 1986 to 2000
  - —Computers: × 500
  - —Networks: × 340,000
- 2001 to 2010
  - —Computers: × 60
  - —Networks: × 4000

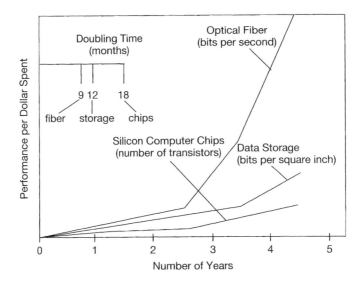

Figure 3.9   Speed increase for "laboratory-level" networks. Copyright © 2002 University of Chicago and The University of Southern California. All Rights Reserved.

ular mail) to the dozen or so supercomputers that may need access to the information in order to run a certain algorithm.

Computation

The next most common resource on a grid is obviously computing cycles provided by the processors on the grid. The processors can vary in speed, architecture, software platform, and storage apparatus. There are efforts underway to develop very high-speed supercomputers. Whereas clustering is a common approach at the TFLOPS speeds, grid computing can also play a role in these initiatives by refining architectures that link remote computers into an assembly of loosely or tightly coupled processors. At the business level, grid computing is expected by the industry to be more practical than cluster-based supercomputing.

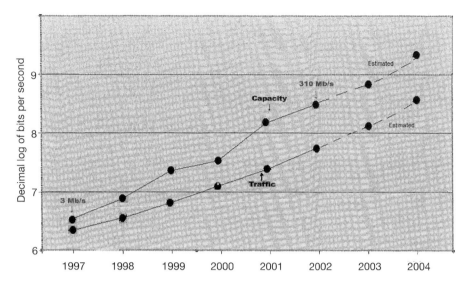

Figure 3.10   Speed increase for the Internet backbone.

Related to supercomputing, the High-End Computing Revitalization Task Force's program under the Defense Advanced Research Projects Agency recently recommended that the United States acquire a multihundred-teraflops system as a cornerstone of a new national supercomputing center [135] (Cray, IBM and Sun Microsystems are working on building the first petaFLOPS computer with novel hardware to be completed by 2010.)

Storage

The next most common resource used in a grid is data storage. In a grid environment, a file or data base can span several physical storage devices and processors, bypassing size restrictions often imposed by file systems that are preembedded with operating systems. Storage capacity available to an application can be increased by making use of the storage on multiple processors with a unifying file system. Each processor on the grid usually provides some quantity of storage for grid use. Storage can be "primary storage," "secondary storage," or "tertiary storage." Memory directly attached to a processor has fast access capabilities but is volatile; this kind of memory is used to cache data to serve as temporary storage for running applications. "Secondary storage" is generally implemented in hard disk drives, such as with RAID (redundant array of inexpensive drives). "Tertiary storage" is generally implemented in near-real-time accessible media such as tape or other permanent storage media. Many grid systems use mountable networked file systems, such as Network File System (NFS), Distributed File System (DFS), or General Parallel File System (GPFS). Special grid database software can "federate" a group of individual data bases and files to form a larger, more inclusive data base.

A **k-hyperperfect number** (aka *hyperperfect number*) (introduced in 1975 by Minoli and Bear) is a number n for satisfying the equality $n = 1 + k(\sigma(n) - n - 1)$, where $\sigma(n)$ is the divisor function (i.e., the sum of all positive divisors of n). A number is perfect iff it is 1-hyperperfect.

The first few k-hyperperfect numbers are 6, 21, 28, 301, 325, 496, ... , with the corresponding values of k being 1, 2, 1, 6, 3, 1, ... The first few k-hyperperfect numbers that are not perfect are 21, 301, 325, 697, ...

It can be shown that if $k > 1$ is an odd integer and $p = (3k + 1)/2$ and $q = 3k + 4$ are prime numbers, then p^2q is k-hyperperfect; McCraine has conjectured that all k-hyperperfect numbers for odd $k > 1$ are of this form, but the hypothesis has not been proven so far. Furthermore, it can be proven that if $p \neq q$ are odd primes and k is an integer such that $k(p + q) = pq - 1$, then pq is k-hyperperfect. It is also possible to show that if $k >$ and $p = k + 1$ is prime, then for all $i > 1$ such that $q = p^i - p + 1$ is prime, $n = p^{i-1}q$ is k-hyperperfect (see examples.)

k	Values of i
16	11, 21, 127, 149, 469, ...
22	17, 61, 445, ...
28	33, 89, 101, ...
36	67, 95, 341, ...
42	4, 6, 42, 64, 65, ...
46	5, 11, 13, 53, 115, ...
52	21, 173, ...
58	11, 117, ...
72	21, 49, ...
88	9, 41, 51, 109, 483, ...
96	6, 11, 34, ...

k	Known k-hyperperfect numbers
100	3, 7, 9, 19, 29, 99, 145, ...
1	6, 28, 496, 8128, 33550336, ...
2	21, 2133, 19521, 176661, 129127041, ...
3	325, ...
4	1950625, 1220640625, ...
6	301, 16513, 60110701, 1977225901, ...
10	159841, ...
11	10693, ...
12	697, 2041, 1570153, 62722153, 10604156641, 13544168521, ...
18	1333, 1909, 2469601, 893748277, ...
19	51301, ...
30	3901, 28600321, ...
31	214273, ...
35	306181, ...
40	115788961, ...
48	26977, 9560844577, ...
59	1433701, ...
60	24601, ...
66	296341, ...
75	2924101, ...
78	486877, ...
91	5199013, ...
100	10509080401, ...
108	275833, ...
126	12161963773, ...
132	96361, 130153, 495529, ...
136	156276648817, ...
138	46727970517, 51886178401, ...
140	1118457481, ...
168	250321, ...
174	7744461466717, ...
180	12211188308281, ...
190	1167773821, ...
192	163201, 137008036993, ...
198	1564317613, ...
206	626946794653, 54114833564509, ...
222	348231627849277, ...
228	391854937, 102744892633, 3710434289467, ...
252	389593, 1218260233, ...
276	72315968283289, ...
282	8898807853477, ...
296	444574821937, ...
342	542413, 26199602893, ...
348	66239465233897, ...
350	140460782701, ...
360	23911458481, ...
366	808861, ...
372	2469439417, ...
396	8432772615433, ...
402	8942902453, 813535908179653, ...
408	1238906223697, ...
414	8062678298557, ...
430	124528653669661, ...
438	6287557453, ...
480	1324790832961, ...
522	723378252872773, 106049331638192773, ...
546	211125067071829, ...
570	1345711391461, 5810517340434661, ...
660	13786783637881, ...
672	142718568339485377, ...
684	154643791177, ...
774	8695993590900027, ...
810	5646270598021, ...
814	31571188513, ...
816	31571188513, ...
820	1119337766869561, ...
968	52335185632753, ...
972	289085338292617, ...
978	60246544949557, ...
1050	64169172901, ...
1410	80293806421, ...
2772	95295817, 124035913, ...
3918	61442077, 217033693, 12059549149, 60174845917, ...
9222	404458477, 3426618541, 8983131757, 13027827181, ...
9828	432373033, 2797540201, 3777981481, 13197765673, ...
14280	848374801, 2324355601, 4390957201, 16498569361, ...
23730	2288948341, 3102982261, 6861054901, 30897836341, ...
31752	4660241041, 7220722321, 12994506001, 52929885457, 60771359377, ...
55848	15166641361, 44783952721, 67623550801, ...
67782	18407557741, 18444431149, 34939858669, ...
92568	50611924273, 64781493169, 84213367729, ...
100932	50969246953, 53192980777,

Figure 3.11 Example of a computationally intensive mathematical problem that can be addressed with grid computing.

Articles on Hyperperfect Numbers

- Daniel Minoli, Robert Bear, *Hyperperfect Numbers*, PME Journal, Fall 1975, pp. 153-157.
- Daniel Minoli, *Sufficient Forms For Generalized Perfect Numbers*, Ann. Fac. Sciences, Univ. Nation. Zaire, Section Mathem; Vol. 4, No. 2, Dec 1978, pp. 277-302.
- Daniel Minoli, *Structural Issues For Hyperperfect Numbers*, Fibonacci Quarterly, Feb. 1981, Vol. 19, No. 1, pp. 6-14.
- Daniel Minoli, *Issues In Non-Linear Hyperperfect Numbers, Mathematics of Computation*, Vol. 34, No. 150, April 1980, pp. 639-645.
- Daniel Minoli, *New Results For Hyperperfect Numbers*, Abstracts American Math. Soc., October 1980, Issue 6, Vol. 1, pp. 561.
- Daniel Minoli, W. Nakamine, *Mersenne Numbers Rooted On 3 For Number Theoretic Transforms*, 1980 IEEE International Conf. on Acoust., Speech and Signal Processing.
- Judson S. McCranie, *A Study of Hyperperfect Numbers*, Journal of Integer Sequences, Vol. 3 (2000), http://www.math.uwaterloo.ca/JIS/VOL3/mccranie.html

Books with Hyperperfect Numbers Information

- Daniel Minoli, Voice over MPLS, McGraw-Hill, New York, NY, 2002, ISBN 0071406158 (p.114-134)

Figure 3.11 *Continued.*

More advanced file systems on a grid can automatically duplicate sets of data to provide redundancy, increased reliability, and improved performance. Certain applications may require synchronous replication of data files; in this case, the "speed of light," namely, the propagation delay to/from the data storage device, may adversely impact the functioning of the application. This situation can be addressed and/or ameliorated by placing the data closer to the processing point. An intelligent grid scheduler can select the appropriate storage devices to hold a job's data, based, for example, on usage patterns or replication needs. Jobs can then be scheduled closer to the data, preferably on the processors that have direct SAN access to the storage devices holding the requisite data [147].

Storage[2] is increasingly recognized as a distinct resource, one that is best thought of separately from the computer systems (hosts) that are its consumers and beneficiaries. Such storage is often shared by multiple hosts and is acquired and managed independently from them. This is in contrast to the historical view (host-attached storage) that storage is an intrinsic part of a computer system, that is, a "peripheral." This trend toward shared storage recognizes the critical value of the information entrusted to the storage system, as well as the fact that storage represents a significant portion of the investment in a typical computing environment.

Through much of the history of computing, storage has been seen as an intrinsic part of computer systems. Although storage once was regarded as a "peripheral," more recently, it has come to be thought of as a storage subsystem, but still uniquely associated with a computer. The principal exceptions to this have been mainframe computer complexes and computer clusters in which a modest number of cooperat-

---

[2]The rest of this subsection is loosely based on the Storage Networking Industry Association (SNIA) SNIA Shared Storage Model [140], used with permission of SNIA.

ing computer systems share a common set of storage devices. The key enabling technology for shared storage is networking technology that can provide high bandwidth, reliable connectivity, and significant geographic scope at a cost that makes shared storage an attractive alternative to the historical host-attached storage model.

Because the traditional computing model associates storage uniquely with a computer system, a computing environment with many computer systems has many storage and storage management environments to maintain and operate—one per computer system.

As business has become more dependent upon computing, it has also become more dependent upon data. Although a failed processor can usually be replaced and operations continued almost immediately after the replacement, a failed storage resource requires replacement, typically followed by time-consuming restoration of data, all too often with some loss of recent changes to that data, which requires recovery action before operations can continue. As a result, storage and the disciplines of caring for data and the storage media on which it resides have gown in visibility and importance.

In addition, the fraction of the purchase price of a computer system that is represented by the storage component has grown over time to the point that now the cost of the storage component of a computer system is often in the vicinity of half of the total price. Beyond the purchase price of storage, the total cost of owning storage has become a significant part of the cost of maintaining the computing environment. In other words, the acquisition cost is a small portion of the total cost of ownership of storage over its lifetime.

In responding to these trends, the IT community has come to view storage as a resource that should be purchased and managed independently of the computer systems that it serves. The IT community has also increasingly come to view storage as a resource that should be shared among computer systems. These changes allow more focused attention on storage that is expected to lead to reduced costs, higher levels of service, and more flexibility through the sharing of the storage resource.

Although shared storage environments can bring many benefits, they also present a number of challenges, particularly, interworking. The Storage Network Industry Association (SNIA) Technical Council has developed a framework that captures the functional layers and properties of a storage system, regardless of the underlying design, product, or installation. Much like the Open Systems Interconnection (OSI) 7-layer model in conventional networking, the SNIA Shared Storage Model may be used to describe common storage architectures graphically, while exposing what services are provided, where interoperability is required, and the pros and cons of each potential architecture; see Figure 3.12. SNIA was formed to communicate the benefits of this new paradigm, and to provide a forum for computer vendors, storage vendors, and the IT community to address its challenges together.

The SNIA Shared Storage Model supports the following kinds of components:

- **Interconnection network**—the network infrastructure that connects the elements of the shared storage environment. This network may be a network that

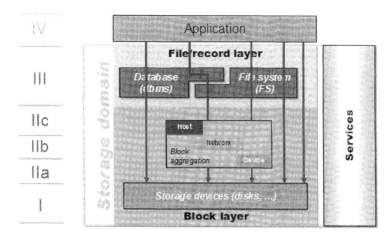

The SNIA Shared Storage Model is a layered one. The
figure shows a picture of the stack with a numbering
scheme for the layers. Roman numerals are used to avoid
confusion with the ISO and IETF networking stack
numbers. The layers are as follows:
- IV. Application
- III. File/record layer
  - IIb. Database
  - IIa. File system
- II. Block aggregation layer, with three
  function placements:
  - IIc. Heat
  - IIb. Network
  - IIa. Device
- I. Storage devices

Figure 3.12   The SNIA Shared Storage Model. Copyright © 2001, 2003 Storage Networking Industry Association. Used with permission of SNIA.

is primarily used for storage access, or one that is also shared with other uses.
The important requirement is that it must provide an appropriately rich, high-
performance, scalable connectivity upon which a shared storage environment
can be based. The physical-layer network technologies that are used (or have
been used) for this function include Fibre Channel, Fast- and Gigabit-Ether-
net, Myrinet, the VAX CI network, and ServerNet. Network protocols that
are used at higher layers of the protocol stack also cover a wide range, including SCSI FCP, TCP/IP, VI, CIFS, and NFS. Redundancy in the storage network allows communication to continue despite the failure of various components; different forms of redundancy protect against different sorts of failures.
Redundant connections within an interconnect may enable it to continue to
provide service by directing traffic around a failed component. Redundant
connections to hosts and/or storage enable the use of multipath I/O to tolerate
interface and connection failures; multipath I/O implementations may also
provide load balancing among alternate paths to storage. An important topol-

ogy for multi-path I/O uses two completely separate networks to ensure that any failure in one network cannot directly affect the other.

- **Host computer**—a computer system that has some or all of its storage needs supplied by the shared storage environment. In the past, such hosts were often viewed as external to the shared storage environment, but we take the opposite view. A host typically attaches to a storage network with a host–bus adapter (HBA) or network interface card (NIC). These are typically supported by associated drivers and related software; both hardware and software may be considered part of the shared storage environment. The hosts attached to a shared storage environment may be largely unaware of each other, or they may explicitly cooperate in order to exploit shared storage environment resources. Most commonly, this occurs across subsets of the hosts ("clusters"). One of the advantages of separating hosts from their storage devices in a shared storage world is that the hosts may be of arbitrary and differing hardware architecture and run different versions and types of operating system software.

- **Physical storage resource**—a nonhost element that is part of the shared storage environment, and attached to the storage network. Examples include disk drives, disk arrays, storage controllers, array controllers, tape drives and tape libraries, and a wide range of storage appliances. (Hosts are not physical storage resources.) Physical storage resources often have a high degree of redundancy, including multiple network connections, replicated functions, and data redundancy via RAID and other techniques, all to provide a highly available service.

- **Storage device**—a special kind of physical-storage resource that persistently retains data.

- **Logical storage resource**—a service or abstraction made available to the shared storage environment by physical storage resources, storage management applications, or combination thereof. Examples include volumes, files, and data movers.

- **Storage management**—functions that observe, control, report, or implement logical storage resources. Typically, these functions are implemented by software that executes in a physical storage resource or host.

These storage constructs can avail themselves of grid technology, especially when the storage needs to reside in distinct locations (for example, for administrative, business continuity, or disaster recovery reasons.)

Scientific Instruments

Grids, particularly an intergrid, can provide shared access to expensive scientific equipment or interconnect dispersed equipment into a larger overall scientific tool. We will not discuss this issue further, since our emphasis is on commercial applications.

Software and Licenses

Two issues are of interest to organizations: application software and licensed software.

Regarding application software, the most basic approach is to use the grid environment to allow the application to run on an available processor on the grid (rather than running locally.) Further along the transition trajectory, one can modify the application to segment its work in such a way that the separate parts can execute in parallel on different grid processors. As noted in Table 3.3, "scalability" is a measure of how efficiently the multiple processors on a grid are used. However, there may be limits to scalability.

Regarding licensed software, there may be software of interest to the organization that may be too expensive to install on a large set of processors. For example, a Fortune 500 company might have, say, 1500 servers; replicating the software on all these servers may be way too expensive and/or inefficient from a budgetary standpoint. In a grid environment, specific business software may be installed on a few designated grid processors. In this instance, the jobs requiring this software are routed to the particular processors on which this software happens to be installed. When the licensing fees are significant line items in the IT budget, this approach can save significant expenses for the organization.

Some software license arrangements permit the software to be physically installed on all of the processors an organization may own, but may limit the number of sites that can simultaneously use the software. With a grid-based license management software, the grid system keeps track of how many concurrent copies of the software are being used and ascertains that no more than that number of copies are executing at any given time.

3.4 BASIC CONSTITUENT ELEMENTS—SERVICE VIEW

Of late, a service view of grid computing has taken hold. Figures 1.5 and 1.6 in Chapter 1 depicted a relatively simple and intuitive view of a layered model that can help the reader understand the basic underlying concepts. Naturally, a standardized layered model, like the well-known and widely used Open Systems Interconnection Reference Model, is needed if interoperable systems are to be built. Standards such as the OGSA provide the necessary stable framework. OGSA is a proposed grid service architecture based on the integration of grid and Web services concepts and technologies. The OGSI specification is a companion specification to OGSA that defines the interfaces and protocols to be used between the various services in a grid environment; it is the mechanism that provides the interoperability between grids designed using OGSA. Key constructs for the architecture are functional blocks, protocols, (network-enabled) grid services (by implementing end-to-end protocols), APIs, and software development kits (SDKs). Web services have emerged in the past few years as a standards-based approach for developing and accessing network applications. The service view and OGSA will be discussed in detail in Chapters 4 and 5. This section provides some basic supporting information.

Some concepts that should be understood are:

- Service-oriented architectures (SOA)
- Simple Object Access Protocol (SOAP)
- Web services standards
- Web Services Description Language (WSDL)
- Web Services Inspection Language (WSIL)
- Universal Description, Discovery, and Integration (UDDI)
- .NET
- Web Services Resource Framework (WSRF)

The fundamental concept behind OGSA is that it is a service-oriented architecture comprised of constituent *grid services,* that are defined as special Web services (more on this below) that provide a set of well-defined interfaces that follow specific conventions [119]. A SOA, defines how two computing entities interact to enable one entity to perform a unit of work on behalf of another entity. The unit of work is referred to as a service, and the service interactions are defined using a description language. Each interaction is self-contained and loosely coupled, so that each interaction is independent of any other interaction [15]. Business applications are typically designed to automate business processes, but often without necessarily embodying in them the ability to adapt themselves to changing business needs; modifying and/or updating business processes and information flows in this environment is rather challenging. This is because business applications have traditionally been written as single, monolithic, and all-inclusive aggregates, making updates and changes to these applications expensive and time-consuming. In a SOA environment, applications are assembled as a collection of services, each of which represents separate and discrete functions or features. As business needs change, services can be added, deleted, or updated as needed, to evolve as the business needs it [86].

The protocol independence of SOA means that different consumers can use services by communicating with the service in different ways. Ideally, there should be a management layer between the providers and consumers to ensure flexibility in reference to implementation protocols [15]. It should be noted that many early Web services projects focused on repurposing already-proven request/response Web architectures, in which transactional support was either explicitly coded into the application layer using classic Web techniques such as HTTP sessions or cookies, or explicitly avoided by having the application simply provide read-only access to back-end business systems [75]. Chapter 4 expands on these concepts.

Preston Gralla explains SOA as follows [38]. In a SOA environment, software components can be exposed as services on the network, and so can be reused for different applications and purposes. In SOA, developing new applications can be a matter of mix-and-match: decide on the application that one needs, find out the existing components that can help build that application, glue them all together, and one is done. SOA is an increasingly popular concept, but it has been around since the mid-1980s. SOA has really not taken off because there has been no standard middleware or application programming interfaces that would allow it to take root. There were attempts to build them, such as the Distributed Computing Environment (DCE) and Common Object Request Broker Architecture (CORBA), but neither caught on, and SOA languished as an interesting concept, but with no significant real-world applications. Then Web services (see below) came along and gave SOA a new opportunity. The Web services underlying architecture works well with the concept of SOA, so much so, in fact, that some analysts and software makers believe that the future of Web services rests with SOA. SOA is an architecture that publishes services in the form of an XML interface; it is really no different from a traditional Web service architecture in which Universal Description, Discovery and Integration is used to create a searchable directory of Web services. In fact, UDDI is the solution of choice for enterprises that want to make available software components as services internally on their networks, using the SOA paradigm. Most Web services implementations are point-to-point, where one has an intimate knowledge of the platform to which one is interested in connecting. That means that the Web service is not made available publicly on the network, and cannot be "discovered"; in a sense, it is hard-coded in the point-to-point connection. In an SOA implementation, information about the Web service and how to connect to it is published in a UDDI-built directory, so that Web service can be easily discovered and used in other applications and implementations. But although the basic Web services architecture fits into the SOA concept, there are still roadblocks to setting them up. Notable among them are security, identity management issues, and management problems—having software that will be able to track and manage hundreds or dozens of Web services and their development and deployment. Software is just becoming available to do that. On the security side, the issues still have not been solved.

Web services (sometimes called application services or simply services) refers to a developing distributed computing environment that has a foundation on simple Internet-based standards to enable heterogeneous distributed computing. Web services define a technique for describing software components to be accessed, methods for accessing these components, and discovery methods that enable the identification of relevant service providers. Web services are programming-language-, programming-model-, and system-software-neutral [114]. A Web service comprises content, or process, or both, with an open programmatic interface. Some examples include currency converters, stock quotes, and dictionaries. More complex examples include travel planners and procurement workflow systems. A Web service has the following characteristics [85]:

- It is an Internet-based application that performs a specific task and complies with a standard specification.
- It is an executable that can be expressed and accessed using XML and XML messaging.
- It can be published, discovered, and invoked dynamically in a distributed computing environment.

- It is platform- and language-independent.

Web services have created a new communication pathway between applications, enabling them to talk to each other and exchange information in a platform-neutral, language-independent way. Originally, these services were designed to reduce costs and facilitate application integration. Web services have now also become a new platform for information providers. Extensive data is now available through Web services, from real-time stock quotes to information about local vehicular traffic. The constituent technologies of Web services—SOAP and WSDL—have been implemented in production environments for several years and the tools to build, test, and deploy Web services have matured significantly. In-depth knowledge of these key technologies was a prerequisite in early days; today, with advanced developer tools, a Web service can be created/accessed very rapidly by developers without needing a background in Web services technologies such as SOAP or WSDL [18].

Web services standards are being defined within the W3C and other standards bodies and form the basis for major new industry initiatives such as IBM (Dynamic e-Business), Microsoft (.NET), and Sun Microsystems (Sun ONE). Web services are small units of code that are independent of operating systems and programming languages. They are designed to handle a limited set of tasks. Web services make it easy to communicate between discrete applications; applications are able to access Web services via standard Web formats with no need to know how the Web service itself is implemented. They also make it possible for developers to reuse existing capabilities and services instead of writing new ones. Web services utilize XML-based communicating protocols. Web services use the standard web protocols Hypertext Transfer Protocol, XML (eXtensible Markup Language[3]), SOAP, WSDL, and UDDI. HTTP is the World Wide Web standard for communication over the Internet. HTTP is standardized by the World Wide Web Consortium (W3C). XML is a well-known standard for storing, carrying, and exchanging data. XML is standardized by the W3C. SOAP-based Web services are becoming the most common implementation of SOA (however, there also are

---

[3]XML provides an essential mechanism for transferring data between services in an application- and platform-neutral format. However, it is not well suited to large datasets with repetitive structures, such as large arrays or tables. Furthermore, many legacy systems and valuable data sets exist that do not use the XML format. The GGF was working at press time to define an XML-based language, the Data Format Description Language (DFDL), for describing the structure of binary and character-encoded (ASCII/Unicode) files and data streams so that their format, structure, and metadata can be exposed. DFDL endeavors to describe existing formats in an actionable manner that makes the data in its current format accessible through generic mechanisms. The DFDL description [which is saved in a (logically) separate file from the data itself] provides a hierarchical description that structures and semantically labels the underlying bits. It plans to capture how bits are to be interpreted as parts of low-level data types (integers, floating point numbers, strings); how low-level types are assembled into scientifically relevant forms such as arrays; how meaning is assigned to these forms through association with variable names and metadata such as units; and how arrays and the overall structure of the binary file are parameterized based on array dimensions, flags specifying optional file components, etc. Further, if the data file contains repetitive structures, such as large arrays or tables, such a description can be very concise [33]. IT planners should track future developments in this arena.

non-Web implementations) [15]. SOAP is a lightweight platform- and language-neutral communication protocol that allows programs to communicate via standard Internet HTTP. SOAP is standardized by the W3C. WSDL is an XML-based language used to define Web services and to describe how to access them. WSDL is a suggestion by Ariba, IBM, and Microsoft for describing services for the W3C XML activity on XML protocols. UDDI is a directory service with which businesses can register and search for Web services. UDDI is a public registry, where one can publish and inquire about Web services [16].

Web services are services (normally including some combination of programming and data, but possibly including human resources as well) that are made available from a business's Web server for Web users or other Web-connected programs. Providers of Web services are generally known as application service providers (ASPs). Web services range from such major services as storage management and customer relationship management down to much more limited services such as the furnishing of a stock quote and the checking of bids for an auction item. The accelerating creation and availability of these services is a major Web trend. Users can access some Web services through a P2P arrangement rather than by going to a central server. Some services can communicate with other services and this exchange of procedures and data is generally enabled by a class of software known as middleware. Besides the standardization and wide availability to users and businesses of the Internet itself, Web services are also increasingly enabled by the use of XML as a means of standardizing data formats and exchanging data. As Web services proliferate, concerns include the overall demands on network bandwidth and, for any particular service, the effect on performance as demands for that service rise. A number of new products have emerged that enable software developers to create or modify existing applications that can be "published" (made known and made accessible) as Web services [15]. In its use of Web services, grid services equate *PortType* to *class*; *operation* to *method*; *service* to *object instance.* Grid services add *properties, PortType inheritance* (via an extension to WSDL), *Factory pattern* for creation of new objects and services, and a base *set of classes.*

As noted, SOAP, WSDL, WS-Inspection, and UDDI are of particular import to OGSA/OGSI [114]; see Figure 3.13. These protocols and constructs are also revisited in the chapters that follow, particularly Chapter 4.

SOAP provides a mechanism of messaging between a service requestor and a service provider. It is a mechanism for formatting a Web service invocation, a simple enveloping process for XML payloads that defines a remote procedure call convention and a messaging convention. SOAP payloads are independent of the underlying transport protocol and can be carried on HTTP, File Transfer Protocol (FTP), or Java Messaging Service (JMS).

SOAP is a way for a program running in one kind of operating system (such as Windows 2003) to communicate with a program in the same or another kind of an operating system (such as Linux) by using the World Wide Web's HTTP/XML as the mechanisms for information exchange. Because Web protocols are installed and available for use by all major operating system platforms, HTTP and XML provide a solution to the question of how programs running under different operating sys-

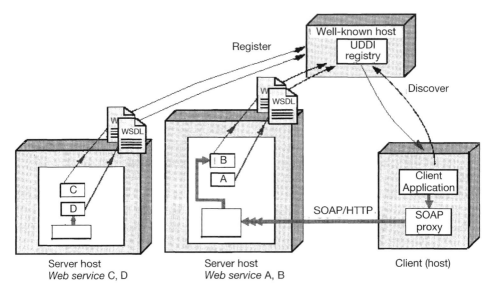

A Web service is a software system identified by a URI
whose public interfaces and bindings are defined and described
using XML. Its definition can be discovered by other
software systems. These other systems may then interact with the
Web service in a manner prescribed by its definition, using
XML-based messages conveyed by Internet protocols.
The Web Services Description Language (WSDL) is the
de-facto XML-based standard for describing Web services.
The Simple Object Access Protocol (SOAP) over HTTP is
the XML-based standard network protocol for exchanging
messages between Web services (W3C-given definitions).

Figure 3.13 Relationship among key protocols and constructs.

tems in a network can communicate with each other. SOAP specifies how to encode
an HTTP header and an XML file so that a program in one computer can invoke a
program in another computer and transact information. It also specifies how the
called program can return a response. SOAP was developed by Microsoft, Develop-
Mentor, and Userland Software and has been proposed as a standard interface to the
IETF. A point of consideration is that SOAP-based programs typically can readily
get through firewall servers that filter out packet sequences (requests) other than
those for known applications (through the designated-port mechanism); otherwise,
some firewall script changes may be needed. Since HTTP requests are usually allowed through firewalls, programs using SOAP are able to communicate with programs anywhere [15].

The Web Services Description Language is an XML mechanism for describing
Web services as a set of endpoints operating on messages. These messages contain
either document-oriented (messaging) or remote procedure call payloads. Service
interfaces are defined abstractly in terms of message structures and sequences of

message exchanges. Service interfaces are bound to a tangible network protocol and data-encoding format to define an endpoint. Related endpoints are bundled to define services. WSDL is extensible to allow description of endpoints and the concrete representation of their messages for a variety of different message formats and network protocols.

The Web Services Inspection Language consists of the XML language along with related conventions for locating service descriptions published by a service provider. A WSIL document can contain a collection of service descriptions (e.g., a URL to a WSDL document) and links (e.g., a URL to another WISL document) to other sources of service descriptions. A service provider creates a WSIL document and makes the document network accessible. Service requestors use standard Web-based access mechanisms (e.g., HTTP GET) to retrieve this document and discover what services the service provider advertises.

UDDI is an XML-based registry for businesses worldwide to list themselves on the Internet. UDDI's ultimate goal is to streamline online transactions by enabling companies to find one another on the Web and make their systems interoperable for e-commerce. UDDI is often compared to a telephone book's white, yellow, and green pages. The registry allows businesses to list themselves by name, product, location, or the Web services they offer. Microsoft, IBM, and Ariba spearheaded UDDI. The registry now includes 100 companies. The UDDI specification utilizes World Wide Web Consortium (W3C) and IETF standards such as XML, HTTP, and Domain Name System (DNS) protocols. It has also adopted early versions of the proposed SOAP messaging guidelines for cross-platform programming. UDDI entered its public beta-testing phase in late 2000. Each of the three founder companies now operates a registry server that is interoperable with servers from other members. As information goes into a registry server, it is shared by servers in the other businesses (additional companies were expected to be acting as operators of the UDDI Business Registry at a future time.) UDDI registration is open to companies worldwide, regardless of their size [15].

Above, we also mentioned .NET. .NET is Microsoft's Internet and Web strategy launched in 2000. .NET is an Internet- and Web-based infrastructure that delivers software as Web services and is a framework for universal services. It is a server-centric computing model. Initially, Windows 2000 and Windows XP comprised the backbone of .NET; as time goes by the .NET infrastructure was expected to be integrated into all Microsoft operating systems and desktop and server products (the .NET plan includes a new version of the Windows operating system, a new version of Office, and a variety of new development software for programmers to build Web-based applications) [16]. .NET is based on Web standards such as HTTP, the communication protocol between Internet applications; XML, the format for exchanging data between Internet applications; SOAP, the standard format for requesting Web services; and UDDI (described above), the standard to search and discover Web services. Web services provide data and services to other applications. The .NET framework is a common environment for building, deploying, and running Web services and Web applications. The .NET framework contains common class libraries like ADO.NET, ASP.NET, and Windows Forms to provide advanced standard services that can be integrated into a variety of computer systems. The .NET framework is language-neutral. Currently, it supports C++, C#, Visual Basic, JScript (the Microsoft version of JavaScript), and COBOL. Web services are the main building blocks in the Microsoft .NET programming model [16].

We also made reference to the Web Services Resource Framework (WSRF). The effective merging of grid and Web services that occurred in the early 2000s has lead to the WSRF, a series of OASIS-developed[4] specifications for performing grid computing on top of Web services [173]. The purpose of the WSRF Technical Committee ("TC") in OASIS is to define a generic and open framework for modeling and accessing stateful resources using Web services. This includes mechanisms to describe views on the state, to support management of the state through properties associated with the Web service, and to describe how these mechanisms are extensible to groups of Web Services [174]. WSRF includes the WS-ResourceProperties, WS-ResourceLifetime, WS-BaseFaults, and WS-ServiceGroup specifications (see Table 3.6 [174]). WSRF.NET is a project at the University of Virginia that allows the creation of WSRF-compliant Web services using the Microsoft .NET platform [173].

WSRF defines the means by which [174]:

- Web services can be associated with one or more stateful resources (named, typed, state components).
- Service requestors access stateful resources indirectly through Web services that encapsulate the state and manage all aspects of Web-service-based access to the state.
- Stateful resources can be destroyed through immediate or time-based destruction.
- The type definition of a stateful resource can be associated with the interface description of a Web service to enable well-formed queries against the resource via its Web service interface.
- The state of the stateful resource can be queried and modified via Web service message exchanges.

---

[4]OASIS (Organization for the Advancement of Structured Information Standards) is a nonprofit, international consortium whose goal is to promote the adoption of product-independent standards for information formats such as Standard Generalized Markup Language (SGML), XML, and HTML. Currently, OASIS (formerly known as SGML Open) is working to bring together competitors and industry standards groups with conflicting perspectives to discuss using XML as a common Web language that can be shared across applications and platforms. OASIS sponsors XML.org , a nonprofit XML Web portal. The goal of OASIS is not to create structured information standards for XML, but to provide a forum for discussion, to promote the adoption of interoperability standards, and to recommend ways members can provide better interoperability for their users. OASIS has worked with the United Nations to sponsor ebXML, a global initiative for electronic business data exchange. EbXML, whose goal is to make it easier for companies of all sizes and locations to conduct business on the Internet, is currently focusing on the specific needs of business-to-business and Internet security as it relates to XML (http://searchwebservices.techtarget.com/gDefinition/0,294236,sid26_gci527425,00.html).

Table 3.6 WSRF specifications

WS-ResourceProperties

This defines how the data associated with a stateful resource can be queried and changed using Web services technologies. This allows a standard means by which data associated with a WS-Resource can be accessed by clients. The declaration of the WS-Resource's properties represents a projection of or a view on the WS-Resource's state. This projection represents an implied resource type which serves to define a basis for access to the resource properties through Web service interfaces.

WS-ResourceLifetime

This defines two ways of destroying a WS-Resource: immediate and scheduled. This allows designers flexibility to design how their Web services applications can clean up resources no longer needed.

WS-BaseFaults

This defines an XML schema type for a base fault, along with rules for how this fault type is used by Web services. A designer of a Web services application often uses interfaces defined by others. Managing faults in such an application is more difficult when each interface uses a different convention for representing common information in fault messages. Support for problem determination and fault management can be enhanced by specifying Web services fault messages in a common way. When the information available in faults from various interfaces is consistent, it is easier for requestors to understand faults. It is also more likely that common tooling can be created to assist in the handling of faults.

WS-ServiceGroup

This defines a means by which Web services and WS-Resources can be aggregated or grouped together for a domain-specific purpose. In order for requestors to form meaningful queries against the contents of the ServiceGroup, membership in the group must be constrained in some fashion. The constraints for membership are expressed by intension using a classification mechanism. Further, the members of each intension must share a common set of information over which queries can be expressed.

- Endpoint references to Web services that encapsulate stateful resources can be renewed when they become invalid; for example, due to a transient failure in the network.
- Stateful resources can be aggregated for domain-specific purposes.

Additional related specifications have been developed that will be considered by OASIS for the WSRF. These were developed by Computer Associates, Fujitsu, Globus Alliance, Hewlett-Packard, IBM, and the University of Chicago. The motivation for these specifications is that whereas Web service implementations typically do not maintain state information during their interactions, their interfaces must frequently allow for the manipulation of state, that is, data values that persist across and evolve as a result of Web service interactions [174]. For example, an online airline reservation system must maintain state concerning flight status, reservations made by specific customers, and the system itself: its current location, load, and performance. Web service interfaces that allow requestors to query flight status, make reservations, change reservation status, and manage the reservation system must necessarily provide access to this state. In the Web Services Resource Framework, we model state as stateful resources and codify the relationship between Web services in terms of an implied resource pattern.

The concepts introduced in this section will be revisited and used in the chapters that follow.

CHAPTER 4

Standards Supporting
Grid Computing: OGSI

In recent years, grid computing has attracted the attention of the technical community with the evolution of on-demand and autonomic computing, as discussed earlier in the book. The business community is also starting to consider its potential merits at this juncture. For any kind of new technology, corporate and business decision makers typically seek answers to a set of questions, including "Are there firm standards to support the technology and its widespread deployment?" As the reader is well aware by now, grid computing is a process of coordinated resource sharing and problem solving in dynamically established, multiinstitutional, virtual organizations, and/or in a computing utility environment [3, 49]. The grid computing paradigm based on open standards can also be utilized to define a "portable" form of outsourcing (call it "open source outsourcing"), in which service providers can be painlessly replaced "at will." A vision for grid computing is as follows:

> IBM's ultimate vision for Grid is a utility model over the Internet, where clients draw on computer power much as they do now with electricity. With more than 60% of IT budgets dedicated to maintenance and integration—a percentage that continues to rise—the need to reduce complexity and management demands is a pressing one. [118]

Quite a bit more remains to be done at the technical level, however, to make this vision a true reality. We noted in the preceding chapters that the absence of standards has been a retarding factor in the recent past with regard to widespread commercial deployment of grid computing. The following quote is representative of the recent situation:

> Much of Grid Computing is undiscovered country, and many groups are turning their attention to the emerging open standards. In many ways, the discussions about Grid Services parallel those around Internet and XML standards in the mid-1990s. [143]

The fundamental purpose of a computing grid is to make use of broadly distributed computing power across any kind of network (including a company's own

A Networking Approach to Grid Computing. By Daniel Minoli   **101**
ISBN 0-471-68756-1 © 2005 John Wiley & Sons, Inc.

computing power); but without standards, one is actually limiting, rather than extending, one's ability to "harvest" and utilize spare computer power on remote computers [73]. An array of heterogeneous resources comprise a grid, and, hence, it is nearly a mandatory necessity that these resources interact and behave in a well-defined and consistent manner [50]. Without industry-wide standards, it is a technical challenge to achieve highly effective interactions among resources, especially when these belong to different (virtual) organizations.

In 1987, in a Bellcore/Telcordia Special Report, in a section called "Network for a Computing Utility," we stated that protocols and standards were critical:

> The proposed service provides the entire apparatus to make the concept of the Computing Utility possible. . . . [S]ecurity and accounting . . . are much more complex in the distributed, public (grid) environment. . . . This service is basically feasible once a transport and switching network with strong security and accounting (chargeback) capabilities is deployed. A high degree of intelligence in the network is required . . . a physical network is required . . . security and accounting software is needed . . . protocols and standards will be needed to connect servers and users, as well as for accounting and billing. These protocols will have to be developed before the service can be established. . . . [6]

Lately, the Global Grid Forum has indeed started a number of architecture standardization efforts in order to provide the required improved software interoperability, security (confidentiality, integrity, and availability), resource definition, resource discovery, policy, and grid manageability. The Global Grid Forum is a community-initiated forum of researchers and practitioners working on grid computing, and a number of working groups are producing technical specs, documenting user experiences, and writing implementation guidelines. The need for open standards that define these interactions and foster interoperability between components supplied from different sources has been the motivation for the Open Grid Service Architecture/Open Grid Services Infrastructure (OGSA/OGSI) milestone documentation published by the Forum [50]. The OGSI documentation was published in 2002 and a draft version of the OGSA was published late in 2003. As of press time, both OGSA and OGSI were still "works in progress." Efforts are also underway in the GGF to document "best practices," implementation guidelines, and ancillary standards for grid technologies. More than two dozen working groups at the GGF were defining grid standards in areas such as applications and programming models, data management, security, performance, scheduling, and resource management. The Globus Toolkit™ is an open architecture, open standards, commercial-grade tool for building computational grids; it is a widely cited, solid reference implementation of the OGSA/OGSI standards.

As noted in Chapter 3, OGSA is a service-oriented architecture (SOA). OGSI defines mechanisms for creating, managing, and exchanging information among *grid services.* A grid service is a Web service that conforms to a set of conventions (interfaces and behaviors) that define how a client interacts with a grid capability (Web services were also briefly discussed in Chapter 3) [84]. Specifically, the

Table 4.1   OGSA/OGSI documents

- *Anatomy of the Grid.* This architecture white paper by Ian Foster et al. defines the field of grid computing. The document includes a description all of a grid's constituent parts and what they do.

- *Physiology of the Grid.* This white paper by Ian Foster et al. explains how grid computing can be supported using Web services. It presents details about OGSA and grid semantics (i.e., services). Along with the *Anatomy of the Grid,* these two papers provide a detailed overview of grid computing, and are inputs to the OGSA and OGSI specifications.

- *Open Grid Services Architecture (OGSA),* GWD-R (draft-ggf-ogsa-ogsa-011), Draft. Editors: I. Foster et al., September 23, 2003.

- *Open Grid Services Infrastructure (OGSI),* Version 1.0, Editors: S. Tuecke et al., June 27, 2003.

OGSI specifications define the standard interfaces and behaviors of a grid service, building on a Web services base [36]. This approach provides a common and open standards-based mechanism to access various grid services using existing industry standards such as SOAP, XML, and WS-Security [17].

In this chapter, we drill down on grid standards and standardization activities. At press time, there was only a short list of grid computing standards: the just-cited draft Open Grid Services Architecture, along with its companion implementation standard, the Open Grid Services Infrastructure. Here, we first take a high-level view of OGSI (and OGSA, by implication), and then we proceed with a more detailed assessment; because of the relationship between the two documents, OGSA also gets some coverage in this chapter. In Chapter 5, we focus more extensively on OGSA itself[1]. For a more comprehensive description of these concepts, the reader should consult the key references listed in Table 4.1. The purpose of this chapter is to highlight the standardization progress and *not* to provide a comprehensive specification and/or description. Also, note that some of the details may change over time (but one hopes that the overall framework is stable, after these many years of research and funding); hence, the reader should always consult the latest GGF documentation, after acquiring a basic understanding through the material presented herewith.

In Chapter 3, we also made reference to WSRF. The recent practical confluence of grid and Web services that occurred in the early 2000s led to the development by OASIS of WSRF, which is a specification for performing grid computing on top of Web services [173]. This is another level of useful standardization that will further foster the introduction of "open-source" grid computing services on the part of GSPs.

---

[1]This ordering is dictated by the chronological development of these documents; pedagogically, the reverse order is ostensively better.

4.1 INTRODUCTION

The architecture for grid computing is defined in the Open Grid Services Architecture that describes the overall structure and the services to be provided in grid environments [17]. Figure 4.1 depicts the network's role in supporting a (standardized) grid. Figure 4.2 is the reference diagram that illustrates the OGSA. The companion implementation standard, the OGSI, is a formal specification of the concepts described by the OGSA; it specifies a set of service primitives that define a nucleus of behavior common to all Grid Services [17]. OGSI, in effect, is the base infrastructure on which the OGSA is built, as illustrated pictorially in Figure 4.3.

As just noted, OGSA is a distributed interaction and computing architecture that is based around the grid service concept, assuring interoperability on heterogeneous systems. As a result, different types of systems can communicate and share information. OGSA allows system composition to perform a specific task, or solve a challenging problem, by using distributed resources over the interconnecting network [119]. The grid architecture is now being developed based on Internet protocols (for example, communication, routing, file transfer, name resolution, etc.) and services. The grid architecture defined in OGSA leverages the emerging Web services to define the WSDL interfaces that are relevant to the grid environment [119]. We introduced WSDL in Chapter 3; as noted there, WSDL is an XML-based language used to describe the services that a business offers, and to provide a mechanism for individuals and businesses to access these services in an on-line fashion (WSDL is derived from Microsoft's SOAP and IBM's Network Accessible Service Specification Language). WSDL is used in the context of UDDI. As noted in Chapter 3, UDDI is an XML-based global registry for businesses that enables these businesses to list themselves and their services on the Internet (additional details are

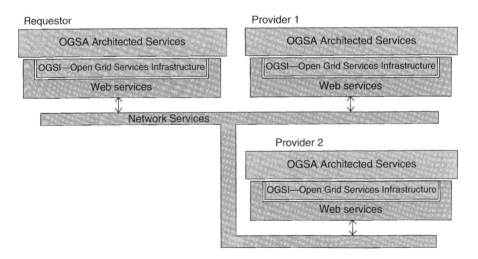

Figure 4.1 Networking role.

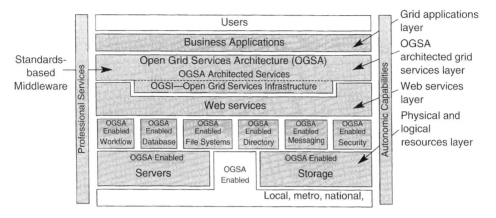

Figure 4.2 Basic functional model for grid environment.

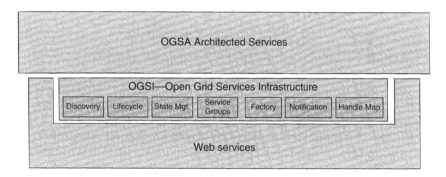

Figure 4.3 OGSA reliance on OGSI.

provided later in the chapter). These conventions and other OGSI mechanisms associated with grid service creation and discovery provide for the controlled, fault-resilient, and secure management of the distributed and often long-lived "state" that is commonly required in advanced distributed applications[2] [84]. Building on Web services standards (see relevant footnotes in Chapter 3), OGSA takes the view that a *grid service is simply a Web service that conforms to a particular set of conventions;* that is, grid services are defined in terms of standard WSDL with some (minor) extensions. With this approach, OGSA is driving the hosting environment to accept modifications or additions for supporting the repertoire of grid services [119]. To further clarify what a grid service is, note that an OGSI-compliant grid service defines a subclass of Web services whose ports all inherit capabilities from a standard grid service port (so a grid service is a Web service that conforms to a set

---

[2]"State" is "service data," that is a collection of XML elements encapsulated as service data elements

of conventions that provide for controlled, fault-resilient, and secure management of stateful services) [31].

The running of an individual service (for example, an information query) is called a *service instance*. Services and service instances can be "lightweight" and transient, or they can be long-term tasks that require "heavy-duty" support from the grid. Services and service instances can be dynamic or interactive, or they can be batch processed. Service can run at scheduled times, or they can run at arbitrary times [143]. As seen in Figure 4.3, grid services include:

- Discovery
- Lifecycle
- State management
- Service groups
- Factory
- Notification
- Handle map

These are defined later on in the chapter, in Section 4.5.

A "layering" approach is used to the extent possible in the definition of a grid architecture because it is advantageous for higher-level functions to use common lower-level functions [44]. With standards, IT staff at Fortune 500 companies will have a predictable way to find, identify, and utilize new grid services as they become available. Additionally, OGSA will provide for interoperability between grids that may have been built using different underlying tools [17]. Hence, the going-forward grid-deployment "formula" is as follows: open standards and protocols lead to the development of services, and services are the foundation blocks of the grid. Services allow users to carry out tasks on the grid. Grid functionality can include the following, among others [143]: information queries, network bandwidth allocation, data management/extraction, processor requests, managing data sessions, and balance workloads.

As noted, the GGF is comprised of a set of working groups that are developing standards and best practices for distributed computing ("grids" and "metacomputing") efforts, including those specifically aimed at very large data sets, high-performance computing, and P2P. GGF represents a merger of three technical communities: those in North America (originally called "Grid Forum"), Asia–Pacific, and the European Grid Forum (eGrid). GGF has become a key point of coordination, information exchange, and collaboration for staff involved in large-scale R&D programs in the U.S., Europe, Canada, and Asia–Pacific. Major industry players are getting involved in the Global Grid Forum and we are seeing increasing and significant collaboration with industry groups such as the Peer-to-Peer Working Group and the New Productivity Initiative. The GGF creates an opportunity to reduce the costs of these programs, to accelerate their progress, and to promote and ensure common practices and interoperability between large-scale "metacomputing" or "grid" systems [33]. IBM and other industry leaders, researchers, and representa-

Table 4.2 GGF groups

Group	Status*
Grid Policy Architecture	Approved RG
IPv6	Approved WG
Open Grid Service Architecture Authorization	Approved WG
Authority Recognition	Approved RG
Job Submission Description Language	Approved WG
Grid File System	Pending WG
Proposed New Groups	
Astronomy—RG	Pending BOF
Business Grid—RG	Pending BOF
CDOLM—WG	Pending BOF
Grid API—WG	Pending BOF
Grid Federations—RG	Pending BOF
Metadata Management—WG	Pending BOF
Persistent Archives	Pending BOF
Ubiqutous Computing—RG	Pending BOF
Workflow Management—WG	Pending BOF

*Group status as of press date.
RG = Research Group.
WG = Working Group.

tives from a variety of grid software vendors are actively involved in the work of the GGF [17]. The development of OGSA specifications receives support from IBM, the U.S. Department of Energy, the National Science Foundation, and NASA's Information Power Grid program, among others [36]. It is hoped that the OGSI will form the basis of a number of open and more functional grid implementations. GGF sponsors the Global Grid Forum Workshop. By press time, nine workshops had been held.[3] OGSA-related GGF groups include (also see Table 4.2) [36]:

- The Open Grid Services Architecture Working Group (OGSA-WG)
- The Open Grid Services Infrastructure Working Group (OGSI-WG)
- The Open Grid Service Architecture Security Working Group (OGSA-SEC-WG)
- Database Access and Integration Services Working Group (DAIS-WG)

Table 4.3 identifies key documents and/or specifications that have been produced by the GGF. Since OGSA builds on Web services, it likely will incorporate

---

[3]For example, the 11th Global Grid Forum Workshop (GGF11, June 2004) was held in Honolulu. GGF10 took place in Berlin in March 2004. GGF8 (June 2003) was held in Seattle, Washington, June 24–27, 2003. The meeting gathered about 700 grid stakeholders, practitioners, and experts. The 7th Global Grid Forum Workshop (GGF7, March 2003) GGF7 was held in Tokyo, Japan, March 4–7, 2003. The meeting gathered about 780 grid stakeholders, practitioners, and experts for the first GGF held in the Asia–Pacific region.

Table 4.3 Recent GGF documents

Document	Title	Document type	Group	Author(s)
GFD.1	GGF Document Series	Community practice	GFSG	C. Catlett
GFD.2	GGF Structure	Community practice	GFSG	C. Catlett, I. Foster, W. Johnston
GFD.3	GGF Management	Community practice	GFSG	C. Catlett, I. Foster, W. Johnston
GFD.4	Ten Actions When Superscheduling	Informational	SRM	J. Schopf
GFD.5	Advanced Reservation API	Experimental	SRM	V. Sander, A. Roy
GFD.6	Attributes for Communication between Scheduling Instances	Informational	SRM	U. Schwiegelshohn, R. Yahyapour
GFD.7	A Grid Monitoring Architecture	Informational	ISP	B. Tierney, R. Aydt, D. Gunter, W. Smith, M. Swany, V. Taylor, R. Wolski
GFD.8	A Simple Case Study of a Grid Performance System	Informational	ISP	R. Aydt, D. Gunter, W. Smith, M. Swany, B. Tierney, V. Taylor
GFD.9	Overview of Grid Computing Environments	Informational	APME	G. Fox, M. Pierce, D. Gannon, M. Thomas
GFD.10	Grid User Services Common Practices	Informational	APME	J. Towns, J. Ferguson, D. Frederick, G. Myers
GFD.11	Grid Scheduling Dictionary of Terms and Keywords	Informational	SRM	M. Roehrig, W. Ziegler, P. Wieder
GFD.12	Security Implications of Typical Grid Computing Usage Scenarios	Informational	SEC	M. Humphrey, M. Thompson
GFD.13	Grid Database Access and Integration: Requirements and Functionalities	Informational	DATA	M. P. Atkinson, V. Dialani, L. Guy, I. Narang, N.W. Paton, D. Pearson, T. Storey, P. Watson
GFD.14	Services for Data Access and Data Processing on Grids	Informational	DATA	V. Raman, I. Narang, C. Crone, L. Haas, S. Malaika, T. Mukai, D. Wolfson, C. Baru

Table 4.3 *Continued*

Document	Title	Document type	Group	Author(s)
GFD.15	Open Grid Services Infrastructure	Recommendation	ARCH	S. Tuecke, K. Czajkowski, I. Foster, J. Frey, S. Graham, C. Kesselman, T. Maguire, T. Sandholm, D. Snelling, P. Vanderbilt
GFD.16	Global Grid Forum Certificate Policy Model	Community practice	SEC	R. Butler, T. Genovese
GFD.17	CA-based Trust Issues for Grid Authentication and Identity Delegation	Informational	SEC	M. Thompson, D. Olson, R. Cowles, S. Mullen, M. Helm
GFD.18	An Analysis of the UNICORE Security Modal	Informational	SEC	T.Goss-Walter, R.Letz, T. Kentemich, H.-C Hoppe
GFD.19	Job Description for GGF Steering Group Members	Informational	GFSG	J. Schopf, P. Clarke, B. Nitzberg, C. Catlett
GFD.20	GridFTP: Protocol Extensions to FTP for the Grid	Recommendation	DATA	W. Allcock
GFD.21	GridFTP Protocol Improvements	Experimental	DATA	I. Mandrichenko

specifications defined within the W3C, IETF, OASIS, and other standards organizations. A first (prototype) grid service implementation that follows GGF specs was demonstrated in early 2002, at the Globus Toolkit tutorials held at Argonne National Laboratory; the Globus Toolkit Version 3.0 is based on the OGSI standard. Other commercial products are also under development.

In the material that follows, we provide some (additional) motivations for standardization. We then look at architectural elements. Following this discussion, we look at OGSA/OGSI from a pragmatic perspective, and then from a more formal perspective.

4.2 MOTIVATIONS FOR STANDARDIZATION

We opened this chapter by noting the reasons for pursuing industry and/or regulatory (de jure) standards. In this area, most standards are of the "industry type." When we use the word *standard* throughout this book, we mean, multivendor, public-forum, or accredited-standard-organization-based agreements.[4] The lack of standards has meant that companies, developers, and organizations have had to develop and support grid technology using proprietary techniques and solutions, thereby limiting its deployment potential. An effective grid relies on making use of computing power, whether via a LAN, over an extranet, or through the Internet. To use computing power efficiently, one needs to support a gamut of computing platforms; also; one needs a flexible mechanism for distributing and allocating the work to individual clients [73].

The time is now ripe for grid standards. Telecommunications standards were developed (in critical-mass fashion) in the 1980s, Internet standards were developed (in critical-mass fashion) in the 1990s, and, hopefully, grid computing standards will developed (in critical-mass fashion) in the 2000s.

A closed, proprietary environment limits the ease with which one can distribute work, who can become a service provider, who can be a service requester, and how one can find out about the available grid resources. The reader can grasp the limitations of a nonstandard approach by considering any of the better-known grid computing projects on the Internet. For example, consider distributed.net. This organization is a "loosely knit" group of computer users located all over the world, that takes up challenges and run projects that require a lot of computing power. It solves these by utilizing the combined idle processing cycles of their members' computers.[5] To illustrate the point about standards, to become a service provider for the distributed.net grid, one must download a specific client that is capable of processing the work units from the corresponding servers. However, with a distributed.net client installed, one can only process work units supplied by distributed.net [73]. Furthermore, distributed.net service providers can only process those work units supplied by distributed.net. For example, if distributed.net wanted to allow their service providers (those people with the distributed.net client installed) to process SETI@Home work units, it would be problematic. distributed.net would have to re-

[4]The fact that a company such as Cisco (as an example) may have an internal "standard" to paint their boxes teal is not what we are focusing on here.

[5]Distributed.net (Distributed Computing Technologies Inc.) was founded in 1997. Now the project has grown to encompass thousands of users around the world. distributed.net's computing power has grown to become equivalent to that of more than 160,000 PII 266 MHz computers working 24 hours a day, 7 days a week, 365 days a year. Distributed.net counts five victories so far. Their first victory was announced on October 19, 1997, indicating that they had found the correct solution for RSA Labs' RC5-32/12/7 56-bit secret-key challenge. Confirmed by RSA Labs, the key 0x532B744CC20999 presented the plaintext message for which they had been searching for 250 days. Their second victory was announced on February 24, 1998, confirming that they had found the correct solution for the RSA Labs' DES-II-1 56-bit secret-key challenge. The solution key was 76 9E 8C D9 F2 2F 5D EA and was found after 40 days of work. Their third victory was on January 19, 1999, when they found the correct key to the RSA Labs DES-III contest, with the help of the Electronic Frontier Foundation's "Deep Crack" customized DES cracker. The correct key was 92 2C 68 C4 7A EA DF F2. Their fourth victory was on 16 January 2000 when they received the winning key for the CSC contest from a SPARC in the United States after searching for 62 days. Their fifth victory was on July 14, 2002 when they completed the RC5-64 project, after 1,757 days of working. The key was 0x63DE7DC154F4D039. (From the distributed.net FAQ.)

deploy their service provider functionality. They would also have to redesign many of their discovery and distribution systems to allow different work units to be deployed to service providers, and they would need to update their statistical analysis on completed units to track it all properly [73].

As this example illustrates, standards are critical to making the computing utility concept a reality. On the other hand, a corporate user just looking to secure better utilization of its platforms and internal resources could start with a vendor-based solution and then move up to a standards-based solution in due course. Some specific areas where a lack of grid standards limit deployment are [73]:

- **Data management.** For a grid to work effectively, there is a need to store information and distribute it. Without a standardized method for describing the work and how it should be exchanged, one quickly encounters limits related to the flexibility and interoperability of the grid.

- **Dispatch management.** There are a number of approaches that can be used to handle brokering of work units and to distribute these work units to client resources. Again, not having a standard method for this restricts the service providers that can connect to the grid and accept units of work from the grid; this also restricts the ability of grid services users to submit work.

- **Information services.** Metadata[6] about the grid service helps the system to distribute information. The metadata is used to identify requesters (grid users), providers, and their respective requirements and resource availability. Again, without a standard, one can only use specific software and solutions to support the grid applications.

- **Scheduling.** As covered in Chapter 3, work must be scheduled across the service providers to ensure they are kept busy. To accomplish this, information about remote loads must be collected and administered. A standardized method of describing the grid service enables grid implementations to specify how work is to be scheduled.

- **Security.** Without a standard for the security of a grid service and for the secure distribution of work units, one runs the risk of distributing information to the "wrong" clients. Although proprietary methods can provide a level of security, they limit accessibility.

- **Work unit management.** Grid services require management of the distribution of work units to ensure that the work is uniformly distributed over the

---

[6]Metadata is a definition or description of data. In data processing, metadata is definitional data that provides information about, or documentation of, other data managed within an application or environment. For example, metadata would document data about data elements or attributes (name, size, data type, etc.), data about records or data structures (length, fields, columns, etc.), and data about data (where it is located, how it is associated, ownership, etc.). Metadata may include descriptive information about the context, quality, condition, or characteristics of the data. For example, the data of a newspaper story is the headline and the story, whereas the metadata describes who wrote it, when and where it was published, and what section of the newspaper it appears in. Metadata can help us determine who content is for and where, how, and when it should appear. (This footnote based on [82]; also see [83].)

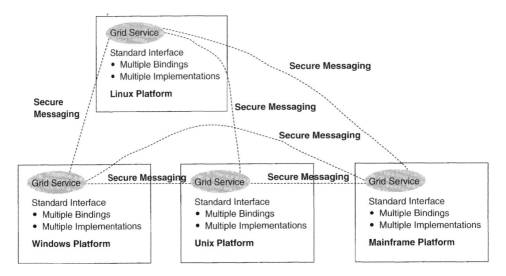

Figure 4.4 Example of a service-oriented architecture.

service providers. Without a standard way of advertising and managing this process, efficiencies are degraded.

Looking from the perspective of the grid applications developer, a closed environment is similarly problematic because to make use of the computing resources across a grid, the developer must utilize a specific tool kit or environment to build, distribute, and process work units. The closed environment limits the choice of grid-resident *platforms* on which work units can be executed and, at the same time, also limits how one uses and distributes work and/or requests to the grid [73]. It also means that one cannot combine or adopt other grid solutions for use within an organization's grid without redeploying the grid software.

The generic advantages of the standardized approach are well known, since they apply across any number of disciplines. In the context of grid computing, they all reduce to one basic advantage: the extension and expansion of the resources available to the user for grid computing. From an end user's perspective, standardization translates into the ability to purchase middleware and grid-enabled applications from a variety of suppliers in an off-the-shelf, shrink-wrapped fashion. Figure 4.4 depicts an example of the environment that one aims to achieve.

For example, standard APIs enable application *portability;* without standard APIs, application portability is difficult to accomplish (different platforms access protocols in different ways). Standards enable cross-site *interoperability;* without standard protocols, interoperability is difficult to achieve. Standards also enable the deployment of a *shared infrastructure* [44].[7] Use of the OGSI standard, therefore, provides the following benefits [73]:

---

[7]This material is licensed for use under the terms of the Globus Toolkit Public License.

- **Increased effective computing capacity.** When the resources utilize the same conventions, interfaces, and mechanisms, one can transparently switch jobs among grid systems, both from the perspective of the server as well as from the perspective of the client. This allows grid users to use more capacity and allows clients a more extensive choice of projects that can be supported on the grid. Hence, with a gamut of platforms and environments supported, along with the ability to more easily publish the services available, there will be an increase in the effective computing capacity.

- **Interoperability of resources.** Grid systems can be more easily and efficiently developed and deployed when utilizing a variety of languages and a variety of platforms. For example, it is desirable to mix service-provider components, work-dispatch tracking systems, and systems management; this makes it easier to dispatch work to service providers and for service providers to support grid services.

- **Speed of application development.** Using middleware (and/or toolkits) based on a standard expedites the development of grid-oriented applications supporting a business environment. Rather than spending time developing communication and management systems to help support the grid system, the planner can, instead, spend time optimizing the business/algorithmic logic related to the processing the data.

For useful applications to be developed, a rich set of grid services (the OGSA architected services) need to be implemented and delivered by both open source efforts (such as the Globus project) and by middleware software companies. In a way, OGSI and the extensions it provides for Web services are necessary but insufficient for the maturation of the service-oriented architecture; the next required step is that these standards be fully implemented and truly observed (in order to provide portability and interoperability) [3, 33, 39, 50, 84, 114]. Figure 4.5 depicts a simple environment to put the network-based services in context.

4.3 ARCHITECTURAL CONSTRUCTS

The previous section described the use of standards in a grid environment. This section, based in part on reference [44], provides some basic machinery for the construction of a grid-oriented architecture. The sections that follow actually describe the OGSA/OGSI architecture itself.

4.3.1 Definitions

A service-oriented grid architecture is *descriptive:* it provides a common vocabulary for use when describing grid systems. A grid architecture is *normative:* it provides guidance and identifies key areas in which services are required. A grid architecture is also *prescriptive:* it defines standard "Intergrid" protocols and application programmer interfaces (APIs) to facilitate creation of interoperable grid systems and portable applications.

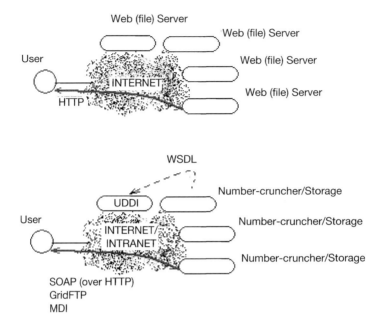

Figure 4.5 Network-based grid services.

The challenge (also known in grid computing circles as The Programming Problem) is, how does the planner develop robust, secure, long-lived, well-performing applications for dynamic, heterogeneous grids? "Grid applications" are diverse (data, collaboration, computing, sensors, and so on); hence, it seems unlikely there is one single solution to this problem. Most applications have been written "from scratch," with or without grid services. Application-specific libraries have been shown to provide significant benefits. What is needed is [44]:

- Abstractions and models to add to speed and robustness of development
- Tools to ease application development and diagnose common problems
- Code/tool sharing to allow reuse of code components developed by others

The evolving grid architecture aims at addressing these and other issues. Some important definitions used in the development of the architecture definition are:

- Resource: An entity that is to be shared; e.g., computers, storage, data, software
- Network protocol: A formal description of message formats and a set of rules for message exchange across a variety of subsystems such as SANs, LANs, MANs, WANs, and Global Area Networks (GANs)
- Network-enabled service: Implementation of protocols that define a set of capabilities

- API: A specification for a set of routines to facilitate application development. Examples include GSS API (Generic Security Service API) and MPI (Message Passing Interface). A protocol can have multiple APIs (e.g., TCP/IP APIs include BSD sockets, Winsock, and System V streams). The protocol provides interoperability: programs using different APIs can exchange information and one does not need to know the remote APIs. An API can have multiple protocols. For example, MPI provides portability: any correct program compiles and runs on a platform. MPI does not provide interoperability: all processes must link to same SDK (e.g., MPICH and LAM versions of MPI).

- Software Development Kit (SDK): A particular instantiation of an API (it may consists of libraries and tools). Examples of SDKs include MPICH and Motif Widgets.

- Syntax: Rules for encoding information. Examples include XML, Condor ClassAds, Globus RSL, X.509 certificate format [Request For Comments (RFC) 2459], Cryptographic Message Syntax (RFC 2630), and ASN.1. Syntaxes are distinct from protocols in the sense that a syntax may be used by many protocols (e.g., XML), and be useful for other purposes. Syntaxes may be layered (e.g., Condor ClassAds uses XML, which uses ASCII.)

4.3.2 Protocol Perspective

A protocol-oriented view of a grid architecture emphasizes the following [44]:

- Development of grid protocols and services
  - ○ Protocol-mediated access to remote resources
  - ○ New services, for example, resource brokering
  - ○ Mostly (extensions to) existing protocols
- Development of grid APIs and SDKs
  - ○ Interfaces to grid protocols and services
  - ○ Facilitate application development by supplying higher-level abstractions
- The model is the Internet, which has been hugely successful

A protocol and a service have a complementary, dualistic relationship. A well-defined protocol provides a clearly defined service; a well-defined service can be supported by a clearly defined protocol. Protocols, services, APIs, and SDKs will, ideally, be largely self-contained.

Some key upper-layer protocols and constructs of interest that were described in the previous chapter are included in Table 4.4 [43]. The GGF's approach has been to propose a set of core services as basic infrastructure. These core services have been used to construct high-level, domain-specific solutions. The design principles are: keep participation cost low, enable local control, provide support for adaptation, and use the "IP hourglass" model of many applications using a few core services to support many fabric elements (e.g., OSs). Figure 1.6 in Chapter 1 provided

Table 4.4 Key protocols/constructs of interest (upper layers)

SOAP (Simple Object Access Protocol)	Transport mechanism that is independent of the underlying platform and protocol. For example, two disparate processes can communicate without the intimate knowledge of systems and platforms on which both of them are running.
UDDI (Universal Description, Discovery and Integration)	Repository that stores the descriptions of Web services.
WSDL (Web Services Definition Language)	A language that provides a way of describing the specific interfaces of Web services and APIs, and is used by UDDI.
XML (eXtensible Markup Language)	A meta-language used to describe grammatical descriptions of objects and describing data structures in an open manner. It is similar in appearance to HTML, is platform-neutral, and can be used to represent both documents and data.
WSIL (WS-Inspection Language)	An XML-based format utilized to facilitate the discovery and aggregation of Web service descriptions in a simple and extensible fashion.

a working model used by the Globus Project. Figure 4.6 expands on this by showing the APIs and SDKs involved. As the figure shows, protocols, services, and APIs occur at each level. Middleware such as Globus Toolkit can help planners move in the direction of standards. Globus Toolkit has emerged as the de facto standard for several important *connectivity, resource,* and *collective* protocols, as shown in Figure 4.6. It should be noted, however, that the going-forward graphical representation of the functional hierarchy is the one shown in Figure 4.2.

The guiding principle is that that each distinct programming environment should not be required to implement the grid-supporting protocols and services from scratch. Rather, such environments should be able to share common code that implements core functionality. The code ought to be robust, well-architected, and self-consistent. Also, the code ought to be "open source," with broad industry input. The emerging OGSA architecture enhances Web services to accommodate requirements of the grid. OGSA defines the semantics of a grid service instance including service instance creation, naming, lifetime management, and communication protocols. The services that are included in the standardization are: *Discovery, Lifecycle, State Management, Service Groups, Factory, Notification,* and *Handle Map* (these are discussed in Section 4.5.)

Web services address discovery and invocation of persistent services. Web services provide interface to persistent states of entire enterprises. In grid environments, one must also support transient service instances that are created/destroyed dynamically; hence, there is a need for interfaces to the states of distributed activities. It follows that there are crucial implications for how services are managed, named, discovered, and used. The creation of a new grid service instance involves the creation of a new process in the hosting environment that has the primary re-

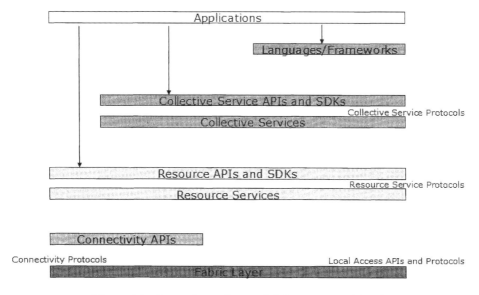

Figure 4.6 Protocols, services, and APIs occur at each level.

sponsibility for ensuring that the services it supports adhere to defined grid service semantics. Multiple grid service instances may correspond to the same grid service interface [119]. OGSA enables application programs and application users to create transient services, as well as to discover and evaluate the properties of existing/available grid services. The OGSA Factory, Registry, Grid Service, and Handle Map interfaces address the creation of transient grid service instances, the service discovery, and characterization in a VO [119].

In a typical grid-enabled environment, one needs to capture users' inputs, discover a grid service (or several services, as needed), create a grid service instance, invoke the grid service instance, and display the results. Often, this is facilitated with tool kits, such as the Globus Toolkit (also see Chapter 6). Tool kits are implementations of the OGSA Development Framework (OGSADF). OGSADF is a mechanism to actualize the grid service definition of OGSA (interface) through the run-time hosting environment.

We expand below some of the concepts that we introduced in Chapter 3.

WSDL Use For Web Services. The material that follows in this subsection provides a short tutorial on WSDL; this subsection is based on reference [16].

WSDL is a language that provides a way of describing the specific interfaces of Web services and APIs. Practically, WSDL can be perceived as a document written in XML. The WSDL document describes a Web service; it specifies the location of the service and the operations (or methods) the service exposes. In other words, it is an XML language for describing the syntax of Web service interfaces and their locations. The WSDL specification calls it "an XML format for describing network services as a set of endpoints operating on messages containing either document-oriented or procedure-oriented information." WSDL 1.1 was submitted as a W3C Note by Ariba, IBM, and Microsoft for describing services for the W3C XML Activity on XML Protocols in early 2001 (a W3C Note is made available by the W3C for discussion only; publication of a Note by W3C indicates no endorsement by W3C or the W3C Team, or any W3C Members). The first Working Draft of WSDL 1.2 was released by W3C in 2002.

A WSDL document defines a Web Service using these major elements:

Element	Defines
<portType>	The operations performed by the Web service. An abstract set of operations supported by one or more endpoints.
<message>	The messages used by the Web service. An abstract definition of the data being communicated.
<types>	The data types used by the Web service. Provides information about any complex data types used in the WSDL document. When simple types are used, the WSDL document does not need this section.
<binding>	The communication protocols used by the Web service. Describes how the operation is invoked by specifying concrete protocol and data format specifications for the operations and messages.
<port>	Specifies a single endpoint as an address for the binding, thus defining a single communication endpoint.
<service>	Specifies the port address(es) of the binding. The service is a collection of network endpoints or ports.

A WSDL document has a definitions element that contains the types, message, portType, binding, and service elements as described in the table above. The main structure of a WSDL document looks like this:

```
<definitions>
<types>
  definition of types . . .
</types>

<message>
  definition of a message . . .
</message>

<portType>
  definition of a port . . .
</portType>

<binding>
  definition of a binding . . .
</binding>

</definitions>
```

A WSDL document can also contain other elements such as extension elements and a service element that makes it possible to group together the definitions of several Web services in one single WSDL document.

WSDL Services. A service definition element supports the following attributes (it defines one or more services):

- Name is optional.
- targetNamespace is the logical namespace for information about this service. WSDL documents can import other WSDL documents, and setting target-Namespace to a unique value ensures that the namespaces do not clash.
- xmlns is the default namespace of the WSDL document, and it is set to http://schemas.xmlsoap.org/wsdl/. All the WSDL elements such as <definitions>, <types> and <message> reside in this namespace.
- xmlns:xsd and xmlns:soap are standard namespace definitions that are used for specifying SOAP-specific information as well as data types.
- xmlns:tns stands for this namespace.

WSDL Ports. The <portType> element is the most important WSDL element. It defines a Web service, the operations that can be performed, and the messages that are involved. The <portType> element can be compared to a function library (or a module or class) in a traditional programming language.

WSDL Messages. The <message> element defines the data elements of an operation. Each message can consist of one or more parts. The parts can be compared to the parameters of a function call in a traditional programming language.

WSDL Types. The <types> element defines the data type that are used by the Web service. For maximum platform neutrality, WSDL uses XML Schema syntax to define data types.

WSDL Bindings. Binding is an operation that occurs when the service requestor invokes or initiates an interaction with the service at runtime, using the binding details in the service description to locate, contact, and invoke the service [43]. The <binding> element defines the message format and protocol details for each port.

WSDL Example. Below is a simplified fraction of a WSDL document. In this example the "portType" element defines "glossaryTerms" as the name of a "port," and "getTerm" as the name of an "operation." The "getTerm" operation has an "input message" called "getTermRequest" and an "output message" called "getTermResponse." The "message" elements define the "parts" of each message and the associated data types. Compared to traditional programming, "glossaryTerms" is a function library, "getTerm" is a function with "getTermRequest" as the input parameter, and "getTermResponse" as the return parameter.

```
<message name="getTermRequest">
  <part name="term" type="xs:string"/>
</message>

<message name="getTermResponse">
  <part name="value" type="xs:string"/>
</message>
<portType name="glossaryTerms">
  <operation name="getTerm">
    <input message="getTermRequest"/>
    <output message="getTermResponse"/>
  </operation>
</portType>
```

Web Services Inspection Language (WSIL). WSIL is a simple, lightweight mechanism for Web service discovery. WSIL is an XML document format designed to facilitate the discovery and aggregation of Web service descriptions in a simple and extensible fashion. Created by IBM and Microsoft and released in late 2001, WSIL is notable because of its simpler document-based approach; compared with UDDI, it is more lightweight and leverages existing Web architectures better. WSIL's model is a decentralized one that is document-based, and leverages the existing Web infrastructure already in place (e.g., WSDL) [153].

According to the introduction to the specification [149, 152], WSIL provides an XML format for assisting in the inspection of a site for available services and a set of rules for how inspection-related information should be made available for consumption. A WISL document provides a means for aggregating references to preexisting service-description documents that have been authored in any number of formats. These inspection documents are then made available at the point of offering for the service as well as through references that may be placed within a content medium such as HTML. Specifications have been proposed to describe Web services at different levels and from various perspectives. It is the goal of WSDL to describe services at a functional level. What has not yet been provided by these proposed standards is the ability to tie together, at the point of offering for a service, these various sources of information in a manner that is simple to create and to use; the WSIL specification addresses this need by defining an XML grammar that facilitates the aggregation of references to different types of service-description documents, and then provides a well-defined pattern of usage for instances of this grammar. By doing this, the WSIL specification provides a means by which to inspect sites for service offerings. Repositories already exist in which descriptive information about Web services has been gathered together. The WS-Inspection specification provides mechanisms with which these existing repositories can be referenced and utilized, so that the information contained in them need not be duplicated if such a duplication is not desired [149, 152].

Universal Description, Discovery, and Integration (UDDI). UDDI is a standard Web service description format and Web service discovery protocol. A

UDDI registry can contain metadata for any type of service, with "best practices" already defined for services described by WSDL. By organizing Web services into groups associated with categories or business processes, UDDI allows more efficient search and discovery of Web services. The UDDI specification defines a four-tier hierarchical XML schema that provides a model for publishing, validating, and invoking information about Web services [85]. XML was chosen because it offers a platform-neutral view of data and allows hierarchical relationships to be described in a natural way. UDDI uses standards-based technologies, such as common Internet protocols (TCP/IP and HTTP), XML, and SOAP. There are two types of UDDI registries: *public* UDDI registries that serve as aggregation points for a variety of businesses to publish their services, and *private* UDDI registries that serve a similar role within organizations.

It should be noted that UDDI implements a service discovery using a centralized model of one or more repositories containing information on multiple business entities and the services they provide. An analogy for UDDI would be the telephone Yellow Pages: multiple businesses are grouped and listed with a description of the goods or services they offer and how to contact them. The specification provides a high level of functionality through SOAP by specifically requiring an infrastructure to be deployed with substantial overhead and costs associated to its use. The UDDI schema aims at providing a more business-centric perspective (as noted, UDDI describes an online electronic registry that provides an information structure where various business entities register themselves and the services they offer through their WSDL definitions) [149, 152]. On the other hand, WSIL approaches service discovery in a decentralized fashion—service description information can be distributed to any location using a simple extensible XML document format. Unlike UDDI, it does not concern itself with business-entity information, nor does it specify a particular service-description format. WSIL works under the assumption that one is already familiar with the service provider, and relies on other service-description mechanisms such as the WSDL, discussed above [153]. WSIL complements, rather then competes with, UDDI.

A UDDI registry consists of the following data structure types [85]:

- **businessEntity.** The top-level XML element in a business UDDI entry. businessEntity captures the data partners require to find information about a business service, including its name, industry or product category, geographic location, and optional categorization and contact information. It includes support for "yellow pages" taxonomies to search for businesses by industry, product, or geography.
- **businessService.** The logical child of a businessEntity data structure as well as the logical parent of a bindingTemplate structure. businessService contains descriptive business service information about a group of related technical services, including the group name, a brief description, technical service description information, and category information.
- **bindingTemplate.** The logical child of a businessService data structure. bindingTemplate contains data that is relevant for applications that need to in-

voke or bind to a specific Web service. This information includes the Web service URL and other information describing hosted services, routing and load balancing facilities, and references to interface specifications.

- **tModel.** Descriptions of specifications for Web services or taxonomies that form the basis for technical fingerprints. tModel's role is to represent the technical specification of the Web service, making it easier for Web service consumers to find Web services that are compatible with a particular technical specification. Web service consumers can easily identify other compatible Web services based on the descriptions of the specifications for Web Services in the tModel structure. For example, to send a business partner's Web service an RFP, the invoking service must know not only the location/URL of the service, but what format the RFP should be sent in, what protocols are appropriate, what security is required, and what form of a response will result after sending the RFP.

Simple Object Access Protocol (SOAP). SOAP is a lightweight, XML-based protocol for exchanging information in a decentralized, distributed environment. SOAP supports different styles of information exchange, including:[8]

- Information exchange modeled after the Remote Procedure Call. This type of exchange allows for request–response processing, in which an endpoint receives a procedure-oriented message and replies with a correlated response message.
- Information exchange modeled on a message-oriented mechanism. This type of exchange supports organizations and applications that need to exchange business or other types of documents; a message is sent but the sender may not expect or wait for an immediate response.

SOAP has the following features:

- Protocol independence
- Language independence
- Platform and operating system independence
- Support for SOAP XML messages incorporating attachments (using the multipart MIME structure)

A SOAP message consists of (i) a SOAP envelope that encloses two data structures, (ii) the SOAP header and the SOAP body, and, (iii) information about the namespaces used to define them. The header is optional; when present, it conveys information about the request defined in the SOAP body. For example, it might contain transactional, security, contextual, or user profile information. The body contains a Web service request or reply to a request in XML format.

---

[8]The rest of this short section is based on reference [85].

The SOAP specification provides a standard way to encode requests and responses. SOAP messages, when used to carry Web service requests and responses, can conform to the WSDL definition of available Web services. WSDL can define the SOAP message used to access the Web services, the protocols over which such SOAP messages can be exchanged, and the Internet locations where these Web services can be accessed. The WSDL descriptors can reside in UDDI or other directory services, and they can also be provided via configuration or other means such as in the body of SOAP request replies.

The specification describes the structure and data types of message payloads using XML schema. The way that SOAP is used for the message and response of a Web service is:

- The SOAP client uses an XML document that conforms to the SOAP specification and that contains a request for the service.
- The SOAP client sends the document to a SOAP server, and the SOAP servlet running on the server handles the document using, for example, HTTP or HTTPS.
- The Web service receives the SOAP message, and dispatches the message as a service invocation to the application providing the requested service.
- A response from the service is returned to the SOAP server, again using the SOAP protocol, and this message is returned to the originating SOAP client.

SOAP provides a way to leverage the industry investment in XML. Also, since SOAP is typically defined over "firewall-friendly" protocols such as HTTP and SMTP, the industry investment in firewall technology is leveraged as well. Thus, by defining SOAP as an essential part of Web services, the industry will likely enjoy volume production use of Web services far sooner than if other strategies had been employed.

4.3.3 Going From "Art" To "Science"

At press time, there was an effort to move grid technology from "art" to "science" in order to achieve widespread commercialization of the technology. The material that follows describes the "art" (various grid solutions) from which proponents are now in the process of building out a "science" (OGSA/OGSI). The following mechanisms are needed (also refer to Figure 4.7) [44]:

- *Fabric Layer Protocols and Services.* This mechanism includes the plethora of resources that may be shared (discrete computers, file systems, archives, metadata catalogs, networks, sensors, etc.) The goal is to impose few constraints on low-level technology within a grid; namely, connectivity and resource-level protocols form the "neck of the hourglass"—there are many applications on top, many resources at the bottom, and a few key specifications in the middle.

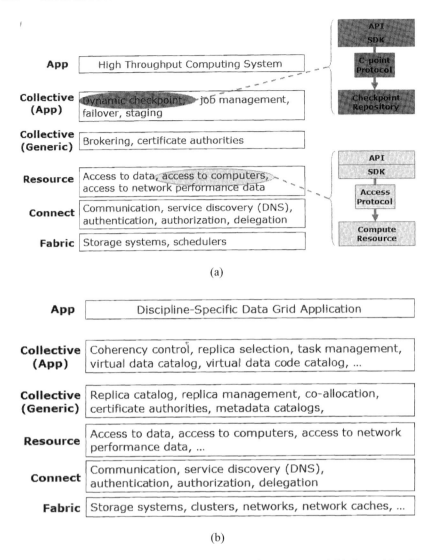

Figure 4.7 Examples of (a) high-throughput computing system and (b) data grid architecture.

- *Connectivity Layer Protocols and Services.* These mechanisms focus on communications (Internet protocols: IP, DNS, routing, etc.), and on security. The Globus Toolkit GSI is an example of "art" with uniform authentication, authorization, and message-protection mechanisms in a multiinstitutional setting. It provides single sign-on, delegation, and identity mapping using public key technology, along with a Secure Sockets Layer (SSL), X.509, and GSS-API [167].

- *Resource Layer Protocols and Services.* These address the following:
  - ○ Remote allocation, reservation, monitoring, and control of compute resources. The Globus Toolkit Grid Resource Allocation Management (GRAM) is an example.
  - ○ High-performance data access and transport. The Globus Toolkit GridFTP protocol (FTP extensions) is an example.
  - ○ Access to structure and state information. The Globus Toolkit Grid Resource Information Service (GRIS) is an example.
- *Collective Layer Protocols and Services.* These address the following:
  - ○ Index servers (also known as metadirectory services) that provide custom views on dynamic resource collections assembled by a community
  - ○ Resource brokers for resource discovery and allocation (e.g., Condor Matchmaker)
  - ○ Replica catalogs
  - ○ Replication services
  - ○ Coreservation and coallocation services
  - ○ Workflow management services

Table 4.5 depicts some examples of grid programming technologies that comprise the "art" of grid as we seek to move up to the "science" [44].

4.4   WHAT IS OGSA/OGSI? A PRACTICAL VIEW

As should be clear by now, OGSA aims at addressing standardization (for interoperability) by defining the basic framework of a grid application structure. Some of the mechanisms employed in the standards formulation of grid computing were described in the previous section and in Chapter 3. In essence, the OGSA standard defines what grid services are, what they should be capable of, and what technologies they are based on. OGSA, however, does not go into specifics of the technicalities of the specification; instead, the aim is to help classify what is and is not a grid system [73]. It is called an *architecture* because it is mainly about describing and building a well-defined set of interfaces from which systems can be built, based on open standards such as WSDL [143].

The objectives of OGSA are to [3, 33, 39, 50, 84, 114]:

- Manage resources across distributed heterogeneous platforms.
- Support QoS-oriented Service Level Agreements (SLAs). The topology of grids is often complex; the interactions between/among grid resources are almost invariably dynamic. It is critical that the grid provide robust services such as authorization, access control, and delegation.
- Provide a common base for autonomic management. A grid can contain a plethora of resources, along with an abundance of combinations of resource

Table 4.5 Examples of grid programming technologies

MPICH-G2: Grid-enabled message passing (Message Passing Interface)
- CoG Kits, GridPort: Portal construction, based on N-tier architectures
- Condor-G: workflow management
- Legion: object models for grid computing
- Cactus: Grid-aware numerical solver framework

Portals
- N-tier architectures enabling thin clients, with middle tiers using grid functions
  - Thin clients = web browsers
  - Middle tier = e.g., Java Server Pages, with Java CoG Kit, GPDK, GridPort utilities
  - Bottom tier = various grid resources
- Numerous applications and projects, e.g.,
  - Unicore, Gateway, Discover, Mississippi Computational Web Portal, NPACI Grid Port, Lattice Portal, Nimrod-G, Cactus, NASA IPG Launchpad, Grid Resource Broker

High-Throughput Computing and Condor
- High-throughput computing
  - Processor cycles/day (week, month, year?) under nonideal circumstances
  - "How many times can I run simulation X in a month using all available machines?"
- Condor converts collections of distributively owned workstations and dedicated clusters into a distributed high-throughput computing facility
- Emphasis on policy management and reliability

Object-Based Approaches
- Grid-enabled CORBA
  - NASA Lewis, Rutgers, ANL, others
  - CORBA wrappers for grid protocols
  - Some initial successes
- Legion
  - University of Virginia
  - Object models for grid components (e.g., "vault" = storage, "host" = computer)

Cactus: Modular, portable framework for parallel, multidimensional simulations

Construct codes by linking
- Small core: management services
- Selected modules: Numerical methods, grids and domain decomps, visualization and steering, etc.
- Custom linking/configuration tools
- Developed for astrophysics, but not astrophysics specific

---

configurations, conceivable resource-to-resource interactions, and a litany of changing state and failure modes. Intelligent self-regulation and autonomic management of these resources is highly desirable.

- Define open, published interfaces and protocols for the interoperability of diverse resources. OGSA is an open standard managed by a standards body.

- Exploit industry standard integration technologies and leverage existing solutions where appropriate. The foundation of OGSA is rooted in Web services, for example, SOAP and WSDL, are a major part of this specification.

OGSA's companion OGSI document consists of specifications on how work is managed, distributed, and how service providers and grid services are described. The Web services component is utilized to facilitate the distribution and the management of work across the grid. Because Web services offer a transparent method of communication between hosts (irrespective of the underlying language or platform), one can utilize these services to transfer work, to describe resources and configuration information, and to communicate and dispatch grid information. WSDL provides a simple method of describing and advertising the Web services that support the grid's application [73]. Summarizing these observations, OGSA is the blueprint, OGSI is a technical specification, and Globus Toolkit is an implementation of the framework.

OGSA describes and defines a Web-services-based architecture composed of a set of interfaces and their corresponding behaviors to facilitate distributed resource sharing and accessing in heterogeneous dynamic environments. A set of services based on open and stable protocols can hide the complexity of service requests by users or by other elements of a grid. Grid services enable *virtualization;* virtualization, in turn, can transform computing into a ubiquitous infrastructure that is more akin to an electric or water utility, as envisioned in the opening paragraph of this chapter [143].

OGSA relies on the definition of grid services in WSDL, which, as noted, defines, for this context, the *operations names, parameters,* and their *types* for grid service access [119]. Based on the OGSI specification, a grid service instance is a Web service that conforms to a set of conventions expressed by the WSDL as service interfaces, extensions, and behaviors [49]. Because the OGSI standard is based on a number of existing standards (XML, Web services, WSDL), it is an open and standards-based solution. This implies that, in the future, grid services can be built that are compatible with the OGSI standard, even though they may be based on a variety of different languages and platforms [73].

Specifically, the grid service interface (see Table 4.6 and [84, 114]) is described by WSDL, which defines how to use the service. A new tag, gsdl, has been added to the WSDL document for grid service description. The UDDI registry and WSIL document are used to locate grid services. The transport protocol SOAP is used to connect data and applications for accessing grid services. All services adhere to specified grid service interfaces and behaviors. Grid service interfaces correspond to portTypes in WSDL used in current Web services solutions [119].

The interfaces of grid services address discovery, dynamic service-instance creation, lifetime management, notification, and manageability; the conventions of grid services address naming and upgrading issues. The standard interface of a grid service includes multiple bindings and implementations ("implementations" include Java and procedural/object-oriented computer programming languages). Grid services, such as the ones just cited, can, therefore, be deployed on different hosting environments, even different operating systems. OGSA also provides a grid security mechanism to ensure that all the communications between services are secure.

Table 4.6 Proposed OGSA grid service interfaces*

Port type	Operation	Description
GridService	FindServiceData	Query a variety of information about the grid service instance, including basic introspection information (handle, reference, primary key, home handle map: terms to be defined), richer per-interface information, and service-specific information (e.g., service instances known to a registry). Extensible support for various query languages.
	SetTermination Time	Set (and get) termination time for grid service instance
	Destroy	Terminate grid service instance.
Notification-Source	SubscribeTo-NotificationTopic	Subscribe to notifications of service-related events, based on message type and interest statement. Allows for delivery via third-party messaging services.
Notification-Sink	Deliver Notification	Carry out asynchronous delivery of notification messages.
Registry	RegisterService UnregisterService	Conduct soft-state registration of grid service handles. Deregister a grid service handle.
Factory	CreateService	Create new grid service instance.
Handle Map	FindByHandle	Return grid service reference currently associated with supplied grid service handle.

*Interfaces for authorization, policy management, manageability, and likely other purposes remain to be defined.

The definition of standard service interfaces and the identification of the protocol(s) are addressed in current OGSA specifications [119].

Service capabilities (that is, the services offered by a particular company or organization) are widely used in existing Web services solutions. Likewise, grid services are characterized by the capabilities they afford. A grid service capability could be comprised of computational resources, storage resources, networks, programs, databases, and so on. A grid service implements one or more interfaces, where each interface defines a set of method operations that is invoked by constructing a method call through, method signature adaptation using SOAP [119].

Like the majority of Web services, OGSI services use WSDL as a service description mechanism. There are two fundamental requirements for describing Web services based on the OGSI [49]:

1. The ability to describe interface inheritance—a basic concept with most of the distributed object systems.
2. The ability to describe additional information elements with the interface definitions.

The WSDL 1.1 specification lacks the two abilities just enumerated in its definition of portType. Hence, at press time, OGSI was utilizing an extension WSDL called GWSDL (Grid-extension to WSDL); however, there was a consensus among OGSI Work Group members to eventually use the WSDL 1.2 specification (when WSDL 1.2 reaches the recommendation stage, it may eliminate the need to use GWSDL) [49]. The WSDL 1.2 working group has agreed to support the just-listed features through portType inheritance and an open content model for portTypes. As an interim decision, OGSI has developed a new schema for portType definition (extended from normal WSDL 1.1 schema portType Type) under the new GWSDL namespace definition. Another noteworthy aspect of OGSI is the naming convention adopted for the portType operations and the lack of support for operator overloading. In these cases, OGSI follows the same conventions as described in the suggested WSDL 1.2 specification [49].

From a near-term implementation perspective, the Globus Toolkit is the primary solution that supports the new standards of the OGSA/OGSI system (we cover some details of this in Chapter 6.) IBM is also deploying a version of WebSphere, the Web development platform, that makes use of grid technology to help spread the load of requests for a Web application [73].

4.5 OGSA/OGSI SERVICE ELEMENTS AND LAYERED MODEL

4.5.1 Key Aspects

This section provides a more detailed view of OGSA/OGSI. Keep in mind that the key principle of OGSA is that all grid resources—both logical and physical—are modeled as services. There are two main logical components of OGSA: (i) the Web-services-plus-OGSI layer, and (ii) the OGSA-architected services layer. Four main layers comprise the OGSA architecture, as shown in Figure 4.2 [3, 33, 39, 50, 84, 114]:

- **Grid applications layer.** This layer is the user-visible layer. It supports user applications. Eventually, a "rich" set of grid-architected services is expected to be developed.

- **OGSA-architected grid services layer.** Services in this layer include: Discovery, Lifecycle, State management, Service Groups, Factory, Notification, and Handle Map. These services are based on the Web services layer. The GGF was working at press time to define many of these architected grid services in areas such as program execution, data services, and core services. Some are already defined (and implementations have already appeared.)

- **Web Services layer, plus the OGSI extensions that define grid services.** The OGSI specification defines grid services and builds on standard Web services technology. OGSI exploits the mechanisms of Web services such as XML and WSDL to specify standard interfaces, behaviors, and interaction for all grid resources. OGSI extends the definition of Web services to provide capabilities for dynamic, stateful, and manageable Web services that are required to model the resources of the grid.

- **Physical and logical resources layer.** The concept of resources is central to OGSA and to grid computing in general. Resources, as discussed in Chapter 3, comprise the capabilities of the grid. *Physical* resources include servers, storage, and network. Above the physical resources are *logical resources.* Logical resources provide additional function by virtualizing and aggregating the resources in the physical layer. General-purpose middleware such as file systems, database managers, directories, and workflow managers provide these abstract services on top of the physical grid.

The GGF OGSA working group found it necessary to augment core Web services functionality to address grid services requirements. OGSI extends Web services by introducing interfaces and conventions in two main areas [3, 33, 39, 50, 84, 114]:

1. "Interfaces." One needs to take into consideration the dynamic and potentially transient nature of services in a grid: particular service instances may come and go as work is dispatched, as resources are configured and provisioned, and as system state changes. Therefore, grid services need interfaces to manage the creation, destruction, and life-cycle management of these dynamic services.
2. "State." Grid services typically have attributes and data associated with them. This is similar in concept to the traditional structure of objects in object-oriented programming: objects have behavior and data. Likewise, Web services were found to be in need of being extended to support state data associated with grid services. Basic Web services are stateless (e.g., add, subtract). Most real-world applications involve stateful transactions [e.g., query (sd2), get-data (row3-row17)]. State is linked to a "handle" or sessionID as a parameter. Protocols such as SOAP, SMTP, and FTP use state mechanisms (sessionID, packet headers, TCP sockets, respectively).

Consistent with these two observations, OGSI introduces an interaction model for grid services. The interaction model provides a uniform way for software developers to model and interact with grid services by providing interfaces for discovery, life cycle, state management, creation and destruction, event notification, and reference management (these services were depicted in Figure 4.3.) Below, we list interfaces and conventions that OGSI introduces [3, 33, 39, 50, 84, 114].

- **Factory.** A mechanism (interface) that provides a way to create new grid services. Factories may create temporary instances of limited function, such as a scheduler creating a service to represent the execution of a particular job; or they may create longer-lived services such as a local replica of a frequently used data set. Not all grid services are created dynamically; for example, some services might be created as the result of an instance of a physical resource in the grid, such as a processor, storage, or network device.

- **Life cycle.** A mechanism architected to prevent grid services from consuming resources indefinitely without requiring a large-scale distributed "garbage collection" scavenger. Every grid service has a termination time set by the service creator or factory. Because grid services may be transient, grid service instances are created with a specified lifetime. The lifetime of any particular service instance can be negotiated and extended, as required, by components that are dependent on or manage that service. In turn, a client with appropriate authorization can use termination time information to check the availability (lease period) of the service; the client can also request to extend the current lease time by sending a keep-alive message to the service with a new termination time. If the service accepts this request, the lease time can be extended to the new termination time requested by the client. This soft-state life cycle is controlled by appropriate security and policy decisions of the service, and the service has the authority to control this behavior (for example, a service can arbitrarily terminate a service or can extend its termination time even while the client holds a service reference) [49].

- **State management.** As previously noted, grid services can have "state." OGSI specifies a framework for representing this state, called service data, and a mechanism for inspecting or modifying that state, named Find/SetServiceData. Furthermore, OGSI requires a minimal amount of state in service data elements that every grid service must support, and requires that all services implement the Find/SetServiceData portType.

- **Service groups.** Service groups are collections of grid services that are indexed (using service data described above) for some specific purpose. For example, they might be used to collect all the services that represent the resources in a particular cluster node within the grid.

- **Notification.** Services interact with one another by exchanging messages based on service invocation. The state information (the service data described above) that is modeled for grid services changes as the system runs. Many interactions between grid services require dynamic monitoring of changing state. Notification applies a traditional publish/subscribe paradigm to this monitoring. Grid services support an interface (NotificationSource) to permit other grid services (NotificationSink) to subscribe to changes. The internal state of a grid service can keep track that this grid service has received one or zero messages. This reliable message delivery mechanism guaranteed by the internal state can build business-oriented transactions [119]. In a transient stateful service, OGSA provides a mechanism to capture the state information associated with any operation that fails. If an operation fails, the keep-alive messages cease if there is no service client for invoking this running service instance. Then the grid service instance automatically times out and frees the computing resources associated with this service instance [119].

- **Handle Map.** This deals with service identity. When Factories are used to create a new instance of a Grid Service, the Factory returns the identity of the newly instantiated service. This identity is composed of two parts: a Grid Service Handle (GSH) and a Grid Service Reference (GSR). A GSH provides a reference the grid service indefinitely; GSR can change within the grid services lifetime. The Handle Map interface provides a way to obtain a GSR given a GSH. The user application invokes *create Grid Service* requests on the Factory interface to create a new service instance. The newly created service instance associated with the grid service interface will be automatically allocated computing resources. Meanwhile, an initial lifetime of the instance can be specified before the service instance is created. The newly created service instance will keep the user credentials for performing further interactions with other systems over the Internet. The newly created grid service instance will be automatically assigned a globally unique name called the GSH, which is used to distinguish this specific service instance from other grid service instances [119].

These enhancements are specified in OGSI. As the OGSI specification was finalized and implementations began to appear, some standards organizations became interested in incorporating a portion of the functionality outlined in OGSI within appropriate Web services standards; hence, over time, it is expected that much of the OGSI functionality will be incorporated in Web services standards [3, 33, 39, 50, 84, 114].

4.5.2 Ancillary Aspects

Drilling down an additional level of detail, one can further categorize grid-architected services into four categories, as shown in Figure 4.8:

- Grid core services
- Grid program execution services
- Grid data services
- Domain-specific services

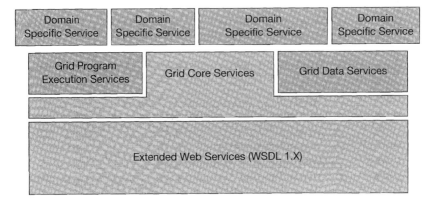

Figure 4.8 The structure of OGSA architected services.

Grid Core Services. Figure 4.9 shows that the grid core services are composed of four main types of services:

1. Service management
2. Service communication
3. Policy management
4. Security

Unlike the OGSI functions that are largely implemented as extensions to basic Web Services protocols and an interaction model, these core services are actually implemented as grid services (upon the OGSI base). These services are considered core primarily because it is expected that they will be broadly exploited by most higher-level services implemented either in support of program execution or data access, or as domain-specific services [3, 33, 39, 50, 84, 114].

Service management. Service management provides functions that manage the services deployed in the distributed grid. It automates a variety of installation, maintenance, monitoring, and troubleshooting tasks within a grid system. Service management includes functions for provisioning and deploying the system compo-

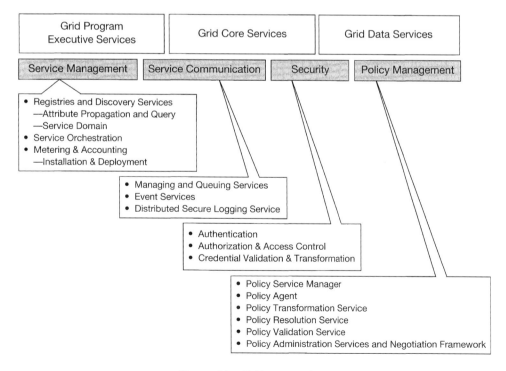

Figure 4.9   Grid core services.

nents; it also includes functions for collecting and exchanging data about the operation of the grid. This data is used for both "online" and "offline" management operations, and includes information about faults, events, problem determination, auditing, metering, accounting, and billing [3, 33, 39, 50, 84, 114].

Service communication. This includes a gamut of functions that support the basic methods for grid services to communicate with each other. These functions support several communication models that may be composed to enable effective inter-service communication, including queued messages, publish–subscribe event notification, and reliable distributed logging [3, 33, 39, 50, 84, 114]. As previously noted, grid services can be published to a UDDI registry, or WSIL documents; the UDDI registry becomes a central place to store such information about and locations for grid services that enables publishing and searching of trading partners' businesses and their grid services. Also, as previously noted, there are two types of UDDI registries: private and public. Application developers and/or service providers can publish the grid services to the public UDDI registries operated by IBM, Microsoft, HP, or SAP. If one wants to publish one's own private or confidential grid services, one can use a private UDDI registry. As an alternative, for testing purposes or for small-scale integration, a developer can publish the company's grid services to WSIL documents, since WSIL enables grid services discovery, deployment, and invocation without the need for a UDDI registry. WSIL provides the means for aggregating references of preexisting service description documents that have been authored in any number of formats; these inspection documents are then made available on a Web site [119]. Figure 4.10 illustrates an example grid service deployment and publishing diagram. The Remote Procedure Call servlet of SOAP and the real implementation of the grid services can be deployed on an application server. All the invocation messages will be captured by the SOAP Remote Procedure Call servlet that routes the messages to the corresponding grid service [119].

Policy services. These create a general framework for creation, administration, and management of policies and agreements for system operation. Policy services include policies governing security, resource allocation, and performance, as well as an infrastructure for "policy-aware" services to use policies to govern their operation. Policy and agreement documents provide a mechanism for the representation and negotiation of terms between service providers and their clients (either user requests or other services); terms include specifications, requirements, and objectives for function, performance, and quality that the suppliers and consumers exchange and that they can then use to influence their interactions [3, 33, 39, 50, 84, 114].

Security services. Security services support, integrate, and unify popular security models, mechanisms, protocols, and technologies in a way that enables a variety of systems to interoperate securely. These security services enable and extend core

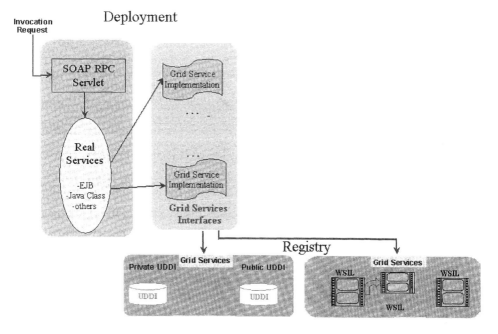

Figure 4.10 An example grid service deployment and publishing diagram.

Web services security protocols and bindings and provide service-oriented mechanisms for authentication, authorization, trust policy enforcement, credential transformation, and so on [3, 33, 39, 50, 84, 114].

Grid Program Execution Services. Grid program execution services are depicted in Figure 4.11. Mechanisms for job scheduling and workload management implemented as part of this class of services are central to grid computing and the ability to virtualize processing resources. Although OGSI and core grid services are generally applicable to any distributed computing system, the grid program execution service class is unique to the grid model of distributed task execution that supports high-performance computing, parallelism, and distributed collaboration [3, 33, 39, 50, 84, 114].

Grid Data Services. Grid data services are also depicted in Figure 4.11. These interfaces support the concept of data virtualization and provide mechanisms related to distributed access to information of many types including databases, files, documents, content stores, and application-generated streams. Services that comprise the grid data services class complement the computing virtualization conventions specified by program execution services (OGSA placing data resources on an equivalent level with computing resources). Grid data services will exploit and vir-

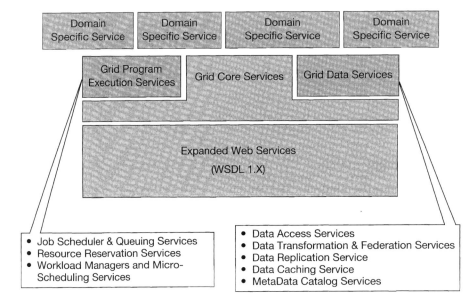

Figure 4.11 Grid program execution services and grid data services.

tualize data using placement methods like data replication, caching, and high-performance data movement to give applications required QoS access across the distributed grid. Methods for federating multiple disparate, distributed data sources may also provide integration of data stored under differing schemas such as files and relational databases [3, 33, 39, 50, 84, 114].

Domain-Specific Services. The three categories discussed above (grid core services, grid program execution services, and grid data services) represent areas of active work by GGF research or working groups. Over time, as these services mature, domain-specific services can also be specified. Domain-specific services will make use of the functionality that these services supply. It is critical that the GGF working groups are concentrating on specifying a broad set of useful grid services that software vendors and developers can then begin to implement.

4.5.3 Implementations of OGSI

As the core of the grid service architecture, OGSI needs to be hosted on a delivery platform that supports Web services. Vendors probably will not compete by offering a wide range of implementations of OGSI. Instead, as part of the "fabric" of Web services implementations, vendors that offer OGSI implementations will likely directly use existing open source implementations provided by organizations like Globus, and/or they will integrate implementations with their hosting platform products like WebSphere, WebLogic, Apache, or .NET [50]. However, grid-architected services provide some opportunities for vendors and organizations to compete and differentiate themselves. This competition will create an "economy" of grid software providers whose innovation will help drive the acceptance of standards like OGSI/OGSA, and this will allow customers to build systems out of interoperable components. Areas of functionality in grid program execution and data services will require innovation and novel approaches, and these may well speed the market acceptance of grid solutions and provide market opportunities for vendors. In Figure 4.12, one notices that grid core services are likely to see a mix of open source reference implementations and vendor-provided "value added" implementations. The bulk of technologies in this area will likely be commoditized, but areas like policy and security could provide vendors a chance to differentiate themselves [3, 33, 39, 50, 84, 114]. Implementations in grid program execution and data services are expected to consist largely of value-added products. These areas represent business opportunities for vendors to integrate leading middleware offerings within the OGSA framework and allow a rich "ecosystem" of grid solutions to develop. Although OGSI/OGSA is novel in the respect that it extends Web Services, it is not clear that software vendors will be able to differentiate themselves based on

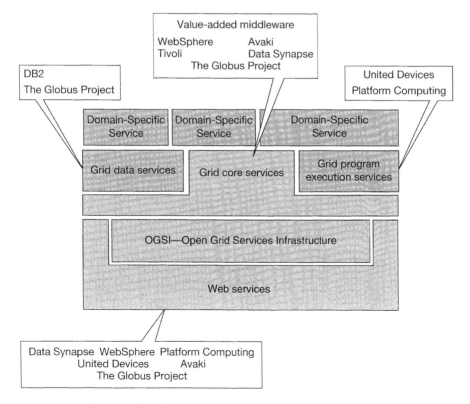

Figure 4.12 Grid program execution and data services hosting.

the quality of their core services implementations; such differention will likely be based on the business creativity and/or import of their domain-specific implementations [50].

For OGSA to grow in acceptance it needs to be implemented on multiple hosting platforms. The Globus Toolkit 3 (GT3) historically was the first full-scale implementation of the OGSI standard (see Chapter 6 for a more extensive discussion of this topic). GT3 was developed by the Globus Project, a research and development project focused on enabling the application of grid concepts to scientific, engineering, and commercial computing. It is expected that many of the OGSI implementations will be delivered via the open source development model and that existing reference implementations (GT3) will be used unmodified in appropriate hosting environments [50]. GT3 is written in Java language using the J2EE framework; however, nothing limits OGSI from being implemented in other programming languages and hosted in other environments (the term "hosting environment" is used to denote the server in which one or more grid service implementations run). Figure 4.13 shows that a Java implementation of OGSI can be hosted on any of several J2EE environments (such as JBOSS, WebSphere, or BEA Weblogic). However, alternative platforms such as a traditional C or C++ environment or C# and Microsoft .NET are other possible hosting environments [50]. Ideally, a small number of core implementations of OGSI (perhaps one per hosting platform) will be jointly developed by the industry and used in many products [50].

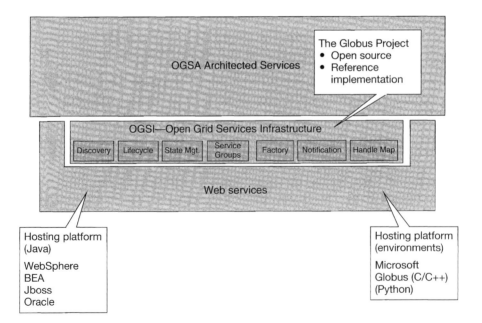

Figure 4.13 OGSI and web services hosting.

4.6   WHAT IS OGSA/OGSI? A MORE DETAILED VIEW

This section[9] provides a more detailed view of OGSI based on the OGSI specification itself. For a more comprehensive description of these concepts, the reader should consult the specification.

4.6.1   Introduction

The OGSA [114] integrates key grid technologies [3, 96] (including the Globus Toolkit) with Web services mechanisms [148] to create a distributed system framework based on the OGSI. A *grid service instance* is a (potentially transient) service that conforms to a set of conventions, expressed as WSDL interfaces, extensions, and behaviors, for such purposes as lifetime management, discovery of characteristics, and notification. Grid services provide for the controlled management of the distributed and often long-lived state that is commonly required in sophisticated distributed applications. OGSI also introduces standard factory and registration interfaces for creating and discovering grid services.

OGSI defines a component model that extends WSDL and XML schema definition to incorporate the concepts of

- Stateful Web services
- Extension of Web services interfaces
- Asynchronous notification of state change
- References to instances of services
- Collections of service instances
- Service state data that augment the constraint capabilities of XML schema definition

The OGSI specification (V1.0 at press time) defines the minimal, integrated set of extensions and interfaces necessary to support definition of the services that will compose OGSA. The OGSI V1.0 specification proposes detailed specifications for the conventions that govern how clients create, discover, and interact with a grid service instance. That is, it specifies (1) how grid service instances are named and referenced; (2) the base, common interfaces (and associated behaviors) that all grid services implement; and (3) the additional (optional) interfaces and behaviors associated with factories and service groups. The specification does *not* address how grid services are created, managed, and destroyed within any particular hosting environment. Thus, services that conform to the OGSI specification are not necessarily portable to various hosting environments, but any client program that follows the

---

[9]This section is based on *Open Grid Services Infrastructure (OGSI)* [84], copyright © Global Grid Forum (2003). This document and translations of it may be copied and furnished to others, and derivative works that comment on or otherwise explain it or assist in its implementation may be prepared, copied, published, and distributed, in whole or in part, without restriction of any kind, provided that the above copyright notice and this paragraph are included on all such copies and derivative works

conventions can invoke any grid service instance conforming to the OGSI specification (of course, subject to policy and compatible protocol bindings).

The term *hosting environment* is used in the OGSI specification to denote the server in which one or more grid service implementations run. Such servers are typically language or platform specific; examples include native Unix and Windows processes, J2EE application servers, and Microsoft .NET.

4.6.2 Setting the Context

GGF calls OGSI the "base for OGSA." Specifically, there is a relationship between OGSI and distributed object systems and also a relationship between OGSI and the existing (and evolving) Web services framework. One needs to examine both the client-side programming patterns for grid services and a conceptual hosting environment for grid services. The patterns described in this section are enabled but not *required* by OGSI.

4.6.2.1 Relationship to Distributed Object Systems. A given grid service implementation is an addressable and potentially stateful instance that implements one or more interfaces described by WSDL portTypes. Grid service factories can be used to create instances implementing a given set of portType(s). Each grid service instance has a notion of identity with respect to the other instances in the distributed grid. Each instance can be characterized as state coupled with behavior published through type-specific operations. The architecture also supports introspection in that a client application can ask a grid service instance to return information describing itself, such as the collection of portTypes that it implements.

Grid service instances are made accessible to (potentially remote) client applications through the use of a grid service handle and a grid service reference (GSR). These constructs are basically network-wide pointers to specific grid service instances hosted in (potentially remote) execution environments. A client application can use a grid service reference to send requests, represented by the operations defined in the portType(s) of the target service description directly to the specific instance at the specified network-attached service endpoint identified by the grid service reference.

In many situations, client stubs and helper classes isolate application programmers from the details of using grid service references. Some client-side infrastructure software assumes responsibility for directing an operation to a specific instance that the GSR identifies.

The characteristics introduced above (stateful instances, typed interfaces, global names, etc.) are frequently also cited as fundamental characteristics of *distributed object-based systems.* There are, however, also various other aspects of distributed object models (as traditionally defined) that are specifically not required or prescribed by OGSI. For this reason, OGSI does not adopt the term distributed object model or distributed object system when describing these concepts, but instead uses the term "open grid services infrastructure," thus emphasizing the connections that are established with both Web services and grid technologies.

Among the object-related issues that are not addressed within OGSI are implementation inheritance, service instance mobility, development approach, and hosting technology. The grid service specification does not require, nor does it prevent, implementations based upon object technologies that support inheritance at either the interface or the implementation level. There is no requirement in the architecture to expose the notion of implementation inheritance either at the client side or at the service provider side of the usage contract. In addition, the grid service specification does not prescribe, dictate, or prevent the use of any particular development approach or hosting technology for grid service instances. Grid service providers are free to implement the semantic contract of the service description in any technology and hosting architecture of their choosing. OGSI envisions implementations in J2EE, .NET, traditional commercial transaction management servers, traditional procedural Unix servers, and so forth. It also envisions service implementations in a wide variety of both object-oriented and nonobject-oriented programming languages.

4.6.2.2 Client-Side Programming Patterns. Another important issue is

how OGSI interfaces are likely to be invoked from client applications. OGSI exploits an important component of the Web services framework: the use of WSDL to describe multiple protocol bindings, encoding styles, messaging styles (RPC versus document oriented), and so on, for a given Web service. The Web Services Invocation Framework (WSIF) and Java API for XML RPC (JAX-RPC) are among the many examples of infrastructure software that provide this capability.

Figure 4.14 depicts a possible (but not required) client-side architecture for OGSI. In this approach, a clear separation exists between the client application and

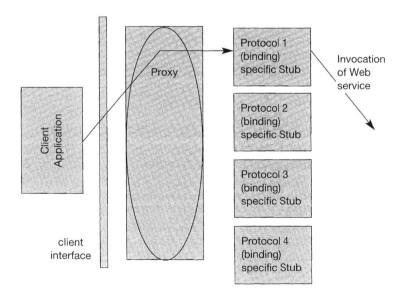

Figure 4.14 Possible client-side runtime architecture.

the client-side representation of the Web service (proxy), including components for marshaling the invocation of a Web service over a chosen binding. In particular, the client application is insulated from the details of the Web service invocation by a higher-level abstraction: the client-side interface.

Various tools can take the WSDL description of the Web service and generate interface definitions in a wide range of programming-language-specific constructs (e.g., Java interfaces and C#). This interface is a front end to specific parameter marshaling and message routing that can incorporate various binding options provided by the WSDL. Further, this approach allows certain efficiencies, for example, detecting that the client and the Web service exist on the same network host, therefore avoiding the overhead of preparing for and executing the invocation using network protocols.

Within the client application runtime, a *proxy* provides a client-side representation of remote service instance's interface. Proxy behaviors specific to a particular encoding and network protocol (*binding,* in Web services terminology) are encapsulated in a *protocol-specific (binding-specific) stub*. Details related to the binding-specific access to the grid service instance, such as correct formatting and authentication mechanics, happen here; thus, the application is not required to handle these details itself.

It is possible, but not recommended, for developers to build customized code that directly couples client applications to fixed bindings of a particular grid service instance. Although certain circumstances demand potential efficiencies gained by this style of customization, this approach introduces significant inflexibility into a system and therefore should only be used under extraordinary circumstances.

The developers of the OGSI specification expect the stub and client-side infrastructure model that we describe to be a common approach to enabling client access to grid services. This includes both application-specific services and common infrastructure services that are defined by OGSA. Thus, for most software developers using grid services, the infrastructure and application-level services appear in the form of a class library or programming language interface that is natural to the caller. WSDL and the GWSDL extensions provide support for enabling heterogeneous tools and enabling infrastructure software.

4.6.2.3 Client Use of Grid Service Handles and References. As noted, a client gains access to a grid service instance through grid service handles and grid service references. A grid service handle (GSH) can be thought of as a permanent network pointer to a particular grid service instance. The GSH does not provide sufficient information to allow a client to access the service instance; the client needs to "resolve" a GSH into a grid service reference (GSR). The GSR contains all the necessary information to access the service instance. The GSR is not a "permanent" network pointer to the grid service instance because a GSR may become invalid for various reasons; for example, the grid service instance may be moved to a different server.

OGSI provides a mechanism, the HandleResolver to support client resolution of a grid service handle into a grid service reference. Figure 4.15 shows a client application that needs to resolve a GSH into a GSR.

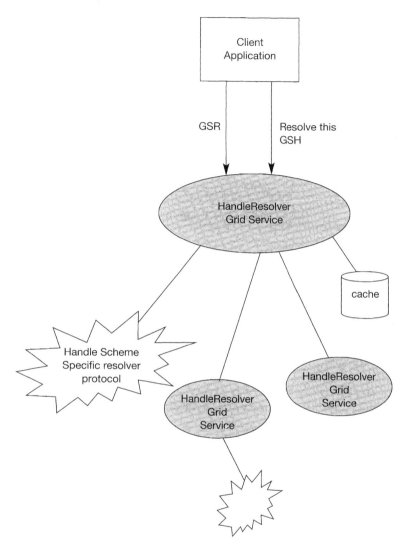

Figure 4.15 Resolving a GSH.

The client resolves a GSH into a GSR by invoking a HandleResolver grid service instance identified by some out-of-band mechanism. The HandleResolver can use various means to do the resolution; some of these means are depicted in Figure 4.15. The HandleResolver may have the GSR stored in a local cache. The HandleResolver may need to invoke another HandleResolver to resolve the GSH. The HandleResolver may use a handle resolution protocol, specified by the particular kind (or scheme) of the GSH to resolve to a GSR. The HandleResolver protocol is specific to the kind of GSH being resolved. For example, one kind of

handle may suggest the use of HTTP GET to a URL encoded in the GSH in order
to resolve to a GSR.

4.6.2.4 *Relationship to Hosting Environment.* OGSI does not dictate a
particular service-provider-side implementation architecture. A variety of approaches are possible, ranging from implementing the grid service instance directly as an operating system process to a sophisticated server-side component model
such as J2EE. In the former case, most or even all support for standard grid service behaviors (invocation, lifetime management, registration, etc.) is encapsulated within the user process; for example, via linking with a standard library. In the
latter case, many of these behaviors are supported by the hosting environment.

Figure 4.16 illustrates these differences by showing two different approaches to
the implementation of argument demarshaling functions. One can assume that, as is
the case for many grid services, the invocation message is received at a network
protocol termination point (e.g., an HTTP servlet engine) that converts the data in
the invocation message into a format consumable by the hosting environment. The
top part of Figure 4.16 illustrates two grid service instances (the oval) associated
with container-managed components (e.g., EJBs within a J2EE container). Here,
the message is dispatched to these components, with the container frequently providing facilities for demarshaling and decoding the incoming message from a format (such as an XML/SOAP message) into an invocation of the component in native programming language. In some circumstances (the oval), the entire behavior
of a grid service instance is completely encapsulated within the component.

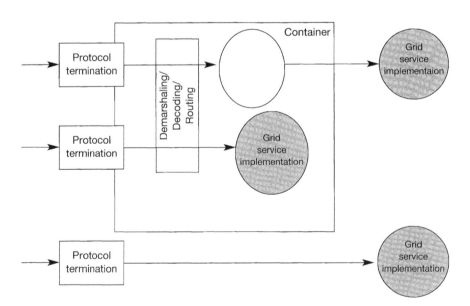

Figure 4.16 Two approaches to the implementation of argument demarshaling functions in
a grid service hosting environment.

In other cases (the oval), a component will collaborate with other server-side executables, perhaps through an adapter layer, to complete the implementation of the grid service behavior. The bottom part of Figure 4.16 depicts another scenario wherein the entire behavior of the grid service instance, including the demarshaling/decoding of the network message, has been encapsulated within a single executable. Although this approach may have some efficiency advantages, it provides little opportunity for reuse of functionality between grid service implementations.

A container implementation may provide a range of functionality beyond simple argument demarshaling. For example, the container implementation may provide lifetime management functions, automatic support for authorization and authentication, request logging, intercepting lifetime management functions, and terminating service instances when a service lifetime expires or an explicit destruction request is received. Thus, one avoids the need to reimplement these common behaviors in different grid service implementations.

4.6.3   The Grid Service

The purpose of the OGSI document is to specify the (standardized) interfaces and behaviors that define a *grid service.* In brief, a grid service is a WSDL-defined service that conforms to a set of conventions relating to its interface definitions and behaviors. Thus, every grid service is a Web service, though the converse of this statement is not true. The OGSI document expands upon this brief statement by

- Introducing a set of WSDL conventions that one uses in the grid service specification; these conventions have been incorporated in WSDL 1.2 [150].
- Defining *service data* that provide a standard way for representing and querying metadata and state data from a service instance
- Introducing a series of core properties of grid service, including:
  - Defining grid service description and grid service instance, as organizing principles for their extension and their use
  - Defining how OGSI models time
  - Defining the grid service handle and grid service reference constructs that are used to refer to grid service instances
  - Defining a common approach for conveying fault information from operations. This approach defines a base XML schema definition and associated semantics for WSDL fault messages to support a common interpretation; the approach simply defines the base format for fault messages, without modifying the WSDL fault message model.
  - Defining the life cycle of a grid service instance

4.6.4   WSDL Extensions and Conventions

As should be clear by now, OGSI is based on Web services; in particular, it uses WSDL as the mechanism to describe the public interfaces of grid services. However, WSDL 1.1 is deficient in two critical areas: lack of interface (portType) extension and the inability to describe additional information elements on a portType (lack of open content). These deficiencies have been addressed by the W3C Web Services Description Working Group [150]. Because WSDL 1.2 is a "work in progress," OGSI cannot directly incorporate the entire WSDL 1.2 body of work. Instead, OGSI defines an extension to WSDL 1.1, isolated to the wsdl:portType element, which provides the minimal required extensions to WSDL 1.1. These extensions to WSDL 1.1 match equivalent functionality agreed to by the W3C Web Services Description Working Group. Once WSDL 1.2 [150] is published as a recommendation by the W3C, the Global Grid Forum is committed to defining a follow-on version of OGSI that exploits WSDL 1.2, and to defining a translation from this OGSI v1.0 extension to WSDL 1.2.

4.6.5   Service Data

The approach to *stateful* Web services introduced in OGSI identified the need for a common mechanism to expose a service instance's state data to service requestors for query, update, and change notification. Since this concept is applicable to any Web service including those used outside the context of grid applications, one can propose a common approach to exposing Web service state data called "serviceData." The GGF is endeavoring to introduce this concept to the broader Web services community.

In order to provide a complete description of the interface of a stateful Web service (i.e., a *grid service*), it is necessary to describe the elements of its state that are externally observable. By externally observable, one means that the state of the service instance is exposed to clients making use of the declared service interface, where those clients are outside of what would be considered the internal implementation of the service instance itself. The need to declare service data as part of the service's external interface is roughly equivalent to the idea of declaring attributes as part of an object-oriented interface described in an object-oriented interface-definition language.

Service data can be exposed for read, update, or subscription purposes. Since WSDL defines operations and messages for portTypes, the declared state of a service must be externally accessed only through service operations defined as part of the service interface. To avoid the need to define serviceData-specific operations for each serviceData element, the grid service portType provides base operations for manipulating serviceData elements by name.

Consider an example. Interface alpha introduces operations op1, op2, and op3. Also assume that the alpha interface consists of publicly accessible data elements of de1, de2, and de3. One uses WSDL to describe alpha and its operations. The OGSI serviceData construct extends WSDL so that the designer can further define the interface to alpha by declaring the public accessibility of certain parts of its state de1, de2, and de3. This declaration then facilitates the execution of operations on the service data of a stateful service instance implementing the alpha interface.

Put simply, the serviceData declaration is the mechanism used to express the elements of the publicly available state exposed by the service's interface. ServiceData elements are accessible through operations of the service interfaces such as those defined in this specification. The private internal state of the service instance is not part of the service interface and is therefore not represented through a serviceData declaration.

4.6.5.1 *Motivation and Comparison to JavaBean Properties.* The

OGSI specification introduces the serviceData concept to provide a flexible, properties-style approach to accessing state data of a Web service. The serviceData concept is similar to the notion of a public instance variable or field in object-oriented programming languages such as Java, Smalltalk, and C++. ServiceData is similar to JavaBean™ properties. The JavaBean model defines conventions for method signatures (getXXX/setXXX) to access properties, and helper classes (BeanInfo) to document properties. The OGSI model uses the serviceData elements and XML schema types to achieve a similar result.

The OGSI specification has chosen not to require getXXX and setXXX WSDL operations for each serviceData element, although service implementers may choose to define such safe get and set operations themselves. Instead, OGSI defines extensible operations for querying (get), updating (set), and subscribing to notification of changes in serviceData elements. Simple expressions are required by OGSI to be supported by these operations, which allows for access to serviceData elements by their names, relative to a service instance. This by-name approach gives functionality roughly equivalent to the getXXX and setXXX approach familiar to JavaBean and Enterprise JavaBean programmers. However, these OGSI operations may be extended by other service interfaces to support richer query, update, and subscription semantics, such as complex queries that span multiple serviceData elements in a service instance.

The serviceDataName element in a GridService portType definition corresponds to the BeanInfo class in JavaBeans. However, OGSI has chosen an XML (WSDL) document that provides information about the serviceData, instead of using a serializable implementation class as in the BeanInfo model.

4.6.5.2 *Extending portType with serviceData.* ServiceData defines a new

portType child element named serviceData, used to define serviceData elements, or SDEs, associated with that portType. These serviceData element definitions are referred to as serviceData declarations, or SDDs. Initial values for those serviceData elements (marked as "static" serviceData elements) may be specified using the staticServiceDataValues element within portType. The values of any serviceData element, whether declared statically in the portType or assigned during the life of the Web service instance, are called serviceData element values, or SDE values.

4.6.5.3 *serviceDataValues.* Each service instance is associated with a collec-

tion of serviceData elements: those serviceData elements defined within the various portTypes that form the service's interface, and also, potentially, additional service-

Data elements added at runtime. OGSI calls the set of serviceData elements associated with a service instance its "serviceData set." A serviceData set may also refer to the set of serviceData elements aggregated from all serviceData elements declared in a portType interface hierarchy.

Each service instance must have a "logical" XML document, with a root element of serviceDataValues that contains the serviceData element values. An example of a serviceDataValues element was given above. A service implementation is free to choose how the SDE values are stored; for example, it may store the SDE values not as XML but as instance variables that are converted into XML or other encodings as necessary.

The wsdl:binding associated with various operations manipulating serviceData elements will indicate the encoding of that data between service requestor and service provider. For example, a binding might indicate that the serviceData element values are encoded as serialized Java objects.

4.6.5.4 SDE Aggregation within a portType Interface Hierarchy.
WSDL 1.2 has introduced the notion of multiple portType extension, and one can model that construct within the GWSDL namespace. A portType can extend zero or more other portTypes. There is no direct relationship between a wsdl:service and the portTypes supported by the service modeled in the WSDL syntax. Rather, the set of portTypes implemented by the service is derived through the port element children of the service element and binding elements referred to from those port elements. This set of portTypes, and all portTypes they extend, defines the complete interface to the service.

The serviceData set defined by the service's interface is the set union of the serviceData elements declared in each portType in the complete interface implemented by the service instance. Because serviceData elements are uniquely identified by QName, the set union semantic implies that a serviceData element can appear only once in the set of serviceData elements. For example, if a portType named "pt1" and portType named "pt2" both declare a serviceData named "tns:sd1," and a portType named "pt3" extends both "pt1 and "pt2," then it has one (not two) serviceData elements named "tns:sd1."

4.6.5.5 Dynamic serviceData Elements.
Although many serviceData elements are most naturally defined in a service's interface definition, situations can arise in which it is useful to add or move serviceData elements dynamically to or from an instance. The means by which such updates are achieved are implementation specific; for example, a service instance may implement operations for adding a new serviceData element.

The grid service portType illustrates the use of dynamic SDEs. This contains a serviceData element named "serviceDataName" that lists the serviceData elements currently defined. This property of a service instance may return a superset of the serviceData elements declared in the GWSDL defining the service interface, allowing the requestor to use the subscribe operation if this serviceDataSet changes, and the findServiceData operation to determine the current serviceDataSet value.

4.6.6   Core Grid Service Properties

This subsection discusses a number of properties and concepts common to all grid services.

4.6.6.1   Service Description and Service Instance.   One can distinguish in OGSI between the *description* of a grid service and an *instance* of a grid service:

- A *grid service description* describes how a client interacts with service instances. This description is independent of any particular instance. Within a WSDL document, the grid service description is embodied in the most derived portType (i.e., the portType referenced by the wsdl:service element's port children, via referenced binding elements, describing the service) of the instance, along with its associated portTypes (including serviceData declarations), bindings, messages, and types definitions.
- A grid service description may be simultaneously used by any number of *grid service instances,* each of which
  - ○ Embodies some state with which the service description describes how to interact
  - ○ Has one or more grid service handles
  - ○ Has one or more grid service references to it

A service description is used primarily for two purposes. First, as a description of a service interface, it can be used by tooling to automatically generate client interface proxies, server skeletons, and so forth. Second, it can be used for discovery, for example, to find a service instance that implements a particular service description, or to find a factory that can create instances with a particular service description.

The service description is meant to capture both interface syntax and (in a very rudimentary, nonnormative fashion) semantics. *Interface syntax* is described by WSDL portTypes. *Semantics* may be inferred through the name assigned to the portType. For example, when defining a grid service, one defines zero or more uniquely named portTypes. Concise semantics can be associated with each of these names in specification documents, and, perhaps in the future, through Semantic Web or other more formal descriptions. These names can then be used by clients to discover services with desired semantics, by searching for service instances and factories with the appropriate names. The use of namespaces to define these names also provides a vehicle for assuring globally unique names.

4.6.6.2   Modeling Time in OGSI.   The need arises at various points throughout this specification to represent time that is meaningful to multiple parties in the distributed Grid. For example, information may be tagged by a producer with timestamps in order to convey that information's useful lifetime to consumers. Clients need to negotiate service instance lifetimes with services, and multiple services may need a common understanding of time in order for clients to be able to manage their simultaneous use and interaction.

The GMT global time standard is assumed for grid services, allowing operations to refer unambiguously to absolute times. However, assuming the GMT time standard to represent time does *not* imply any particular level of clock synchronization between clients and services in the grid. In fact, no specific accuracy of synchronization is specified or expected by OGSI, as this is a service-quality issue.

Grid service hosting environments and clients should utilize the Network Time Protocol (NTP) or equivalent function to synchronize their clocks to the global standard GMT time. However, clients and services must accept and act appropriately on messages containing time values that are out of range because of inadequate synchronization, where "appropriately" may include refusing to use the information associated with those time values. Furthermore, clients and services requiring global ordering or synchronization at a finer granularity than their clock accuracies or resolutions allow for must coordinate through the use of additional synchronization service interfaces, such as through transactions or synthesized global clocks.

In some cases, it is required to represent both zero time and infinite time. Zero time should be represented by a time in the past. However, infinite time requires an extended notion of time. One therefore introduces the following type in the OGSI namespace that may be used in place of xsd:dateTime when a special value of "infinity" is appropriate.

4.6.6.3   *XML Element Lifetime Declaration Properties.*   Since serviceData elements may represent instantaneous observations of the dynamic state of a service instance, it is critical that consumers of serviceData be able to understand the valid lifetimes of these observations. The client may use this time-related information to reason about the validity and availability of the serviceData element and its value, though the client is free to ignore the information.

One can define three XML attributes that together describe the lifetimes associated with an XML element and its subelements. These attributes may be used in any XML element that allows for extensibility attributes, including the serviceData element.

The three lifetime declaration properties are:

1. ogsi:goodFrom. Declares the time from which the content of the element is said to be valid. This is typically the time at which the value was created.

2. ogsi:goodUntil. Declares the time until which the content of the element is said to be valid. This property must be greater than or equal to the goodFrom time.

3. ogsi:availableUntil. Declares the time until which this element itself is expected to be available, perhaps with updated values. Prior to this time, a client should be able to obtain an updated copy of this element. After this time, a client may no longer be able to get a copy of this element (while still observing cardinality and mutability constraints on this element). This property must be greater than or equal to the goodFrom time.

4.6.7 Other Details

The above description is but a summary of the OGSI specification. The interested reader should refer to reference [84] for a more inclusive discussion.

4.7 A POSSIBLE APPLICATION OF OGSA/OGSI TO NEXT-GENERATION OPEN-SOURCE OUTSOURCING

4.7.1 Opportunities

This section looks briefly at the issue of outsourcing of IT services by an increasing number of large and mid-size companies. In the early 1990s, we published an early book on outsourcing [120] that attempted to make the point that *analytics* were needed to make an informed and defensible decision. Up to then, a lot of the outsourcing deals were made on an emulation mode: "If leading company x in industry A made such a choice, then we at company y in industry A should also follow suit." Now, if we were to write a book on this topic we would emphasize the desire and/or advantage to use standards, in particular grid computing standards, to establish "open-source outsourcing," so that an organization can obtain services in a completely commoditized and competitive manner. Open standards enable a company to easily port its business if it finds that an outsourcer is not delivering the service to the stipulated SLA or financial levels. We invite the reader to read Chapter 5 from the perspective of a pending outsourcing decision and to appreciate, while reading, the opportunity that the grid computing standards afford in this context. Figure 4.17 depicts the target open-source-outsourcing architecture. According to press-time market research, the worldwide outsourcing revenue was $120B in 2003 and was expected to grow to $160B by 2006. Because IT costs usually equate to 6% of the revenue line of companies, $120B equates to a revenue top line of $2T. This means that this is equivalent in the aggregate to the top 30 companies in the United States (Exxon Mobil, Wal-Mart, GM, Ford Motor, General Electric, ChevronTexaco, Chrysler, IBM, Altria, HP, State Farm Insurance, and the next nineteen) outsourcing their entire IT operations.

The material that follows focuses only on the current outsourcing trends/imperatives; we let the reader mentally apply the material of this chapter and the chapter(s) that follow to the issue of outsourcing.

4.7.2 Outsourcing Trends

The material in this subsection, characterizing the market momentum toward outsourcing, is synthesized from Gartner Dataquest (Stamford, CT) sources.

40% of the Fortune 500 companies were expected to have outsourced offshore by the end of 2004. More generally, by 2004, 70% of enterprises will selectively outsource applications using a variety of ASPs, traditional outsourcers, niche applications vendors, and offshore providers. Enterprise buyers are demanding that IT service providers offer a range of global sourcing alternatives, including on-site,

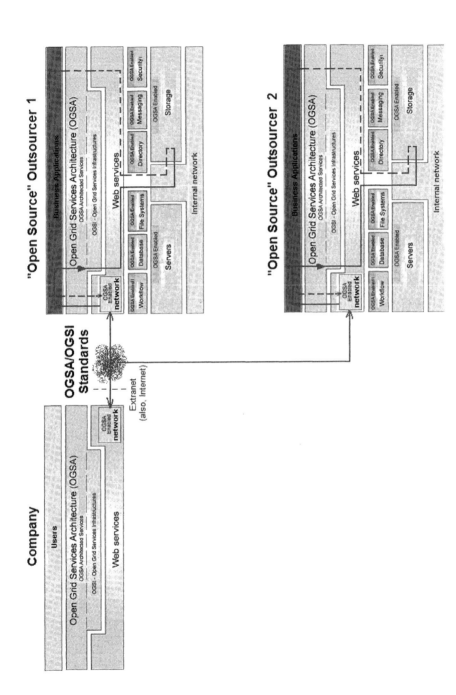

Figure 4.17 Open outsourcing possible with standards.

domestic, nearshore (services delivered from an adjacent or nearby country) and offshore capabilities. Press-time studies show that about 40% of customers reported they are currently outsourcing some aspect of their network operations, typically voice and data. By year-end 2004, one out of every 10 jobs within U.S.-based IT vendors and IT service providers will move to emerging markets, as also will one of every 20 IT jobs within user enterprises. Growth in offshore delivery is expected to be continuous, but moderate compared with the "hype" around the concept in recent years (the revenue growth just cited is a CAGR of 10%). An analysis of IT outsourcing contracts for the past 14 years has shown that the average value of an IT outsourcing contract is $47 million, and the average length of a contract is six years. Enterprises must use structured evaluation and selection criteria or run the risk of engaging with the wrong ASP and/or offering the wrong level of service (here is where open-source outsourcing can be of value).

Reflecting the industry's price sensitivity, cost ranks higher in IT buyer's decision making in financial services than in other industries. For many CIOs, the decision to outsource activities offshore is fiscally calculable: the cost, quality, value, and process advantages are well proven through about 15–20 years of practice starting in the late 1980s. According to observers, offshore outsourcing is becoming a tool for improving service delivery and a source of qualified talent. Most of today's offshore business process outsourcing (BPO) opportunity remains at the level of out-tasking a component of a business process rather than outsourcing an entire business process, and is mostly relegated to back-office transaction processing (and contact centers). In recent years, financial services penned at least 17 large deals, a number also matched by government operations. The telecommunications industry has undertaken 12 deals, transportation has 11, manufacturing and aerospace/defense each have eight, and high-tech has seven.

Observers suggest that enterprises should consider including a 25% acquisition clause into their contract that allows the enterprise to get out of the contract if the service provider is more than 25% acquired by another company. The enterprise should also include a competitive-pricing clause that forces an ASP hosting provider to match a deal it gave to another enterprise if the enterprise with such a clause in the contract could qualify for the volume commitments. Again, open-source outsourcing can be of obvious value.

Enterprises around the world are attempting to focus their investments on their core business processes and are increasingly looking at outsourcing noncore business processes, such as IT with its never-ending overcomplexity. Early adopters of BPO services, primarily large organizations, continue to expand their relationships to include new process areas, and new technology and media are creating opportunities for outsourcing entire lines of products and services, such as online payroll, online benefits administration, online order management, and online transaction processing.

An element in many of the outsourcing initiatives is a focus on IT infrastructure and operations: many of the large initiatives involve substantial consolidation and centralization of IT assets on a global basis. Outsourcing providers promise to meet IT and business needs through new technology and new business models, particularly the on-demand model (akin to the grid computing paradigm) that appears to promise relief from fixed costs.

Although BPO has emerged as one of the fastest growing service opportunities in the financial services market, BPO is not a new service area for financial services: check-processing services have been around for decades, and payment processing showed steady robust growth through the 1990s. What is different now is that BPO is rapidly expanding into areas that were off limits to outsourcing just a few years ago. Increasing acceptance is also driving expansion in the number and scope of deals, which, in turn, increases the market size.

This survey material from Gartner Dataquest documents the market momentum toward outsourcing. The use of an open-source approach, which greatly facilitates portability, will prove to be very advantageous to companies. We encourage IT professionals to explore these opportunities through the machinery afforded by OGSA/OGSI.

CHAPTER 5

Standards Supporting Grid Computing: OGSA

For any kind of new technology, corporate and business decision makers typically seek answers to a set of questions, including "Are there firm standards to support the technology and its widespread deployment?" Any experienced planner is keenly aware of the financial implications of using a technology that does not have standards (or is at least based on a broad-reaching de facto industry standard). In the previous chapter, we discussed OGSI in some detail, it being the original grid standard published by the Global Grid Forum (OGSI defines grid services and the basic mechanisms for creating, managing, and exchanging information between them). A second standard appeared a year later (and was still in initial draft form at press time): the Open Grid Services Architecture (OGSA). As we have discussed up to this point, standards are critical to the commercialization of the Intergrid, just like Internet standards were critical to the commercialization of the Internet in the 1990s (e.g., see [66–68]). These same standards can then be used for enterprise grids, just like browsers are now used for intranet applications. Also, as noted, these standards can be utilized to develop open outsourcing environments.

OGSA specifies the scope of important services required to support grid systems and applications in both e-science and e-business. It identifies a core set of such services that are perceived as being essential for many systems and applications, and it specifies at a high level the functionalities required for and the interrelationships among these core services. These same standards are also very useful in the enterprise grid (intragrid) context. OGSA is a special Web service that provides a set of well-defined interfaces and follows specific conventions. The OGSA document also lists existing technical standards and standard definition activities within GGF, OASIS, W3C, and other standards bodies that speak to required OGSA functionality, and identifies priority areas for further work.

This chapter (based largely on [69][1]) covers the OGSA documentation. Although the OGSA continues to be revised, this write-up is intended to provide a

---

[1]Copyright © Global Grid Forum (2002, 2003). All Rights Reserved. This document and translations of it may be copied and furnished to others, and derivative works that comment on or otherwise explain it or assist in its implementation may be prepared, copied, published and distributed, in whole or in part, without restriction of any kind, provided that the above copyright notice and this paragraph are included on all such copies and derivative works.

A Networking Approach to Grid Computing. By Daniel Minoli
ISBN 0-471-68756-1 © 2005 John Wiley & Sons, Inc.

sense of where this work is going. The purpose of this chapter is to highlight the standardization progress and not to provide a comprehensive normative specification and/or tutorial. The reader should always consult the latest GGF documentation, after acquiring a basic understanding through the material presented herein.

Some of this material parallels some of the information of Chapter 4, but that is the way the OGSI and OGSA documents have been produced by the GGF. Proponents note that

> . . . [I]n theory, utility computing gives managers greater utilization of data-center resources at lower operating costs. At their disposal will be flexible computing, storage and network capacity that can react automatically to changes in business priorities. The data center of the future also will have self-configuring, self-monitoring and self-healing features so managers can reduce today's manual configuration and troubleshooting chores, advocates say. The allure of utility computing is easy to see, but there is no clear road map without a stable set of standards. Getting there requires an open-source standards-based approach that encompasses network gear, servers, software, services and IT governance. Vendors are working to create intelligent devices, management tools and services for utility consumption. [63].

There is plenty of work to do, and standards are a critical initiative in this area, as they are, in reality, in many other arenas (spanning the gamut from WWW to Ethernet, DVDs to HDTV). Hence, our emphasis on this topic, and the motivation for this chapter.

Successful realization of the OGSA vision of a broadly applicable and broadly adopted framework for distributed system integration, virtualization, and management requires the definition of a core set of interfaces, behaviors, resource models, and bindings. The OGSA documentation, developed by the OGSA working group within the GGF, provides a first (but preliminary and incomplete) version of this OGSA definition. Throughout this book as well as in this chapter, the term "resource" is used in its most general sense and can include virtualized physical resources such as processors, storage, memory, and/or virtual resources such as software licenses or data.

5.1 INTRODUCTION

The OGSA, developed within the OGSI working group of the Global Grid Forum, is a proposed enabling infrastructure for grid systems and applications, that is, systems and applications that are concerned with the integration, virtualization, and management of services within distributed, heterogeneous, dynamic "virtual organizations" in industry, e-science, or e-business [58, 114]. Whether confined to a single enterprise or extended to encompass external resource sharing and service provider relationships, one finds that service integration, virtualization, and management in these contexts can be technically challenging because of the need to achieve various end-to-end qualities of service when running on top of different native platforms.

Work on OGSA seeks to address these challenges by defining an integrated set of Web-service-based service definitions designed both to simplify the creation of secure, robust grid systems, and to enable the creation of interoperable, portable, and reusable components and systems via the standardization of key interfaces and behaviors. The purpose of the OGSA document is to summarize current understanding of required OGSA functionality and the appropriate rendering of this functionality into service definitions. More specifically, it presents functionality requirements, a service taxonomy, relationships among the various services, and, finally, more detailed descriptions of specific services.

Activities in OGSA both build on and are contributing to the development of the growing collection of technical specifications that form the emerging Web services architecture [51]. (Indeed, OGSA can be viewed as a particular profile for the application of core WS standards.) Some grid functionality requirements are met by existing or proposed standards. In other cases, grid functionality requirements may require extensions to existing service definitions and/or entirely new service definitions. Where this is the case, the document describes the current state of work underway to define such extensions and/or definitions.

Although the OGSA vision is broad, work to date has focused on the definition of a small set of core semantic elements. In particular, the OGSI specification (discussed in Chapter 4) defines, in terms of WSDL interfaces and associated conventions, extensions and refinements of emerging Web services standards to support basic grid behaviors [52]. OGSI-compliant Web services—what the GGF calls grid services—are intended to form the components of grid infrastructure and application stacks.

OGSI defines essential building blocks for distributed systems, including standard interfaces and associated behaviors for describing and discovering service attributes, creating service instances, managing service lifetime, and subscribing to and delivering notifications. However, it does not define all elements that arise when creating large-scale systems. One may also need to address a wide variety of other issues, both fundamental and domain specific, of which the following are just examples.

How do I establish identity and negotiate authentication?

How is policy expressed and negotiated?

How do I discover services?

How do I negotiate and monitor service-level agreements?

How do I manage membership of, and communication within, virtual organizations?

How do I organize service collections hierarchically so as to deliver reliable and scalable service semantics?

How do I integrate data resources into computations?

How do I monitor and manage collections of services?

Without standardization in each of these (and other) areas, it is hard to build large-scale systems in a standard fashion, achieve code reuse, and achieve interoperability among components—three distinct and important goals. Much of the OGSA document is concerned with defining these services.

GGF's understanding of what is required in OGSA is preliminary and incomplete by their admission. Both the understanding of OGSA's purpose and form, and the details of specific components, are likely to evolve; in the meantime, however, the OGSA document provides a basis for debate and also can serve as input to discussions of priorities for OGSA specification development.

The Global Grid Forum's OGSA Working Group (OGSA-WG) has the following charter and scope:

1. To produce and document the use cases that drive the definition and prioritization of OGSA components, as well as document the rationale for our choices.

2. To identify and outline requirements for, and a prioritization of, OGSA services and components.

3. To identify and outline requirements for, and a prioritization of, hosting environment and protocol bindings that are required for deployment of portable, interoperable OGSA implementations.

4. To identify and outline requirements for, and a prioritization of, models for resources and other important entities.[2]

5. To identify, outline, and prioritize interoperability requirements for the various OGSA components.

6. To define standard OGSA profiles, i.e., sets of OGSA components that meet specific requirements.

7. To define relationships between GGF and other standards bodies activities such as W3C, OASIS, and WSI whose work touches upon OGSA-related issues.

In some cases, work within OGSA-WG may result in the drafting of specifications for OGSA components. However, one expects that the task of completing these specifications will be handled by other working groups.

The OGSA document is intended as a contribution to goals 2, 3, and 4. It is not, in its current form, a final product. However, it does provide a base for understanding grid-based systems and services; hence, our inclusion herein.

5.2 FUNCTIONALITY REQUIREMENTS

The development of the OGSA document has been based on a variety of use case scenarios [55]. The use cases have not been defined with a view to expressing formal requirements (and do not contain the level of detail that would be required for formal requirements), but have provided useful input to the definition process.

---

[2]The resource services focus is the objective of the Common Management Models workgroup at GGF.

Analysis of the use cases, other input from OGSA-WG participants, and other studies of grid technology requirements lead the Working Group to identify important and broadly relevant characteristics of grid environments and applications, along with functionalities that appear to have general relevance to a variety of application scenarios. Although this material does not represent a comprehensive or formal statement of functionality requirements from our use cases, it does provide useful input for subsequent development of OGSA functions. The case scenarios that have been considered include [55]:

- National fusion collaboration
- IT infrastructure and management
- Commercial data centers
- Service-based distributed query processing
- Severe storm prediction
- Online media and entertainment

5.2.1 Basic Functionality Requirements

The following basic functions are universally fundamental:

- *Discovery and brokering.* Mechanisms are required for discovering and/or allocating services, data, and resources with desired properties. For example, clients need to discover network services before they are used, service brokers need to discover hardware and software availability, and service brokers must identify codes and platforms suitable for execution requested by the client [55].
- *Metering and accounting.* Applications and schemas for metering, auditing, and billing for IT infrastructure and management use cases [55]. The metering function records the usage and duration, especially metering the usage of licenses. The auditing function audits usage and application profiles on machines, and the billing function bills the user based on metering.
- *Data sharing.* Data sharing and data management are common as well as important grid applications. Mechanisms are required for accessing and managing data archives, for caching data and managing its consistency, and for indexing and discovering data and metadata.
- *Deployment.* Data is deployed to the hosting environment that will execute the job (or made available in or via a high-performance infrastructure). Also, applications (executable) are migrated to the computer that will execute them.
- *Virtual organizations (VOs).* The need to support collaborative VOs introduces a need for mechanisms to support VO creation and management, including group membership services [58]. For the commercial data center use case [55], the grid creates a VO in a data center that provides IT resources to the job upon the customer's job request. Depending on the customer's request, the grid will negotiate with another grid on a remote commercial data

center and create a VO across the commercial data centers. Such a VO can be used to achieve the necessary scalability and availability.

- *Monitoring.* A global, cross-organizational view of resources and assets for project and fiscal planning, troubleshooting, and other purposes. The users want to monitor their applications running on the grid. Also, the resource or service owners need to surface certain states so that the user of those resources or services may manage the usage using the state information.

- *Policy.* An error and event policy guides self-controlling management, including failover and provisioning. It is important to be able to represent policy at multiple stages in hierarchical systems, with the goal of automating the enforcement of policies that might otherwise be implemented as organizational processes or managed manually. There may be policies at every level of the infrastructure: from low-level policies that govern how the resources are monitored and managed, to high-level policies that govern how business process such as billing are managed. High-level policies are sometimes decomposable into lower-level policies.

5.2.2   Security Requirements

Grids also introduce a rich set of security requirements; some of these requirements are:

- *Multiple security infrastructures.* Distributed operation implies a need to interoperate with and manage multiple security infrastructures. For example, for a commercial data center application, isolation of customers in the same commercial data center is a crucial requirement; the grid should provide not only access control but also performance isolation. For another example, for an online media and entertainment use case, proper isolation between content offerings must be ensured; this level of isolation has to be ensured by the security of the infrastructure.

- *Perimeter security solutions.* Many use cases require applications to be deployed on the other side of firewalls from the intended user clients. Intergrid collaboration often requires crossing institutional firewalls. OGSA needs standard, secure mechanisms that can be deployed to protect institutions while also enabling cross-firewall interaction.

- *Authentication, Authorization, and Accounting.* Obtaining application programs and deploying them into a grid system may require authentication/authorization. In the commercial data center use case, the commercial data center authenticates the customer and authorizes the submitted request when the customer submits a job request. The commercial data center also identifies his/her policies (including but not limited to SLA, security, scheduling, and brokering policies).

- *Encryption.* The IT infrastructure and management use case requires encrypting of the communications, at least of the payload.

- *Application and Network-Level Firewalls.* This is a long-standing problem; it is made particularly difficult by the many different policies one is dealing with and the particularly harsh restrictions at international sites.

- *Certification.* A trusted party certifies that a particular service has certain semantic behavior. For example, a company could establish a policy of only using e-commerce services certified by Yahoo.

5.2.3 Resource Management Requirements

Resource management is another multilevel requirement, encompassing SLA negotiation, provisioning, and scheduling for a variety of resource types and activities:

- *Provisioning.* Computer processors, applications, licenses, storage, networks, and instruments are all grid resources that require provisioning. OGSA needs a framework that allows resource provisioning to be done in a uniform, consistent manner.

- *Resource virtualization.* Dynamic provisioning implies a need for resource virtualization mechanisms that allow resources to be transitioned flexibly to different tasks as required; for example, when bringing more Web servers on line as demand exceeds a threshold.

- *Optimization of resource usage* while meeting cost targets (i.e., dealing with finite resources). Mechanisms to manage conflicting demands from various organizations, groups, projects, and users and implement a fair sharing of resources and access to the grid.

- *Transport management.* For applications that require some form of real-time scheduling, it can be important to be able to schedule or provision bandwidth dynamically for data transfers or in support of the other data sharing applications. In many (if not all) commercial applications, reliable transport management is essential to obtain the end-to-end QoS required by the application.

- *Access.* Usage models that provide for both batch and interactive access to resources.

- *Management and monitoring.* Support for the management and monitoring of resource usage and the detection of SLA or contract violations by all relevant parties. Also, conflict management is necessary; it resolves conflicts between management disciplines that may differ in their optimization objectives (availability goals versus performance goals, for example).

- *Processor scavenging* is an important tool that allows an enterprise or VO to use to aggregate computing power that would otherwise go to waste. How can OGSA provide service infrastructure that will allow the creation of applications that use scavenged cycles? For example, consider a collection of desktop computers running software that supports integration into processing and/or storage pools managed via systems such as Condor, Entropia, and United Devices. Issues here include maximizing security in the absence of strong trust.

- *Scheduling of service tasks.* Long recognized as an important capability for any information processing system, scheduling becomes extremely important and difficult for distributed grid systems. In general, dynamic scheduling is an essential component [55]. Computer resources must be provisioned on-demand to satisfy the need to complete a forecast on time.

- *Load balancing.* In many applications, it is necessary to make sure make sure deadlines are met or resources are used uniformly. These are both forms of load balancing that must be made possible by the underlying infrastructure. For example, for the commercial data center use case, monitoring the job performance and adjusting allocated resources to match the load and fairly distributing end users' requests to all the resources are necessary. For the online media and entertainment use case, the amount of workload is a direct result of how many concurrent online game players are being hosted on a game server. If the game server (server A) is responsible for a 20 square mile area in the game world, and a battle occurred in that area, many players will rush to that area, causing workload on that server to increase. As players enter that area and leave other areas, other servers' workloads will decrease. Hence, when the workload of server A gets above certain threshold, a load balancing routine needs to be triggered to rebalance the resources (i.e., servers). That is, workloads must be redistributed across servers with idle capacity.

- *Advanced reservation.* This functionality may be required in order to execute the application on reserved resources. For example, for the commercial data center use case, the grid decides when to start the request processing based on the customer's request. It interprets the job specification description language in which the request is written and it checks to see if the customer has the right to perform the request.

- *Notification and messaging.* Notification and messaging are critical in most dynamic scientific problems. Notification and messaging are event driven.

- *Logging.* It may be desirable to log processes such as obtaining/deploying application programs because, for example, the information might be used for accounting. This functionality is represented as "metering and accounting."

- *Workflow management.* Many applications can be wrapped in scripts or processes that require licenses and other resources from multiple sources. Applications coordinate using the file system based on events.

- *Pricing.* Mechanisms for determining how to render appropriate bills to users of a grid.

5.2.4 System Properties Requirements

A number of grid-related capabilities can be thought of as desirable system properties rather than functions:

- *Fault tolerance.* Support is required for failover, load redistribution, and other techniques used to achieve fault tolerance. Fault tolerance is particularly important for long running queries that can potentially return large amounts of data, for dynamic scientific applications, and for commercial data center applications.

- *Disaster recovery.* Disaster recovery is a critical capability for complex distributed grid infrastructures. For distributed systems, failure must be considered one of the natural behaviors and disaster recovery mechanisms must be considered an essential component of the design. Autonomous system principles must be embraced as one designs grid applications and should be reflected in OGSA. In case of commercial data center applications if the data center becomes unavailable due to a disaster such as an earthquake or fire, the remote backup data center needs to take over the application systems.

- *Self-healing capabilities* of resources, services and systems are required. Significant manual effort should not be required to monitor, diagnose, and repair faults. There is a need for the ability to integrate intelligent self-aware hardware such as disks, networking devices, and so on.

- *Strong monitoring* for defects, intrusions, and other problems. Ability to migrate attacks away from critical areas.

- *Legacy application management.* Legacy applications are those that cannot be changed, but they are too valuable to give up or to complex to rewrite. Grid infrastructure has to be built around them so that they can continue to be used.

- *Administration.* Be able to "codify" and "automate" the normal practices used to administer the environment. The goal is that systems should be able to self-organize and self-describe to manage low-level configuration details based on higher-level configurations and management policies specified by administrators.

- *Agreement-based interaction.* Some initiatives require agreement-based interactions capable of specifying and enacting agreements between clients and servers (not necessarily human) and then composing those agreements into higher-level end-user structures.

- *Grouping/aggregation of services.* The ability to instantiate (compose) services using some set of existing services is a key requirement. There are two main types of composition techniques: selection and aggregation. Selection involves choosing to use a particular service among many services with the same operational interface. Aggregation involves orchestrating a functional flow (workflow) between services. For example, the output of an accounting service is fed into the rating service to produce billing records. One other basic function required for aggregation services is to transform the syntax and/or semantics of data or interfaces.

5.2.5   Other Functionality Requirements

Although some use cases involve highly constrained environments (that may well motivate specialized OGSA profiles), it is clear that in general grid environments tend to be heterogeneous and distributed:

- *Platforms.* The platforms themselves are heterogeneous, including a variety of operating systems (Unixes, Linux, Windows, and, presumably, embedded systems), hosting environments (J2EE, .NET, others), and devices (computers, instruments, sensors, storage systems, databases, networks, etc.).
- *Mechanisms.* Grid software can need to interoperate with a variety of distinct implementation mechanisms for core functions such as security.
- *Administrative environments.* Geographically distributed environments often feature varied usage, management, and administration policies (including policies applied by legislation) that need to be honored and managed.

A wide variety of application structures are encountered and must be supported by other system components, including the following:

- Both *single-process* and *multiprocess* (both local and distributed) applications covering a wide range of resource requirements.
- *Flows,* that is, multiple interacting applications that can be treated as a single transient service instance working on behalf of a client or set of clients.
- *Workloads* comprising potentially large numbers of applications with a number of characteristics just listed.

5.3 OGSA SERVICE TAXONOMY

As noted above, the purpose of OGSA is to define standard approaches to, and mechanisms for, basic problems that are common to a wide variety of grid systems, such as communicating with other services, establishing identity, negotiating authorization, service discovery, error notification, and managing service collections.

As illustrated in Figure 5.1, the three principal elements of OGSA are the (i) Open Grid Services Infrastructure, (ii) OGSA services, and (iii) OGSA models:

- Building on both grid and Web services technologies, the OGSI defines mechanisms for creating, managing, and exchanging information among entities called grid services (this was discussed at length in the previous chapter). A grid service is a Web service that conforms to a set of conventions (interfaces and behaviors) that define how a client interacts with a grid service. These conventions, and other OGSI mechanisms associated with grid service creation and discovery, provide for the controlled, fault-resilient, and secure management of the distributed and often long-lived state that is commonly required in distributed applications.
- *OGSA services* build on OGSI mechanisms to define interfaces and associated behaviors for various functions not supported directly within OGSI, such as service discovery, data access, data integration, messaging, and monitoring.
- *OGSA models* support these interface specifications by defining models for common resource and service types.

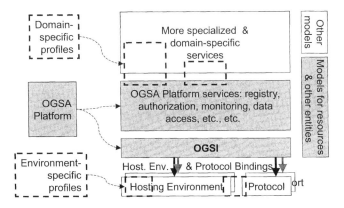

Figure 5.1   OGSA components (shaded) and related profiles (dashed lines).

The GGF anticipates that these OGSA components will be supplemented by a set of environment-specific profiles addressing issues such as the following. The specification mentions these here for completeness, but they are not discussed further.

- *Protocol bindings.* Environment profiles enable interoperability among different grid services by defining common mechanisms for transport and authentication. These issues are not addressed by OGSI, but rather defined as binding properties, meaning that different service implementations may implement them in different ways. For example, "SOAP over HTTP" is a useful grid service transport profile. Another example of such a profile is the proposed GSSAPI profile for security context establishment and message protection using WS-SecureConversation [59] and WS-Trust [60].

- *Hosting environment bindings.* Environment profiles of this sort enable portability of grid service implementations. For example, an "OGSA J2EE Profile" might define standardized Java APIs that allow for portability of grid services among OGSI-enabled J2EE systems. An "OGSA Desktop Grid Profile" could allow for interoperability among systems that allow untrusted (and untrusting) desktop computers to participate in distributed computations. An "OGSA Scientific Linux Profile" could define standard execution environments for computers that run scientific applications, specifying conventions for the locations of key executables and libraries, and for the names of certain environment variables

- *Sets of domain-specific services.* Profiles of this sort define interfaces and models in addition to those defined within OGSA to address the needs of specific application domains. For example, an "OGSA Database Profile" might define a set of interfaces and models for distributed database management; an "OGSA eCommerce Profile" might define interfaces and models for e-commerce applications.

The material that follows expands briefly upon each OGSA element. OGSA defines services that occur within a wide variety of grid systems. One can divide these functions into four broad groups: *core services, data services, program execution services,* and *resource management services.*

5.3.1   Core Services

"Core" services are implementations of functions that are generally used by a wide variety of higher-level services and that implement broadly useful capabilities. Although dependencies between core services and noncore services (by higher-level functions) are likely, there is no known requirement at this time that any particular core service be present in order to implement a grid. The organization of these "core" services into the areas further discussed below is for the convenience of explanation and is not meant to imply dependencies among functions that are grouped together.

5.3.1.1   Service Interaction. This category includes the following subservices: VOs; service group and discovery services; service domain, composition, orchestration, and workflow; and transactions. These subservices are primarily intended to provide interaction mechanisms for the collection of services in the grid. These subservices provide means for services to be registered and to locate each other; they also provide mechanisms for composing multiple lower-level services into aggregations. These subservices also include functions for deploying the software images that implement services in hosting environments and for collecting data about their operation for management, accounting, and billing purposes.

5.3.1.2   Service Management. This category includes functions for managing deployed services. It includes the following subservices: metering and accounting; installation, deployment, and provisoning; fault management; and problem determination. Service management automates and assists with a variety of installation, maintenance, monitoring, and troubleshooting tasks within a grid system. Service management includes functions for provisioning and deploying the system components; it also includes functions for collecting and exchanging data about the operations of the grid. This data is used for both "on-line" and "off-line" management operations and includes information about faults, events, problem determination, auditing, metering, accounting, and billing. Service management may also depend on models and schema that describe dependency relationships between different components, installation and provisioning steps and processes, and external capabilities.

5.3.1.3   Service Communication. This service category includes the following subservices: distributed logging, messaging and queueing, and event. These services provide the basic methods for services to communicate and support several interservice communication models.

5.3.1.4 Security. OGSA security architectural components aim to support, integrate, and unify available security models, mechanisms, protocols, platforms, and technologies, in a way that enables a variety of systems to interoperate securely. Specifically, the security of a grid environment must take into account the security of various aspects involved in a grid service invocation, as depicted in Figure 5.2. As discussed above, a grid service can be accessed over a variety of protocol bindings; given that bindings deal with protocol and message formats, security functions such as confidentiality, integrity, and authentication fall within the scope of bindings and thus are outside the scope of OGSA proper. A supplementary GGF OGSA Security Architecture deals with these goals in a manner consistent with the security model that is currently being defined for the Web services framework; an associated OGSA Security Roadmap document enumerates the security related specifications that will be needed to ensure interoperable implementations of the OGSA Security Architecture.

Each participating endpoint can express the policy it wishes to see applied when engaging in a secure conversation with another endpoint. Policies can specify supported authentication mechanisms, required integrity and confidentiality protection, trust policies, privacy policies, and other security constraints. When invoking grid

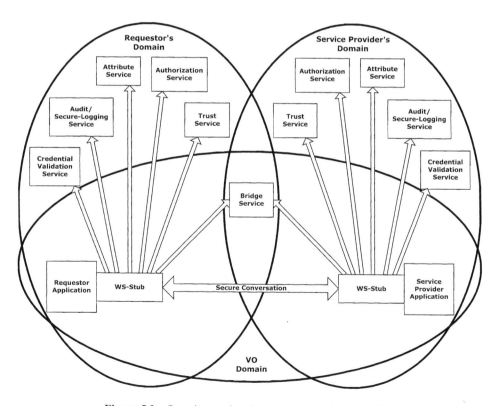

Figure 5.2 Security services in a virtual organization setting.

services dynamically, endpoints may need to discover the policies of a target service and establish trust relationships dynamically. Once a service requestor and a service provider have determined each other's policies, they can establish a secure channel over which subsequent operations can be invoked. Such a channel should enforce the mutual agreed-on qualities of protection, including identification, confidentiality, and integrity. The security model must provide a mechanism by which authentication credentials from the service requestor's domain can be translated into the service provider's domain and vice versa. This translation is required in order for both ends to evaluate their mutual access policies based on the established credentials and the quality of the established channel.

OGSA's security model must address authentication, confidentiality, message integrity, policy expression and exchange, authorization, delegation, single log-on, credential life span and renewal, privacy, secure logging, assurance, manageability, firewall traversal, and security at the OGSI layer. One can expect that existing and evolving standards will be adopted or recognized in the grid security model. Figure 5.2 shows relationships between a requestor, service provider, and many of the security services. Note that both requester and service provider are always subject to the security policies dictated by their respective administrative domains. Furthermore, a VO can have its own security policy that can enable the sharing of the submitted resources, but the associated rights will always be capped by the overruling resource-local policy. For many grid applications, the resource owners and the individual requesters will not "know" each other, as they live in different administrative domains, while their interactions are dynamically discovered and brokered by scheduler services and such. This implies that trust has to be dynamically established through introductions, and the concept of the VO as bridge is seen as an important mechanism to build these dynamic trust relationships. All security interfaces used by a service requestor and service provider need to be standardized within OGSA. Compliant implementations will be able to make use of existing services and defined policies through configuration. Compliant implementations of a particular security-related interface would be able to provide the associated and possibly alternative security services.

5.3.2   Data Services

The scale, dynamism, autonomy, and distribution of data sources in grid environments can result in significant complexity in data access and management. A variety of interfaces need to be defined to assist developers and users in the management of this complexity. In addition to basic data access interfaces and common resource models for storage and data management systems, these interfaces also address the need for transparency, heterogeneity, location, naming, distribution, replicas, ownership, and data access costs. Data virtualization services aimed at providing these transparencies can include federated access to distributed data, dynamic discovery of data sources based on content, dynamic migration of data for workload balancing, and schema management. In implementing such services, one needs to take into account the fact that different data types (e.g., flat file data, streaming media, and relational data) require different approaches to management. Furthermore, different applications require different forms of support; for example, some applications cannot be modified and require transparent access via file systems, whereas other applications need explicit management of data locality and replication.

These considerations suggest a role for variety of potential data management interfaces including:

- Interfaces for data caching (resolving a file handle to a flat file into a data stream)
- Interfaces for data replication
- Interfaces for data access (via mechanisms for accessing a wide range of data types, including flat files, RDBMS, and streaming media)
- Interfaces for file and DBMS services and, possibly, federated data management services that are used as part of a vertical utility grid
- Interfaces for data transformation and filtering; interfaces for schema transformation (allowing different data, service, and policy schema to be reconciled so that the services can interact correctly)
- Interfaces for grid storage services that allow direct access to storage resources/data throughout the grid

5.3.3 Program Execution

Program-execution services enable applications to have *coordinated* access to underlying VO resources, regardless of their physical location or access mechanisms. Figure 5.3 shows the grid services required for program execution. These services include:

- Agreement Factory Service—create agreement services based on domain-specific terms such as job, reservation, and data access terms
- Job Agreement Service—creates, monitors, and controls compute jobs
- Reservation Agreement Service—guarantees that resources are available for running a job
- Data Access Agreement Service—stages the required application and data
- Queuing Service—provides a service that allows administrators to customize and define scheduling policies at the VO level, and/or at the different resource manager levels
- Index Service—allows for the propagation of information between resource managers and the metascheduler

When an application utilizing the grid makes use of more than one physical resource during its execution, program-execution middleware maps the resource requirements of the user application to the multiple physical resources that are required to run that application. Community schedulers are the key to making VO

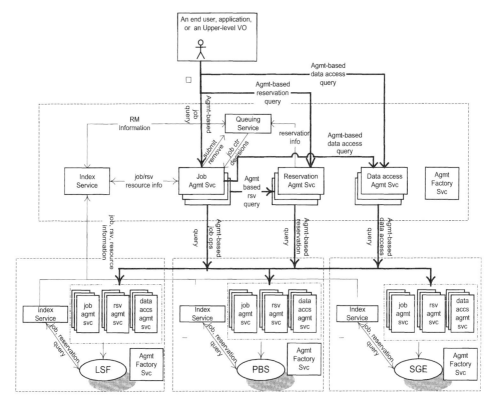

Figure 5.3 Program execution architecture and services.

resources easily accessible to end users, by automatically matching the requirements of a grid application with the available resources while staying within the
conditions that the VO has specified with the underlying resource managers (RMs).

Interoperability is fundamental for a program-execution grid. In order to allow
the higher-order constructs (such as community schedulers) to work with the lower-
level resource managers, there must be agreement on how these entities will interact
with each other, even though the lower-level resource managers might be very different from each other, in function and in interface. In order to meet the requirement
for interoperability, standards are required that define the interfaces through which
resource managers are accessed and managed. Additionally, there is a requirement
that the services representing the resource managers act using standard semantics,
so that the behavior of the resource manager is predictable to the community scheduler.

OGSA-based grid environments may be composed of many different but interacting grid services. Each such service may be subject to different policies governing how to manage the underlying resources. In order to deal with the complexities
of large collections of these services, there must be mechanisms for grid service

management and the allocation of resources for applications. One such mechanism is defined through the proposed WS-Agreement interface [53]. The specification document for WS-Agreement (refer to the specification) describes it as ". . . the ability to create grid services and adjust their policies and behaviors based on organizational goals and application requirements." WS-Agreement defines the Agreement-based Grid Service Management model, which is specified as a set of OGSI-compliant portTypes allowing clients to negotiate with management services in order to manage grid services or other legacy applications. To put it in concrete terms, if a user wishes to submit a computing job to run on a cluster, the user would rely on the grid service client to contact a job management service and negotiate a set of agreements that ensure that the user's job has access to required processors, memory, storage space, and so on.

WS-Agreement defines fundamental mechanisms based on Agreement services, which represent an ongoing relationship between an agreement provider and an agreement initiator. The agreements define the behavior of a delivered service with respect to a service consumer. The agreement will most likely be defined in sets of domain-specific agreement terms (defined in other specifications), since the WS-Agreement specification is focused on defining the abstraction of the agreement and the protocol for coming to agreement, rather than on defining sets of agreement terms.

Referring back to the job submission example, the client might contact a job management service that implements the AgreementFactory interface, with creation parameters that might say, "My job has to have a software license for application X, I would like to have 8 cpus, and I would like to have 4 GB of RAM." If the job management service could not provide the software license, the agreement terms would be rejected; if it could provide all of the terms, an agreement service instance representing the job resources would be created. If, because of available resource constraints, the job management service could not fulfill the terms of the original creation parameters, but could supply either 4 cpus and 4 GB of RAM, or 8 cpus and 2 GB of RAM, the job management service could create an AgreementOffer that included two potential agreements: "app X, 4cpus, 4GB" and "app X, 8cpus, 2GB", one of which the client could choose (because cpus and memory were terms subject to counteroffers). By defining various sets of terms for representing different types of resources available within the VO, community schedulers can be written that can negotiate with resource managers for the use of the underlying resources on behalf of the user community.

Program-execution services utilize WS-Agreement portTypes both for its client interface and its interface to underlying resources, with the goal of allowing hierarchical VO deployment. For example, one community scheduler talks to another community-scheduler-based resource conglomeration. Another key aspect of Agreement Service is to consider the constraints of the service provider based on service-level agreements in a business context and their reflection in underlying resource manager policies. Not all constraints are related to availability or unavailability of resources: some are management policy specified based on pricing, usage, higher priorities, and so on.

Program-execution services can be flexibly composed to offer a spectrum of time-to-result QoS levels, ranging from online/interactive to batch, with specific (or no) turn-around time requirements. *Online/interactive jobs* can be sent by the job agreement service to the resource managers immediately, without getting queued. *Batch jobs* will be queued by the queueing services and dispatched to resource managers later. The jobs with specific turnaround times may use the available reservations made with the resource managers at specific times so that the deadlines can be met. The jobs with no specific turnaround time can be queued at the VO level or the back-end resource manager level until the required resources are available, or back-filled with the existing reservations without delaying the start time of the more time-critical jobs.

Workloads are composite entities and have multiple levels of "execution entities." Workloads are made of jobs, which in turn are made of tasks, which in turn are made of tasklets (see Figure 5.4). Each of these composite entities has a manager. Workload realization (or execution) is the general set of use cases that take workload requests and map them to appropriate resources within the grid that can realize these workloads. Grid services manage and coordinate access and consumption of geographically distributed resources by workloads realized on those resources. Workload realization can be visualized as a mapping between "demand" in the form of workloads and "supply" in the form of the available grid resources. In a fundamental scenario, the system has to map the demand to the supply and provide the mechanism to realize these workloads on the resources. This primary mode can be augmented with other services, mechanisms, and capabilities that provide alternate modes of interaction including optimization of the mapping and scheduling a temporal and topological execution profile. In addition, other services manage and enforce the service-level agreements with the user, and still other services tweak the resources and manage the available capacity to ensure that a desired quality of service is delivered.

Services that belong to the Resource Optimization Framework are focused on the optimization of the supply side of the mapping. This can be done by admission

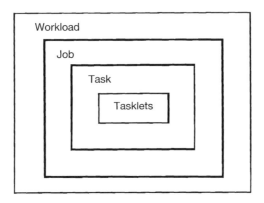

Figure 5.4 Workload, job, task, and tasklet.

control, resource utilization monitoring and metering, capacity projections, resource provisioning and load balancing across equivalent resources, and negotiation with workload optimization and/or management services to migrate workloads onto other resources so as to maximize resource utilization. Services that belong to the Workload Optimization Framework are focused on the demand side of the mapping. These services may queue requests to prevent resource saturation, manage relative priorities in requests, and perform postbalancing by migrating workloads to appropriate resources depending on the potential to violate or be rewarded for missing or exceeding SLAs, respectively. Services in the Resource Optimization Framework are focused on resolving any contentions that the myopic views of the respective resource or workload optimization frameworks may create. These services arbitrate and modulate the primary interactions either in an "in-band" or "out-of-band" manner.

5.3.4 Resource Management

Grid services in this category include:

- *Service orchestration.* These interfaces provide ways to describe and manage the choreography of a set of interacting services.
- *Administration.* Standard interfaces for such tasks as software deployment, change management, and identity management.
- *Provisioning and resource management.* Negotiation of service-level agreements and dynamic resource allocation and redistribution consistent with SLA policy, including mechanisms that allow clients and workflows to acquire access to resources and services at a particular (future) time.
- *Reservation and Scheduling Services.* Reservation services provide the mechanism to make resource reservatios at a particular time duration. Scheduling services provide the mechanism to scheduling tasks according to their priorities
- *Deployment Services.* Deploy necessary software (OS, middleware, application) and data into the hosting environment.

5.4 SERVICE RELATIONSHIPS

Earlier in the chapter, services were classified according to a taxonomy in which two services are related (i.e., put into the same category) if their purpose and functionality are similar; however, this taxonomy does not show the relationships that exist between the services when these services are used in practice. Therefore, one can use a second categorization based on the perspective of a service provider, that is, a person implementing or assembling the various components. Here, one can define types of relationships between services, and organize the services according to these relationships. A new class of services, the *platform services,* is also introduced.

5.4.1 Service Composition

A service composition is a grid service that provides a new set of functions that are derived from, built on, extended from, and/or implemented using functions exposed by other grid services. All services in the composition are *first-class* services (i.e., each individually provides distinct functionality and can, if required, be independent of this service composition or other compositions). An instance of a service composition (representing a specific set of functional and semantic behaviors) includes (or references) instances of all the services that make up this composition. Each composition has an identity that is shared by the individual component services. The instances may be tightly or loosely coupled to the composite service.

In a *tightly coupled composition,* the individual service instances are indistinguishable from the composite that they belong to and are completely subsumed and hidden by the composite. All interactions with these "composed" services are only performed with the composite service. Individual service instances in a tightly coupled composition share the same lifetime and life-cycle characteristics of the composition. In a *loosely coupled composition,* the functionality of the loosely coupled services can be accessed independently of the composite service they support.

Service compositions can be either *pure* or *orchestrated.* Services in a *pure* composition share well-defined common state and/or implement a given state in the composite as an aggregate of the individual states in the services. A pure composition can be established either as an *implementation* composition, meaning that the functionality of the individual services are embedded in a single implemented entity, or as a *managed* composition, meaning that the individual services in the composition are managed by well-defined internal protocols that provide for a shared identity. In an *orchestrated* composition, a master service representing the composition exposes a functionality that is essentially derived by orchestrating a set of loosely coupled services.

OGSA can designate special composite services and define their functions (and the services they compose and compositional methods to be used) similar to other more fundamental services definitions; these composites become *first-class* OGSA services. In other cases, the composition is user defined.

Service compositions can be *heterogeneous,* meaning that services in the composition provide dissimilar functions, or *homogeneous,* meaning that services in the composition are similar in function. A homogeneous composition can be an aggregation of services that are managed as one (more capable) service compared with the individual services. In some cases, the service representing the composition will be the manager for this composite.

An example of a service composition is a job. A job may be composed of other jobs or tasks. Since a job is a composition (i.e., a grid service) it can be managed through well-defined interfaces that are exposed by the service. Another example of service composition is an OGSA batch-job scheduler that provides the same functionality as existing traditional schedulers for batch systems. Such OGSA schedulers represents a combination of functionality like queuing, resource determination, reservation, resource allocation, and so on, where each of these functions can be implemented by specialized services targeted at these functions and can be reused for other usage scenarios. As yet another example, a traditional distributed-resource management cluster (e.g., PBS, LSF) can also be refactored using service composition in which the traditional interfaces into such a cluster become the functionality exposed by the composite service representing this cluster.

5.4.2 Service Orchestration

In addition to identifying specific common services, OGSA describes the common behaviors, attributes, operations, and interfaces needed to allow services to interact with others in a distributed, heterogeneous, grid-enabled environment:

- *Choreography* describes required patterns of interaction among grid services (or, more generally, Web services) and templates for sequences (or more structures) of interactions.
- *Orchestration* describes the ways in which business processes are constructed from Web services and other business processes, and how these processes interact.
- *Workflow* is a pattern of business process interaction, not necessarily corresponding to a fixed set of business processes. All such interactions may be between services residing within a single data center or across a range of different platforms and implementations anywhere.

(Note: For ease of language, in what follows the term "orchestration" refers to "choreography, orchestration, and workflow.")

In the OGSA environment, services, processes, and workflows may both be managed by OGSA and may be vehicles by which management takes place. OGSI capabilities for stateful Web service interaction have much in common with choreography and business process management, but there are technical differences between them. Since one may want grid services to be part of flows, and the flows to be used in grids, it will be important to resolve their relationship. This will enable the GGF to take advantage of emerging work at the OASIS Web Services Distributed Management Group that is addressing both management of web services and management using web services, as well as change management for web services.

OGSI service groups may also play a role in grouping or aggregating services in ways that factor out common usages in grid service/business process interactions; however, it would be best if such factoring took place directly in organizations such as W3C and OASIS that are doing work on these concepts (the role of OGSA is to determine places where existing work will not meet grid architecture needs, rather than to create a competing standard).

For example, since a grid service is a specialized Web service with service data, notifications, life cycles, and service groups, one might define ways in which these concepts can/should be leveraged in grid service interactions that form flows, including ways of monitoring and managing those flows (e.g., using service data and notifications) as well as fault handling. However, there may be mismatches in the life cycles of grid services and flows that will require thought; this is another reason why convergence with mainstream Web service standards is highly desirable. For example, business processes are typically thought to be long lasting, not transient, and are not instance oriented, as grid services are. Grid services standardization people that the view that it is better to cooperate with, rather than compete with, business process industry groups. When addressing some concepts, including quality of service, scheduling, and resource allocation and provisioning, GGF may want to consider their relationships to business process management and service flows. OGSA should have a role in coordinating how, where, and when these aspects get addressed.

Many OGSA services will be constructed using other services; scheduling is one such example. SOAs are designed to permit invocations of services by other services, so capabilities for services built from services are intrinsic to grid/Web services. Whether additional compositional constructs are needed beyond invocations, workflows/choreography, management, and the business transaction/coordination/context (all of which are being addressed by other bodies) has not yet been determined. The notion of composition of stateful services with behavior extended from the services from which the composition is derived is a very important area and one that Web services so far has not fully addressed.

5.4.3 Types of Relationships

OGSA services can be related via "uses relationship," and "extends relationship." In a *"uses* relationship," a first service accesses the interface of a second service to use the functionality provided by this second service. For instance, many services use the handle-resolver service to convert GSHs to GSRs. In an *"extends* relationship," a first service extends the functionality provided by a second service by using portType extensibility. A simple example of this relationship is an event service that extends the OGSI notification functionality; another example is a registry service that extends the service group functionality of OGSI.

5.4.4 Platform Services

OGSA introduces the term *platform services* to denote services that provide functionalities that are basic. Platform services (i) provide underlying functionalities on which other services build, (ii) provide functionalities that are common to (and used by) several high-level services, and (iii) provide functionalities that are designed to be used primarily through the "extends" relationship. The functionality provided by a given platform service is, by definition, present (through extension) in several high-level services; as a consequence, platform service functionalities permeate the high-level services, being pervasive within OGSA. For this reason, they do not fit into (and are not shown as part of) the taxonomy. Platform services form the lower layer of the service relationships, as illustrated in Figure 5.5; however, a platform service may use or extend other platform services (i.e., there is more than one layer of platform services). As OGSA services are organized and categorized, and their

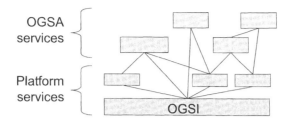

Figure 5.5 The relationships between OGSI, OGSA platform services, and other OGSA services.

functionality is defined in more detail, common functionalities among these services will increasingly appear. These functionalities should/will be redefined as platform services in order to simplify OGSA. As a consequence, as the work on the definition of OGSA progresses, the number of platform services should increase.

The current set of OGSA platform services is as follows:

- OGSI: defines grid services and the basic mechanisms for creating, managing, and exchanging information between them.
- WS-Agreement: provides a set of interfaces that support the negotiation of policies, service-level agreements, reservations, and so on, and maps the related agreements to grid services.
- Common Management Model (CMM): provides the manageability infrastructure for resources in OGSA. CMM defines the base behavioral model for all resources and resource managers in the grid, plus management functionality like relationships and life-cycle management.
- OGSA Data Services (or part of it): provides the basic functionality to manage data in a grid environment.

5.5 OGSA SERVICES

This section provides a more detailed description of required OGSA functionality.

5.5.1 Handle Resolution

As we saw in Chapter 4, OGSI defines a two-level naming scheme for grid service instances based on abstract, long-lived *grid service handles* (GSHs) that can be mapped by HandleMapper services to concrete, but potentially less long lived, *grid service references* (GSRs). These constructs are basically network-wide pointers to specific grid service instances hosted in (potentially remote) execution environments. A client application can use a grid service reference to send requests (represented by the operations defined in the interfaces of the target service) directly to the specific instance at the specified network-attached service endpoint identified by that GSR.

The format of the GSH is a URL, where the schema directive indicates the naming scheme used to express the handle value. Based on the GSH naming scheme, the application should find an associated naming-scheme-specific HandleMapper service that knows how to resolve that name to the associated GSR. This allows different naming scheme implementations to coexist, and to provide different QoS properties through their implementation. OGSI defines the basic GSH format and portType for the HandleMapper service that resolve a GSH to a GSR.

The expectation is that different implementations of the two-level naming scheme with the naming directive will enable features such as transparent service instance migration, fault tolerance through transparent failover, high availability through mirroring, and advanced security through fine-grained access control on the name resolution. All these features come with an associated cost, and their applicability will, therefore, depend on the application itself.

The OGSI Working Group decided to leave the registration of GSHs and associated GSRs undefined for possible standardization elsewhere (the same applies to the authorization of the naming resolution as well as the features that deal with scalability and robustness.) Another unspecified issue is how the bootstrap mechanism should work. In other words, given the handle of the very first service to contact, how would a party find the associated resolution service? Currently, it is left to the implementation, which may decide to use custom configuration data and external naming services; for example, DNS or the Handle System. The handle resolutions require service invocations and could, therefore, affect overall performance. This issue could be addressed by local caching, or possibly by the definition of a generic service that maintains caching information about the GSH–GSR mappings independent from the handle resolution scheme.

5.5.2   Virtual Organization Creation and Management

VOs are a concept that supplies a "context" for operation of the grid that can be used to associate users, their requests, and resources. VO contexts permit the grid resource providers to associate appropriate policy and agreements with their resources. Users associated with a VO can then exploit those resources consistent with those policies and agreements. VO creation and management functions include mechanisms for associating users/groups with a VO, manipulation of user roles (administration, configuration, use, etc.) within the VO, association of services (encapsulated resources) with the VO, and attachment of agreements and policies to the VO as a whole or to individual services within the VO. Finally, creation of a VO requires a mechanism by which the "VO context" is referenced and associated with user requests (this is most likely via a GSH, since a "service" is the likely embodiment of the VO).

5.5.3   Service Groups and Discovery Services

GSHs and GSRs together realize a two-level naming scheme, with HandleResolver services mapping from handles to references; however, GSHs are not intended to contain semantic information and indeed may be viewed for most purposes as opaque. Thus, other entities (both humans and applications) need other means for discovering services with particular properties, whether relating to interface, function, availability, location, policy, or other criteria.

Traditionally, in distributed systems this problem is addressed by creating a third-level "human-readable" or "semantic" name space that is then mapped (bound) to abstract names (in this case, GSHs) via registry, discovery, metadata catalog, or other similar services. It is important that OGSA defines standard functions for managing such name spaces, otherwise services and clients developed by different groups cannot easily discover each other's existence and properties. These functions must address the creation, maintenance, and querying of name mappings. Two types of such semantic name spaces are common—naming by attribute, and naming by path.

Attribute naming schemes associate various metadata with services and support retrieval via queries on attribute values. A registry implementing such a scheme allows service providers to publish the existence and properties of the services that they provide, so that service consumers can discover them. One can envision special-purpose registries being built on the base service group mechanisms provided by the OGSI definition. In other words, an OGSA-compliant registry is a concrete specialization of the OGSI service group.

A ServiceGroup is a collection of entries, where each entry is a grid service implementing the ServiceGroupEntry interface. The ServiceGroup interface also extends the GridService interface. There is a ServiceGroupEntry for each service in the group (i.e., for each group member). Each ServiceGroupEntry contains a serviceLocator for the referred-to service and information (content) about that service. The content element is an XML element advertising some information about the member service. The type of the content element conforms to one of the QName elements in the ContentModelType SDE of the ServiceGroup interface.

It is the content model of the service group definition that suggests the concrete type and specific use of the registry being offered. The content model of the serviceGroupEntry for a given service group is published in the service data of the service group. The content model is the basis on which search predicates can be formed and executed against the service group with the findServiceData operation. In other words, it is the content model that forms the basis of the registry index upon which registry searches can be executed. It is envisioned that many application-specific, special-purpose registries will be developed.

It is also envisioned that many registries will inherit and implement the notificationSource interface so as to facilitate client subscription to register state changes. Again, specific state change subscriptions will be possible through the advertisement of the registry-specific service group content model of the service group on which the registry is built.

As stated earlier, one can envision many application-specific registry implementations being defined. Whether or not one or more general-purpose registry types should be defined and adopted as part of OGSA is to be determined.

Path naming or *directory* schemes (as used, for example, in file systems) represent an alternative approach to attribute schemes for organizing services into a hierarchical name space that can be navigated. The two approaches can be combined, as in LDAP. Directory path naming can be accomplished by defining a PathName Interface that maps strings to GSHs. Thus, a string such as "/data/genomics_dbs/mouse" could map to a service (GSH) that might deliver portions of the mouse genome, and perhaps also do BLAST searches against the mouse genome. Similarly, "/applications/biology/genomics/BLAST" could map to a GSH that has Interfaces for executing BLAST.

The interface will have methods to insert, look up, and delete <string, GSH> pairs, and will in essence be a simple table. It is expected that path_name services will be "chained" together, so that evaluation of a path may involve traversing several path_name services, forming a directed graph. This can be used to link disjoint namespaces into namespace cliques.

5.5.4 Choreography, Orchestration, and Workflow

Over these interfaces OGSA provides a rich set of behaviors and associated operations and attributes for business process management (additional work remains to be done in this area):

- Definition of a job flow, including associated policies
- Assignment of resources to a grid flow instance
- Scheduling of grid flows (and associated grid services)
- Execution of grid flows (and associated grid services)
- Common context and metadata for grid flows (and associated services)
- Management and monitoring for grid flows (and associated grid services)
- Failure handling for grid flows (and associated grid services); more generally, managing the potential transiency of grid services
- Business transaction and coordination services

5.5.5 Transactions

Transaction services are important in many grid applications, particularly in industries such as financial services and in application domains such as supply chain management. However, transaction management in a widely distributed, high-latency, heterogeneous RDBMS environment is more complicated than in a single data center with a single vendor's software. Traditional distributed transaction algorithms, such as two-phase distributed commit, may be too expensive in a wide-area grid, and other techniques such as optimistic protocols may be more appropriate. At the same time, different applications often have different characteristics and requirements that can be exploited when selecting a transaction technique to use. Thus, it is unlikely that there will be a "one size fits all" solution to the transaction problem.

5.5.6 Metering Service

Different grid deployments may integrate different services and resources and feature different underlying economic motivations and models; however, regardless of these differences, it is a quasiuniversal requirement that resource utilization can be monitored, whether for purposes of cost allocation (i.e., charge back), capacity and trend analysis, dynamic provisioning, grid-service pricing, fraud and intrusion detection, and/or billing. OGSA must address this requirement by defining standard monitoring, metering, rating, accounting, and billing interfaces. These interfaces can use or extend those defined within the Common Management Model (CMM) that provides access to basic resource performance and utilization instrumentation, exposed as serviceData. For example, an operating system might publish countervalues corresponding to the state of system activities such as processor utilization, memory usage, disk and tape I/O activity, network usage, and so on. Although interfaces do not provide the values needed for metering and accounting directly (since, for instance, they are not directly related to the consumers), they can provide basic monitoring data to be used by the metering and accounting services.

We address metering in this subsection and rating, accounting, and billing in the sections that follow.

A grid service may consume multiple resources and a resource may be shared by multiple service instances. Ultimately, the sharing of underlying resources is managed by middleware and operating systems. All modern operating systems and many middleware systems have metering subsystems for measuring resource consumption (i.e., monitored data) and for aggregating the results of those measurements. For example, all commercial Unix systems have provisions for aggregating prime-time and nonprime-time resource consumption by user and command.

A metering interface provides access to a standard description of such aggregated data (metering serviceData). A key parameter is the time window over which measurements are aggregated. In commercial Unix systems, measurements are aggregated at administrator-defined intervals (chronological entry), usually daily, primarily for the purpose of accounting. On the other hand, metering systems that drive active workload management systems might aggregate measurements using time windows measured in seconds. Dynamic provisioning systems use time windows somewhere between these two examples.

Several use cases require metering systems that support multitier, end-to-end flows involving multiple services. An OGSA metering service must be able to meter the resource consumption of configurable classes of these types of flows executing on widely distributed, loosely coupled server, storage, and network resources. Configurable classes should support, for example, a departmental charge-back scenario where incoming requests and their subsequent flows are partitioned into account classes determined by the department providing the service. The metering of end-to-end flows in a grid environment is somewhat analogous to the metering of individual processes in a traditional OS. Since traditional middleware and operating systems do not support this type of metering, additional functions must be accommodated by OGSA. In addition to traditional accounting applications, it is anticipated that end-to-end resource consumption measurements will play an important role in dynamic provisioning and pricing grid services.

Finally, in addition to metering resource consumption, metering systems must also accommodate the measurement and aggregation of application-related (e.g., licensed) resources. For example, a grid service might charge consuming services a per-use fee. The metering service must be able to support the measurement of this class of service (resource) consumption. The GGF GESA-WG defines services that use the metering service (we cover this issue in Chapter 7).

5.5.7   Rating Service

A rating interface needs to address two types of behaviors. Once the metered information is available, it has to be translated into financial terms. That is, for each unit of usage, a price has to be associated with it. This step is accomplished by the rating interfaces, which provide operations that take the metered information and a rating package as input and output the usage in terms of chargeable amounts. For example, a commercial UNIX system indicates that 10 hours of prime-time resource and 10 hours on nonprime-time resource are consumed, and the rating package indicates that each hour of prime-time resource is priced at 2 dollars and each hour of non-prime-time resource is priced at 1 dollar, a rating service will apply the pricing indicated in the rating package and translate the usage information into financial information in the terms of 20 dollars of prime-time resource charge, and 10 dollars of nonprime time resource charge.

Furthermore, when a business service is developed, a rating service is used to aggregate the costs of the components used to deliver the service, so that the service owner can determine the pricing, terms, and conditions under which the service will be offered to subscribers.

5.5.8   Accounting Service

Once the rated financial information is available, an accounting service can manage subscription users and accounts information, calculate the relevant monthly charges and maintain the invoice information. This service can also generate and present invoices to the user. Account-specific information is also applied at this time. For example, if a user has a special offer of a 20% discount for his usage of the commercial UNIX system described above, this discount will be applied by the accounting service to indicate a final invoiced amount of 24 dollars.

5.5.9   Billing and Payment Service

Billing and payment service refers to the financial service that actually carries out the transfer of money; for example, a credit card authorization service.

5.5.10   Installation, Deployment, and Provisioning

Computer processors, applications, licenses, storage, networks, and instruments are all grid resources that require installation, deployment, and provisioning (other new resource types will be invented and added to this list.) OGSA affords a framework that allows resource provisioning to be done in a uniform, consistent manner.

5.5.11   Distributed Logging

Distributed logging can be viewed as a typical messaging application in which *message producers* generate *log artifacts,* (atomic expressions of diagnostic information) that may or may not be used at a later time by other independent *message consumers.* OGSA-based logging can leverage the notification mechanism available in OGSI as the transport for messages. However, it is desirable to move logging-specific functionality to intermediaries, or *logging services.* Furthermore, the secure logging of events is required for the audit trails needed to fulfill judiciary and organizational policy requirements, to reconcile security-related inconsistencies, and to provide for forensic evidence both after the fact and in real time. It is expected that standards and implementations for secure logging should be able to considerably leverage on the efforts associated with distributed logging.

Because the invocation of operations on service instances are well defined and typed through their portType definitions, and because it is expected that most implementations will have a model in which the run-time invocation module is generated independent of the application, it follows that there is the opportunity to transparently provide logging of many natural logging-code points (see Figure 5.6). The logging capability includes not only the logging of information about the invocation of the actual service operations on both requestor and service provider, but also the information about the transparent invocations to the all the supporting services that were needed to enable the application service invocation (such as services for discovery, registration, directory, access control, privacy, identity translation, etc.) Most of the required logging and secure logging may be achieved transparent of the application and outside of the application logic, which greatly facilitates an implementation that adheres to the local policy.

Logging services provide the extensions needed to deal with the following issues:

- *Decoupling.* The logical separation of logging artifact creation from logging artifact consumption. The ultimate usage of the data (e.g., logging, tracing, management) is determined by the message consumer; the message producer should not be concerned with this.
- *Transformation and common representation.* Logging packages commonly annotate the data that they generate with useful common information such as category, priority, time stamp, and location. An OGSA logging service should not only provide the capability of annotating data, but also the capability of converting data from a range of (legacy) log formats into a common,

standard canonical representation. Also, a general mechanism for transformation may be required.

- *Filtering and aggregation.* The amount of logging data generated can be large, whereas the amount of data actually consumed can be small. Therefore, it can be desirable to have a mechanism for controlling the amount of data generated and for filtering out what is actually kept and where. Through the use of different filters, data coming from a single source can be easily separated into different repositories, and/or "similar" data coming from different sources can be aggregated into a single repository.

- *Configurable persistency.* Depending on consumer needs, data may have different durability characteristics. For example, in a real-time monitoring application, data may become irrelevant quickly, but be needed as soon as it is generated; data for an auditing program may be needed months or even years after it was generated. Hence, there is a need for a mechanism to create different data repositories, each with its own persistency characteristics. In addition, the artifact retention policy (e.g., determining which log artifacts to drop when a buffer reaches its size limit) should be configurable.

- *Consumption patterns.* Consumption patterns differ according to the needs of the consumer application. For example, a real-time monitoring application needs to be notified whenever a particular event occurs, whereas a postmortem problem determination program queries historical data, trying to find known patterns. Thus, the logging repository should support both synchronous query- (pull-) based consumption and asynchronous push-based (event-driven) notification. The system should be flexible enough that consumers can easily customize the event mechanism—for example, by sending digests of messages instead of each one—and maybe even provide some predicate logic on log artifacts to drive the notifications.

These considerations lead to an architecture for OGSA logging services in which producers talk to *filtering and transformation* services either directly, or indirectly through adapters. Consumers also use this service to create custom message repositories (*baskets*) or look for existing producers and baskets; that is, this service should also function as a factory (of baskets) and a registry (of producers and baskets). There is also a need for a configurable *storage and delivery* service, with which data from different filtering services is collected, stored, and, if required, delivered to interested consumers. Log service also implies the ability to access related log entries (by context) in chronological sequence.

5.5.12   Messaging and Queuing

OGSA extends the scope of the base OGSI Notification Interface to allow grid services to produce a range of event messages, not just notifications that a serviceData element has changed. Several terms related to this work are:

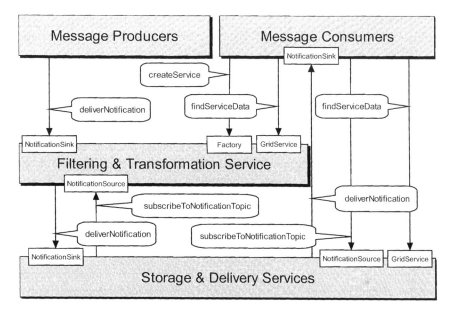

Figure 5.6 Schematic of a messaging service architecture.

- Event—Some occurrence within the state of the grid service or its environment that may be of interest to third parties. This could be a state change or it could be environmental, such as a timer event.

- Message—An artifact of an event, containing information about an event that some entity wishes to communicate to other entities.

- Topic—A "logical" communications channel and matching mechanism to which a requestor may subscribe to receive asynchronous messages and publishers may publish messages.

A message is represented as an XML element with a namespace-qualified QName, and an XML schema-defined complex type. A topic will be modeled as an XML element, describing its internal details, including expected messages associated with the topic. TopicSpaces, or collections of topics will also be modeled. This work also defines:

- An interface to allow any grid service to declare its ability to accept subscriptions to topics and the topics its supports.

- An interface to describe a messaging intermediary (a message broker) that supports anonymous publication and subscription on topics.

- An interface (or set of interfaces) that describe the interface to other messaging services such as a queuing service.

Note that queuing and message qualities of service such as reliability can be considered as both explicit services within an OGSA hosting environment and transport details modeled by the wsdl:binding element in the service description.

5.5.13 Events

Events are generally used as asynchronous signaling mechanisms. The most common form is "publish/subscribe," in which a service "publishes" the events that it exports (makes available to clients). The service may publish the events as reliable or best effort. Clients may then "subscribe" to the event, and when the event is raised, a call-back or message is sent to the client. Once again, the client can usually request either reliable or best effort, though the service may not be able to accept a reliable delivery request. There is also a distinction between the reliability of an event being raised and its delivery to a client. A service my attempt to deliver every occurrence of an event (reliable posting), but not be able to guarantee delivery (best-effort delivery).

An event can be anything that the service decides it will be: a change in a state variable, entry into a particular code segment, an exception such as a security violation or floating point overflow, or the failure of some other expected event to occur.

A second form of event is carried with service method invocations [55]. The basic idea is simple: inside the SOAP message invoking a service method is an Event Interest Set (EIS). The EIS specifies the events in which the caller is interested and a callback associated with each event. If an event named in the EIS is raised during the execution of the method—including outcalls on other grid services—the corresponding call-back is invoked. The EIS is *first class* and can be thought of as a part of the calling context (closure); as such, it can be modified by the callee and propagated down the call chain. This allows the callee to express interest in different events further down the call chain, add himself/herself to a list of subscribers interested in events further down the call chain, or "catch" events and keep them from propagating further up the call chain.

As noted above, events share many properties with notifications in OGSI (see Chapter 4, Section 4.5). However, OGSI notifications are designed to capture state changes in SDE's only, and are not designed as a general event mechanism. Furthermore, notifications are not normally "edge triggered" since they are based on a model that assumes a minimum and maximum time between notifications; for example, notifications that may occur within the minimum time window will not be sent to the client. A complete event framework would allow for event queuing or batching to cover this kind of functionality. OGSA events will build on OGSI notifications.

An event is a representation of an occurrence in a system or application component that may be of interest to other parties. Standard means of representing, communicating, transforming, reconciling, and recording events are important for interoperability. Thus, the OGSA Core should define standard schema for at least certain classes of OGSA events.

A detailed set of services include:

- Standard interface(s) for communicating events with specified QoS. These may be based directly on the Messaging interfaces.
- Standard interface(s) for transforming (mediating) events in a manner that is transparent to the endpoints. This should include aggregation of multiple events into a single event.
- Standard interface(s) for reconciling events from multiple sources.
- Standard interface(s) for recording events. These may be based directly on the Message logging interface(s).
- Standard interface(s) for batching and queuing events.

It is possible that some forms of event services may be built directly on top of OGSI notification interfaces, although more study is needed.

5.5.14   Policy and Agreements

These services create a general framework for creation, administration, and management of policies and agreements for system operation, security, resource allocation, and so on, as well as an infrastructure for "policy aware" services to use the set of defined and managed policies to govern their operation. These services do not actually enforce policies but permit policies to be managed and delivered to resource managers that can interpret and operate on them. Agreements (OGSI-Agreement Specification) provide a mechanism for the representation and negotiation of terms between service providers and their clients (either user requests or other services). These terms include specifications of functional, performance, and quality requirements/objectives that the suppliers and consumers exchange and that they can then use to influence their interactions. The Agreement mechanism provides a general expression framework for these terms but leaves the specification of particular terms to individual disciplines such as performance, security, data quality, and so on. Agreements also contain information on priority, costs, penalties or consequences associated with violation of the "contract" a negotiated agreement represents as well as information on how the service provider and consumer have decided to measure compliance with the agreement.

One can expect that many grid services will use policies to direct their actions. Thus, grids need to support the definition, discovery, communication, and enforcement of policies for such purposes as resource allocation, workload management, security, automation, and qualities of services. Some policies need to be expressed at the operational level, that is, at the level of the devices and resources to be managed, whereas higher-level policies express business goals and SLAs within and across administrative domains. Higher-level policies are hard to enforce without a canonical representation for their meaning to lower-level resources. Thus, business policies probably need to be translated into a canonical form that can then be used to derive lower-level policies that resources can understand. Standard mechanisms are also needed for managing and distributing policies from producers (e.g., administrators, autonomic managers, and SLAs) to endpoints that consume and enforce them (i.e., devices and resources).

These services (interfaces) provide a framework for creating, managing, validating, distributing, transforming, resolving, and enforcing policies within a distributed environment. A detailed set of services include:

- A management control point for policy life cycle: the Policy Service Manager interface
- An interface that policy consumers can use to retrieve required policies: the Policy Service Agent interface
- A way to express that a service is "policy aware": the Policy Enforcement Point interface
- A way to effect change on a resource, for example, by using Common Management Models

The Policy Service Manager controls access to the policy repository. It also controls when notifications of policy changes are sent out so that multiple updates can be made and notifications are only sent after all updates are complete. *The Policy Service Agent* is a management service that other "policy aware" services depend on for delivery of their policies. The agent can provide additional services like understanding time-period conditions so it can inform policy consumers of when policies become active or inactive. Services that consume policies will implement the *Policy Enforcement Point* interface to allow them to be registered with policy agents, participate in the subscription to and notification of policy changes, and to allow policies to be pushed down onto them when needed. These enforcement points will need to interpret the policies and make the necessary configuration changes in the resource(s) they manage by using the Common Management Model.

The OGSA Policy Service provides for a *transformation service* for this purpose and includes a canonical representation of policy in the form of an information model, grammar, and core XML schema (also see Figure 5.7).

A set of secondary validation interfaces can allow automated managers and administrators to act on the same set of policies and validate consistency. An interface is also required for translating policies to and from the canonical form so that consumers that have their own policy formats can plug into the service. Finally, there is a need for run-time resolution of policy conflicts that may require specific application knowledge to determine the cost of violating an agreement and selecting the policy that will have the appropriate impact.

5.5.15 Base Data Services

OGSA data interfaces are intended to enable a service-oriented treatment of data so that data can be treated in the same way as other resources within the Web/grid services architecture. Thus, for example, one can integrate data into registries and co-

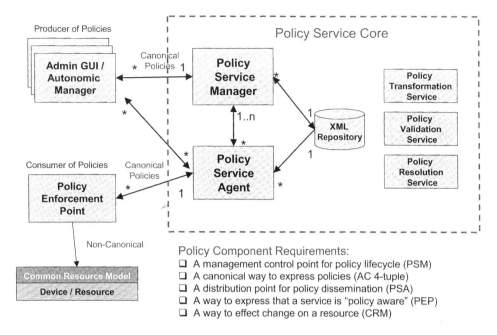

Figure 5.7 A set of potential policy service components.

ordinate operations on data using service orchestration mechanisms. A service-oriented treatment of data also allows us to use OGSI grid service handles as global names for data, manage the lifetime of dynamically created data by using OGSI lifetime management mechanisms, and represent agreements concerning data access via WS-Agreement.

OGSA data services are intended to allow for the definition, application, and management of diverse abstractions—what can be called *data virtualizations*—of underlying data sources. A data virtualization is represented by, and encapsulated in, a *data service,* an OGSI grid service with SDEs that describe key parameters of the virtualization, and with *operations* that allow clients to inspect those SDEs, access the data using appropriate operations, derive new data virtualizations from old, and/or manage the data virtualization. For example, a file containing geographical data might be made accessible as an image via a data service that implements a "JPEG image" virtualization, with SDEs defining size, resolution, and color characteristics, and operations provided for reading and modifying regions of the image. Another virtualization of the same data could present it as a relational database of coordinate-based information, with various specifics of the schema (e.g., table names, column names, types) as SDEs, and SQL as its operations for querying and updating the geographical data. In both cases, the data service implementation is responsible for managing the mapping to the underlying data source.

Four *base data interfaces* (WSDL portTypes) can be used to implement a variety of different data service behaviors:

1. *DataDescription* defines OGSI service data elements representing key para-meters of the data virtualization encapsulated by the data service.

2. *DataAccess* provides operations to access and/or modify the contents of the data virtualization encapsulated by the data service.

3. *DataFactory* provides an operation to create a new data service with a data virtualization derived from the data virtualization of the parent (factory) data service.

4. *DataManagement* provides operations to monitor and manage the data ser-vice's data virtualization, including (depending on the implementation) the data sources (such as database management systems) that underlie the data service.

A data service is any OGSI-compliant Web service that implements one or more of these base data interfaces.

5.5.16   Other Data Services

A variety of higher-level data interfaces can and must be defined on top of the base data interfaces, to address functions such as:

- Data access and movement
- Data replication and caching
- Data and schema mediation
- Metadata management and looking

There are likely to be strong relationships to discovery, messaging, agreement, and coordination functions. Basic data access interfaces allow clients to directly access and manipulate data. A number of such interfaces are required, corresponding to different data types, for example, files, directories, file systems, RDBMS, XML data bases, object data bases, and streaming media. A "file access" service may ex-port interfaces to *read, write,* or *truncate.* GridFTP, an existing data access service, provides mechanisms to *get* and *put* files, and supports third-party transfers.

  Data replication, data caching, and schema transformation subservices are de-scribed below.

Data Replication.   Data replication can be important as a means of meeting per-formance objectives by allowing local computer resources to have access to local data. Although closely related to caching (indeed, a "replica store" and a "cache" may differ only in their policies), replicas may provide different interfaces. Services that may consume data replication are group services for clustering and failover, utility computing for dynamic resource provisioning, policy services ensuring vari-ous qualities of service, metering and monitoring services, and also higher-level workload management and disaster recovery solutions. Each may need to migrate data for computation or to replicate state for a given service.

Work is required to define an OGSA-compliant set of data replication services that, through the use of "adapters," can move data in and out of heterogeneous physical and logical environments without any changes needed to the underlying local data access subsystems. The adapters handle the native "reading" and "writing" of data and the replication software coordinates the run time (recoverability, monitoring, etc.) associated with every data transfer. A central "monitor" sets up and handles communication with the calling service or program and sets up a "subscription–pair" relationship between capture and apply services on a per-replication-request basis to ensure reliability.

Data Caching. In order to improve performance of access to remote data items, caching services will be employed. At the minimum, caching services for traditional flat file data will be employed. Caching of other data types, such as views on RDBMS data, streaming data, and application binaries, are also envisioned. Issues that arise include (but are not limited to):

- Consistency—Is the data in the cache the same as in the source? If not, what is the coherence window? Different applications have very different requirements.
- Cache invalidation protocols—How and when is cached data invalidated?
- Write through or write back? When are writes to the cache committed back to the original data source?
- Security—How will access control to cached items be handled? Will access control enforcement be delegated to the cache, or will access control be somehow enforced by the original data source?
- Integrity of cached data—Is the cached data kept in memory or on disk? How is it protected from unauthorized access? Is it encrypted?

How the cache service addresses these issues will need to available as service data.

Schema Transformation. Schema transformation interfaces support the transformation of data from one schema to another. For example, XML transformations as specified in XSLT.

5.5.17 Discovery Services

Discovery interfaces address the need to be able to organize and search for information about various sorts of entities in various ways. In an OGSA environment, it is normally recommended that entities of whatever type be named by GSHs; thus, discovery services are concerned with mapping from user-specified criteria to appropriate GSHs. Different interface definitions and different implementation behaviors may vary according to how user requests are expressed, the information used to answer requests, and the mechanisms used to propagate and access that information.

5.5.18   Job Agreement Service. The job agreement service is created by the agreement factory service with a set of job terms, including command line, resource requirements, execution environment, data staging, job control, scheduler directives, and accounting and notification terms. The job agreement service provides an interface for placing jobs on a resource manager (i.e., representing a machine or a cluster), and for interacting with the job once it has been dispatched to the resource manager. The job agreement service provides basic matchmaking capabilities between the requirements of the job and the underlying resource manager available for running the job. More advanced job agreement services take into account more advanced job characteristics such as interactive execution, parallel jobs across resource managers, and jobs with requirements based on SLAs.

The interfaces provided by the job agreement service are:

- Manageability interface
  - ○ Supported job terms: defines a set of service data used to publish the job terms supported by this job service, including the job definition (command line and application name), resource requirements, execution environment, data staging, job control, scheduler directives, and accounting and notification terms.
  - ○ Workload status: total number of jobs, statuses such as number of jobs running or pending and suspended jobs.
- Job control: control the job after it has been instantiated. This would include the ability to suspend/resume, checkpoint, and kill the job.

The job agreement service makes use of information in the form of policies that are defined at the VO level, and resource information about available resource managers, queues, host, and job status as provided by the Global Information Service. The job agreement service uses the WS-Agreement Protocol with the underlying resource managers in order to submit jobs to the underlying resource managers, and in order to control the running jobs and to access data.

5.5.19   Reservation Agreement Service

The reservation agreement service is created by the agreement factory service with a set of terms including time duration, resource requirement specification, and authorized user/project agreement terms. The reservation agreement service allows end users or a job agreement service to reserve resources under the control of a resource manager to guarantee their availability to run a job. The service allows reservations on any type of resource (e.g., hosts, software licenses, or network bandwidth). Reservations can be specific (e.g., provide access to host "A" from noon to 5 PM), or more general (e.g., provide access to 16 Linux cpus on Sunday). Once a reservation is made, a job service can send a job to a resource manager that is attached to the provided reservation. Some of the policy decisions made by the reservation service for use with a resource manager include notions of who can make reservations (e.g., administrators only), how many hosts a particular user or user group can reserve at a time, when reservations can be made (i.e., define "blackout periods"), and what types of hosts can be reserved.

The reservation agreement service provides one interface, manageability, which defines a set of service data that describe the details of a particular reservation, including resource terms, start time, end time, amount of the resource reserved, and the authorized users.

The reservation service makes use of information about the existing resource managers available and any policies that might be defined at the VO level, and will make use of a logging service to log reservations. It will use the resource manager adapter interfaces to make reservations and to delete existing reservations.

5.5.20 Data Access Agreement Service

The data access agreement service is created by the agreement factory service with a set of terms, including (but not restricted to) source and destination file path, bandwidth requirements, and fault-tolerance terms (such as retrial times). The data access agreement service allows end users or a job agreement service to stage application or required data.

5.5.21 Queuing Service

The queuing service provides scheduling capability for jobs. Given a set of policies defined at the VO level, a queuing service will map jobs to resource managers based on the defined policies. For example, a queuing service might implement a fair-share policy that makes sure that all users within the VO get reasonable turnaround time on their jobs, rather than being starved out by other users' jobs ahead of them in the queue.

The manageability interface defines a set of service data for accessing QoS terms supported by the queuing services. QoS terms for the queuing service can include whether the service supports on-line or batch capabilities, average turn-around time for jobs, throughput guarantees, the ability to meet deadlines, and the ability to meet certain economic constraints.

The following terms apply to the queuing service:

- Enqueue—add a job to a queue
- Dequeue—remove a job from a queue

5.5.22 Open Grid Services Infrastructure

As we now know, the OGSI defines fundamental mechanisms on which OGSA is constructed. These mechanisms address issues relating to the creation, naming, management, and exchange of information among entities called grid services. A grid service instance is a (potentially transient) service that conforms to a set of conventions (expressed as WSDL interfaces, extensions, and behaviors) for such purposes as lifetime management, discovery of characteristics, and notification.

These conventions provide for the controlled management of the distributed and often long-lived state that is commonly required in distributed applications. OGSI also introduces standard factory and registration interfaces for creating and discovering grid services (OGSI was covered in detail in Chapter 4.) The following list recaps the key OGSI features and briefly discusses their relevance to OGSA.

- *Grid Service descriptions and instances.* OGSI introduces the twin concepts of the grid service description and grid service instance as organizing principles of distributed systems. A grid service description comprises the WSDL (with OGSI extensions) defining the grid service's interfaces and service data (see next item). A grid service instance is an addressable, potentially stateful, and potentially transient instantiation of such a description. These concepts provide the basic building blocks used to build OGSA-based distributed systems. Grid service descriptions define interfaces and behaviors, and a distributed system comprises a set of grid service instances that implement those behaviors, have a notion of identity with respect to the other instances in the system, and can be characterized as state coupled with behavior published through type-specific operations.

- *Service state, metadata, and introspection.* OGSI defines mechanisms for representing and accessing (via both queries and subscriptions) metadata and state data from a service instance (*service data*), as well as providing uniform mechanisms for accessing state. These mechanisms support introspection, in that a client application can ask a grid service instance to return information describing itself, such as the collection of interfaces that it implements.

- *Naming and name resolution.* OGSI defines a two-level naming scheme for grid service instances based on abstract, long-lived *grid service handles* that can be mapped by HandleMapper services to concrete but potentially less-long-lived *grid service references.* These constructs are basically network-wide pointers to specific grid service instances hosted in (potentially remote) execution environments. A client application can use a grid service reference to send requests (represented by the operations defined in the interfaces of the target service) directly to the specific instance at the specified network-attached service endpoint identified by the grid service reference.

- *Fault model.* OGSI defines a common approach for conveying fault information from operations.

- *Life cycle.* OGSI defines mechanisms for managing the life cycle of a grid service instance, including both explicit destruction and soft-state lifetime management functions for grid service instances, and grid service factories that can be used to create instances implementing specified interfaces.

- *Service groups.* OGSI defines a means of organizing groups of service instances.

OGSI does not address how grid services are created, managed, and destroyed within a particular hosting environment. Thus, services that conform to the OGSI specification are not necessarily portable across various hosting environments, but can be invoked by any client that conforms to this specification, subject, of course, to policy and compatible protocol bindings.

Stateful instances, typed interfaces, and global names are frequently also cited as fundamental characteristics of so-called distributed object-based systems. However, various other aspects of distributed object models (as traditionally defined) are specifically not required or prescribed by OGSI. For this reason, one does not use the terms "distributed object model" or "distributed object system" when describing this work, but instead uses the term "open grid services infrastructure," thus emphasizing the connections with both Web services and grid technologies.

Among the object-related issues that are not addressed within OGSI are implementation inheritance, service mobility, development approach, and hosting technology. The grid service specification does not require, nor does it prevent, implementations based upon object technologies that support inheritance at either the interface or the implementation level. There is no requirement in the architecture to expose the notion of implementation inheritance either at the client side or the service-provider side of the usage contract. In addition, the grid service specification does not prescribe, dictate, or prevent the use of any particular development approach or hosting technology for the grid service. For example, there is nothing about OGSI that is Java specific; one can implement OGSI behaviors in C, Python, or other languages.

Grid service providers are free to implement the semantic contract of the service in any technology and hosting architecture of their choosing. One can envision implementations in J2EE, .NET, traditional commercial transaction management servers, traditional procedural UNIX servers, and so on. One can also envision service implementations in a variety of programming languages that would include both object-oriented and nonobject-oriented *alternatives.*

5.5.23 Common Management Model

A fundamental requirement of grid management infrastructure is the ability to define the resources and resource management functions of the system in a standard and interoperable way. The capabilities of the grid management infrastructure rely on the ability to discover, compose, and interact with the resources and the resources managers responsible for them.

The Common Management Model specification defines the base behavioral model for all resources and resource managers in the grid management infrastructure. A mechanism is defined by which resource managers can make use of detailed manageability information for a resource that may come from existing resource models and instrumentation, such as those expressed in CIM, JMX, SNMP, and so on, combined with a set of canonical operations introduced by base CMM interfaces. The result is a manageable resource abstraction that introduces OGSI-compliant operations over an exposed underlying base-content model. CMM does not define yet another manageability (resource information) model.

The CMM specification defines

- The base manageable resource interface, which a resource or resource manager must provide to be manageable
- Canonical lifecycle states—the transitions between the states, and the operations necessary for the transitions that complement OGSI lifetime service data
- The ability to represent relationships among manageable resources (instances and types), including a canonical set of relationship types
- Life cycle metadata (XML attributes) common to all types of managed resources for monitoring and control of service data and operations based on life cycle state
- Canonical services factored out from across multiple resources or domain-specific resource managers, such as an operational port type (start/stop/pause/resume/quiesce)

Additional items that may come within the scope of the CMM specification are

- New data types or metadata to convey semantic meaning of manageability information, such as counter or gauge
- Versioning information
- Metadata to associate a metered usage (unit of measure) with manageability information
- Classification of properties such as metric and configuration
- Registries and locating fine-grained resources
- Managed resource identifier

To summarize, CMM represents the proposed industry standard for representing manageable resources and resource managers within the grid management infrastructure. CMM interfaces are OGSI compliant and are used as the base abstract interfaces from which specific manageable resource types are derived. Standardization of the base management behavior is required in order to integrate the vast number and types of resources and more limited set of resource managers introduced by multiple vendors.

5.6 SECURITY CONSIDERATIONS

The OGSA specification defines requirements for interfaces, behaviors, and models used to structure and achieve interactions among grid services and their clients. Although it is assumed that such interactions must be secured, the details of security are out of scope of the OGSA specification. Instead, security should be addressed in related specifications that define how abstract interactions are bound to specific communication protocols, how service behaviors are specialized via policy-management interfaces, and how security features are delivered in specific programming environments.

5.7 EXAMPLES OF OGSA MECHANISMS IN SUPPORT OF VO STRUCTURES

Grid applications and users typically have a need to create transient services, to discover available services, and to determine the properties of these services. The *Factory, Registry, GridService,* and *HandleMap* interfaces support the creation of transient service instances and the discovery and characterization of the service instances associated with a VO. Registry service is a service instance that supports the Registry interface for registration; it also supports the *GridService* interface's FindServiceData operation for discovery.

These interfaces can be used to construct a variety of VO service structures, such as (see Figure 5.8):

- *Simple hosting environment.* A basic grid environment is comprised of a set of resources located within a single administrative domain and supporting native facilities for service management. These native facilities can include, for example, a Linux cluster, a Microsoft .NET system, or a J2EE application server. Each factory is recorded in the registry; this enables clients to discover available factories. The user interface can be structured as a registry, one or more factories, and a HandleMap service. When a factory receives a client request to create a grid service instance, the factory invokes hosting-specific capabilities to create the new instance. It then assigns the grid service instance a handle, registers the instance with the registry, and makes the handle available to the HandleMap service. The implementations of these various services map directly into local operations [114].

- *Virtual hosting environment.* In more complex environments, the resources associated with a VO can span geographically distributed, heterogeneous "hosting environments." The "virtual hosting environment" can be made accessible to a client through the same interfaces that were used for the hosting environment described above. In this situation, one can have one or more "higher-level" factories that delegate creation requests to lower-level factories. Similarly, one has a higher-level registry that is aware of the higher-level factories and the service instances that these "higher-level" factories have created. The higher-level registry is also aware of any VO-specific policies that regulate the use of VO services. Clients can then utilize the VO registry to discover factories and other service instances associated with the VO. Clients then use the handles returned by the registry to communicate directly with such service instances. The higher-level factories and registries implement standard interfaces; hence, from the perspective of the user, it follows that higher-level factories are indistinguishable from any other factory or registry [114].

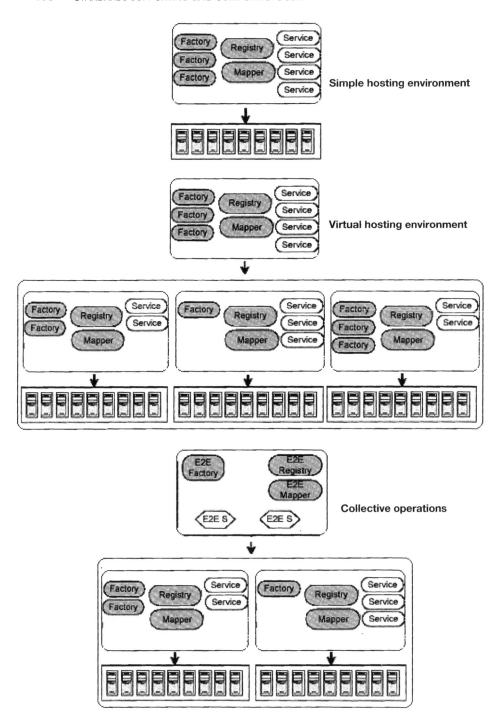

Figure 5.8 Examples of VO structures.

- *Collective operations.* One can also construct a "virtual hosting environment" that provides VO members with more elaborate services. In these environments, the registry keeps track of factories that create higher-level service instances. Such higher-level instances are implemented by utilizing lower-level factories that create multiple service instances. The behaviors of the multiple lower-level service instances are synthesized into that single, higher-level service instance.

Figure 5.9 depicts, for illustrative purposes, the work flow (remote service invocation, lifetime management, and notification) of a data-mining, grid-enabled computation application. In the figure, the "R"s represent local registry services. There will also be a VO registry service that provides information about the location of all depicted services. Four stages are shown in the figure:

Stage 1. The environment encompasses (from left to right) five simple hosting environments: one that runs the user application, a hosting environment that

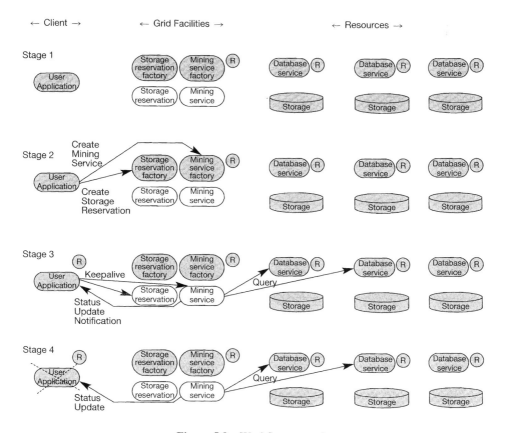

Figure 5.9 Workflow example.

encapsulates computing and storage resources (and that supports two factory services, one for creating storage reservations and the other for creating mining services), and three that support database services.

Stage 2. The user application invokes "create grid service" requests on the two factories in the second hosting environment, requesting the creation of a "data-mining service" and an allocation of temporary storage resources. The "data-mining service" performs the data-mining operation and the temporary storage is utilized for use by that computation. Each request involves mutual authentication of the user and the relevant factory followed by authorization of the request. This is accomplished using an authentication mechanism described in the factory's service description. Each request is successful, resulting in the creation of a grid service instance with an initial lifetime span.

Stage 3. During Stage 2, the new data-mining service instances were also provided with delegated proxy credentials that allow them to perform further remote operations on behalf of the user. In Stage 3, the newly created data-mining service uses its proxy credentials to initiate the requests for data from the two data base services. The intermediate results are placed in local storage. The data mining service also uses notification mechanisms to provide the user application with updates. Simultaneously, the user application generates routine "keepalive" requests to the two grid service instances that it has created.

Stage 4. For some reason, the user application fails. The data-mining computation continues for awhile. Eventually, due to the application failure and to the stoppage of the keep-alive messages, the two grid service instances time out in due course and are terminated. This frees up the storage and computing resources that the applications were consuming.

CHAPTER 6

Grid System Deployment Issues, Approaches, and Tools

This chapter looks at number of issues related to grid deployment and highlights related relevant issues. Security is given particular emphasis. The chapter has two major themes/sections: (i) implementation tools and issues; and (ii) security. We open the chapter with a survey of the functionality and capabilities of Globus Toolkit™, this being a usable grid implementation mechanism. We also cover grid computing environments (GCE), management issues, and generic deployment considerations. Then we cover security, it being a very critical issue. It is not the intent of this chapter to be a checklist or a cookbook formula for deployment, but rather to highlight some topics that play a role in such a process.

6.1 GENERIC IMPLEMENTATIONS: GLOBUS TOOLKIT

As we discussed in Chapters 4 and 5, the Open Grid Services Architecture was modeled (developed) with Globus Toolkit as a point of reference and with Web services as its building blocks. Whereas OGSA is an abstract conceptual architecture, a "concrete" realization mechanism is also needed if one is to achieve widespread penetration of standards-based grid services. One well-known example of the implementation of OGSA/OGSI is the just-named Globus Toolkit maintained by the Globus Alliance.[1] The Globus Toolkit grid-enables a wide range of computing environments. It is a software tool kit addressing key technical issues in the development of grid-enabled environments, services, and applications. As mentioned several times in the previous chapters, Globus Toolkit has been widely adopted as a grid technology solution for scientific and technical computing. Globus Toolkit mechanisms are in use at hundreds of sites and by dozens of major (scientific) grid pro-

---

[1]The Globus Alliance is a partnership of Argonne National Laboratory's Mathematics and Computer Science Division, the University of Southern California's Information Sciences Institute, the University of Chicago's Distributed Systems Laboratory, the University of Edinburgh in Scotland, and the Swedish Center for Parallel Computers. Major partners in the public sector currently include the National Computational Science Alliance, the NASA Information Power Grid project, the National Partnership for Advanced Computational Infrastructure, the University of Chicago, and the University of Wisconsin. Major corporate partners currently include IBM and Microsoft.

A Networking Approach to Grid Computing. By Daniel Minoli
ISBN 0-471-68756-1 © 2005 John Wiley & Sons, Inc.

jects worldwide [114]. The hope is that it will now, as it implements OGSA/OGSI in a cost-effective manner, also stimulate commercial deployment of grid services.

The Globus Alliance is a research and development "project" (The Globus Project™) focused on enabling the application of grid concepts to scientific and engineering computing [129]. The Globus Project started in 1996. The Alliance has and continues to develop fundamental technologies needed to build computational grids with persistent environments that enable software applications to integrate instruments, displays, and computational and information resources that are managed by diverse organizations in widespread locations. Users around the world are building grids and developing grid applications, and Globus Alliance's research targets technical challenges that arise from these activities. Typical research areas include resource management, data management and access, application development environments, information services, and security. Globus Alliance software development has resulted in the Globus Toolkit, a set of services and software libraries to support grids and grid applications. The Globus Toolkit includes software for security, information infrastructure, resource management, data management, communication, fault detection, and portability [129]. The tool kit's open-source software and services have transformed the way on-line resources are shared across organizations [124]. The tool kit allows developers to concentrate (almost exclusively) on the higher-level design of the grid service definitions without having to worry about a specific hosting environment.

The Globus Toolkit is a community-based, open-architecture, open-source set of services and software libraries (available under a liberal open-source license) that supports grids and grid applications. Globus Toolkit has already gone through three generations in the past few years, with the original version in the late 1990s, the next version in early 2000s, and the third version in 2003. It is viewed as the reference software system for generic grid implementation (as contrasted to some domain-specific e-science implementations.) With the publication of OGSI and OGSA, the Globus Toolkit is now moving even more fully toward implementation of standard grid protocols and APIs. As noted, the Globus Toolkit influenced the development of OGSA/OGSI and, in turn OGSA/OGSI is now influencing it [36, 44]. As we saw in previous chapters, OGSI specifies a set of "service primitives" that, rather than stipulating precise services, establish a nucleus of behavior common to all grid/Web services that can be leveraged by meta- and system-level services. Globus Toolkit 3.0 (GT3) uses this specification to provide tools for resource monitoring, discovery, management, security, and file transfer [37, 124].

The official release of the GT3 took place in 2003 (there was an alpha release of the toolkit approximately one year earlier; the release of the Open Grid Services Architecture Development Framework, the core services of GT3, allowed early adopters to start developing OGSA-based services immediately.) GT3 is the first full-scale implementation of the OGSI Version 1.0 that the Globus Project played a key role in defining [124]. Leading grid participants have previously committed to use of GT3 and OGSA; companies include, but are not limited to, Avaki, Cray, Entropia, Hewlett-Packard, IBM, Oracle, Platform Computing, Silicon Graphics, Inc., Sun Microsystems, and Veridian. The introduction of the Globus Toolkit 3.0 with the Open Grid Services Architecture is an important step in moving grid computing beyond the laboratories of academia and research and through the doors of commercial enterprises [124]. The Globus Project is also working with Microsoft to develop a Windows™ version of the Globus Toolkit. Beta versions for Windows XP/2000 platform were released in 2002, with more work under way.

6.1.1 Globus Toolkit Tools and APIs

The Globus Toolkit offers a modular "basket of services and technologies" that enables incremental development of grid-enabled tools and applications. These facilities can be utilized, in the aggregate, to build grids and grid-enabled applications. Globus Toolkit took a layered, modular approach early on; this approach is conducive to openness and standardization. The toolkit defines grid protocols and APIs.

Globus Toolkit provides protocol-mediated access to remote resources for intergrid environments. The Globus Toolkit integrated and extended existing standards and developed a reference implementation. Client and server APIs, SDKs, services, tools, and other capabilities have been defined [119]. As shown in Figure 6.1,

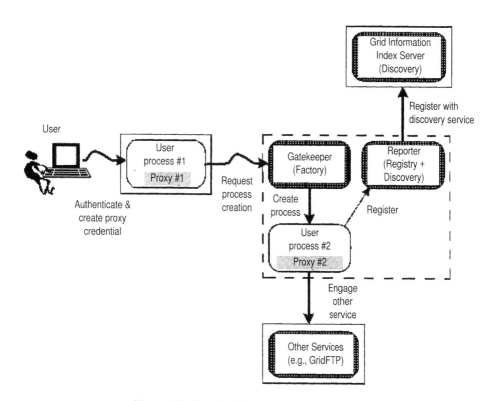

Figure 6.1 Principal Globus Toolkit mechanisms.

Table 6.1   The Globus Toolkit capabilities

- A set of basic facilities needed for grid computing
  - ○ Security: GSI for single sign-on, authentication, authorization, and secure data transfer
  - ○ Resource management: remote job submission and management
    - GRAM protocol and its "gatekeeper" service that provides for secure, reliable, service creation and management (resource layer)
    - Metadirectory service (MDS) that provides for information discovery through soft state registration, data modeling, and a local registry ("GRAM reporter")
  - ○ Data management: secure and robust data movement
    - Data transfer: Grid File Transfer Protocol (GridFTP)
  - ○ Information services: directory services of available resources and their status
    - Grid Resource Information Protocol (GRIP)
  - ○ Collective layer protocols
- APIs to the above facilities
- C bindings (header files) needed to build and compile programs

---

Globus Toolkit components provide the basic elements of an SOA, but with less generality than encompassed in full OGSA. The Globus Toolkit is a set of useful components that can be used either independently or together to develop useful grid applications and programming tools as follows (also see Table 6.1) [129]:

- The *Globus Resource Allocation Manager (GRAM)* provides resource allocation, process creation, monitoring, and management services. GRAM implementations map requests expressed in a resource specification language (RSL) into commands to local schedulers and computers.

- The *Grid Security Infrastructure (GSI)* provides a single-sign-on, run-anywhere authentication service, with support for local control over access rights and mapping from global to local user identities. Smartcard support increases credential security.

- The *Monitoring and Discovery Service (MDS)*[2] is an extensible grid information service that combines data discovery mechanisms with the LDAP (LDAP defines a data model, query language, and other related protocols). MDS provides a uniform framework for providing and accessing system configuration and status information such as computer server configuration, network status, or the locations of replicated datasets.

- *Global Access to Secondary Storage (GASS)* implements a variety of automatic and programmer-managed data movement and data access strategies, enabling programs running at remote locations to read and write local data.

---

[2]MDS-2 supersedes MDS-1, which pioneered the use of GIS concepts but did not address all requirements of interest. MDS-2 provides a configurable information provider component called a Grid Resource Information Service (GRIS) and a configurable aggregate directory component called a Grid Index Information Service (GIIS) [14].

- *Nexus and globus_io* provide communication services for heterogeneous environments, supporting multimethod communication, multithreading, and single-sided operations.

- The *Heartbeat Monitor (HBM)* allows system administrators or ordinary users to detect failure of system components or application processes.

For each component, an API written in the C programming language is provided for use by software developers. Command line tools are also provided for most components, and Java classes are provided for the most important ones. Some APIs make use of Globus servers running on computing resources. Table 6.2 identifies API specifications for software libraries distributed with the Globus Toolkit. In addition to these core services, the Globus Alliance has developed prototypes of higher-level components (resource brokers and resource coallocators) and services.

In addition to these, other components are available that complement, or build on top of, these facilities. For example, the Globus Toolkit provides a rapid-development kit known as Commodity Grid (CoG) that supports technologies such as Java, Python, Web Services, and CORBA. A GUI mechanism available with the Globus Toolkit simplifies the developer's effort for capturing users' inputs, discovering a grid service, creating a grid service instance, invoking a grid service instance, and displaying results. The GUI framework can get a grid service reference (GSR) from a grid service handle (GSH) (discussed in Chapters 4 and 5). Furthermore, the GUI can be customized for supporting an organization's specific grid application. A WSIL document is an example grid service registry used in current Globus Toolkit. The GUI framework can also publish the WSDL document to the grid service registry [119]. Next we compare the Globus Toolkit to two other well-known systems: Condor and Legion.

Condor is a tool for harnessing the capacity of idle workstations for computational tasks [129]. It is well suited for parameter studies and high-throughput computing, where microjobs generally do not need to communicate with each other. Condor and Globus are complementary technologies, as demonstrated by Condor-G, a Globus-enabled version of Condor that uses Globus to handle interorganizational problems like security, resource management for supercomputers, and executable staging. Condor can be used to submit jobs to systems managed by Globus, and Globus tools can be used to submit jobs to systems managed by Condor. The Condor and Globus teams work closely with each other to ensure that the Globus Toolkit and Condor software fit well together [129].

Legion is developing an object-oriented framework for grid applications. The goal of the Legion project is to promote the principled design of distributed system software by providing standard object representations for processors, data systems, and so on. Legion applications are developed in terms of these standard objects. The two technologies are in some respects complementary: Globus focuses on low-level services and Legion on higher-level programming models. There are also significant areas of overlap. It may be useful to note that the Globus Toolkit is being used as the basis for numerous production grid environments (from modest collabo-

Table 6.2   API specifications for software libraries distributed with the Globus Toolkit

Security APIs	globus_gss_assist—simplifies the use of the GSSAPI in the globus environment
	GSS API—the Generic Security Service API C bindings (IETF draft)
Information Service APIs	OpenLDAP—an API for the LDAP protocol used by MDS (developed by the OpenLDAP Project)
Communication APIs	globus_io—provides high-performance I/O with integrated security and a socket-like interface
	globus_nexus—provides multithreaded, asynchronous, thread-safe multiprotocol communication facilities
	globus_nexus_fd—provides NEXUS-based support for file descriptors and timed events. (This API is obsolete as of release 1.1.2; recommend use of globus_io instead.)
Data Access APIs	globus_ftp_control—provides low-level services for implementing FTP client and servers
	globus_ftp_client—provides a convenient way of accessing files on remote FTP servers
	globus_gass_copy—provides a uniform interface for accessing files using a variety of protocols
	globus_gass—provides clients with access to remote files
	globus_gass_transfer—provides an API for clients and servers involved in GASS data transfer
	globus_gass_cache—manages the local GASS cache on a client system
	globus_gass_server_ez—provides a simple set of GASS server capabilities
	globus_gass_server—provides GASS server functionality. (This API is obsolete as of release 1.1.2; recommend use of globus_gass_transfer instead.)
	globus_gass_client—allows clients to get and put remote files via several protocols. (This API is obsolete as of release 1.1.2; recommend use of globus_gass_transfer instead.)
Data Management APIs	globus_replica_catalog—provides an interface to a catalog of data collections, logical files, and physical locations
	globus_replica_management—allows clients to manage files within a file replication system
Resource Management APIs	globus_gram_client—provides remote job submission and management capabilities
	globus_gram_myjob—provides a basic communication mechanism for processes within a GRAM job

Table 6.2 Continued

Resource Management APIs	globus_gram_jobmanager—provides a simple, consistent way to interact locally with a variety of schedulers such as LSF, LoadLeveler, PBS, Condor, etc.
	globus_duroc—provides resource coallocation services for starting distributed jobs
Fault Detection APIs	globus_hbm_client—allows a client process to be monitored by a Heartbeat Monitor system
	globus_hbm_datacollector—allows clients to monitor multiple processes and enables the notification of exceptions
Portability APIs	globus_module—provides a mechanism for activating and deactivating software modules
	globus_libc—provides a portable implementation of libc
	globus_thread—implements threads and synchronization mechanisms
	globus_dc—provides cross-platform data conversion services
	globus_utp—supports the use of timers for monitoring applications and other programs
	globus_list—support for linked lists
	globus_fifo—supports first-in–first-out queues
	globus_hashtable—supports hash tables
	globus_url—supports URL strings
	globus_error—provides an abstract error type for function return codes
	globus_poll—supports polling on I/O channels

rative research projects to huge international scientific ventures), whereas Legion's community of users is smaller and more focused [129].

6.1.2   Details of Key Too Kit Protocols

As the discussion above points out, the Globus Toolkit is built around four key protocols: GSI, GRAM, GRIP, and GridFTP. The Globus Toolkit also utilizes existing Web services technologies that were described in previous chapters, including SOAP, WSDL, and WSIL, to support distributed state management, inspection, discovery, invocation, and asynchronous notifications. Specifically, all the Grid service interfaces are exposed in WSDL format [119]. Some additional information on these protocols is included next; these protocols will continue to exist in GT3.

GSI.   Globus Toolkit implements GSI protocols and APIs, to address grid security functionality. GSI protocols extend existing standard public key protocols: ITU

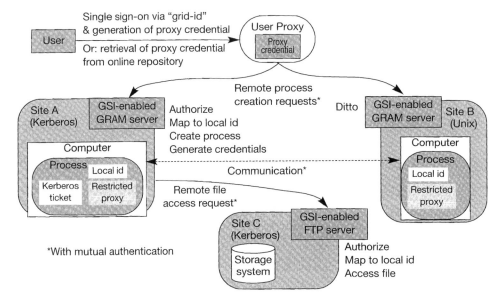

Figure 6.2 GSI in action "Create processes at A and B that communicate and access files at C." Copyright © 2002 University of Chicago and The University of Southern California. All Rights Reserved.

X.509 and IETF Transport Layer Security (TLS-RFC 2246) protocol (the follow-on protocol to SSL). Extensions have been made to X.509 Proxy Certificates and Delegation. GSI also extends standard GSS-API (see below). Figure 6.2 depicts GSI-based operations [44]. GSI utilizes X.509 certificates as the basis for user authentication and defines an X.509 Proxy Certificate to leverage X.509 public key cryptographic techniques capabilities for supporting single sign-on and delegation. GSI utilizes the TLS protocol for authentication.[3]

ITU-T X.509 is a well-accepted security protocol.[4] Users of a public key need to be confident that the associated private key is owned by the correct remote subject (person or system) with which an encryption or digital signature mechanism will be used. This confidence is obtained through the use of public key certificates, which are data structures that bind public key values to subjects. The binding is asserted by having a trusted CA digitally sign each certificate. The CA may base this assertion upon technical means (this is also known as proof of possession through a challenge–response protocol), presentation of the private key, or on an assertion by the subject. A certificate has a limited valid lifetime that is indicated in its signed contents. Because a certificate's signature and timeliness can be independently checked by a client, certificates can be distributed via untrusted communications and server systems, and the certificates can be cached in unsecured storage in certificate-using systems.

---

[3]Other public key-based authentication protocols can also be used with X.509 Proxy Certificates.
[4]These observations on X.509 in the rest of this subsection are based on [42].

ITU-T X.509 (formerly CCITT X.509) or ISO/IEC/ITU 9594-8, which was first published in 1988 as part of the X.500 Directory set of recommendations, defines a standard certificate format. The certificate format in the 1988 standard is called the Version 1 (v1) format. When X.500 was revised in 1993, two more fields were added, resulting in the Version 2 (v2) format (these two fields may be used to support directory access control). The Internet Privacy Enhanced Mail (PEM) RFCs, published in 1993, include specifications for a public key infrastructure based on X.509 v1 certificates. The experience gained in attempts to deploy RFC 1422 made it clear that the v1 and v2 certificate formats are deficient in several respects. Most importantly, more fields were needed to carry information that PEM design and implementation experience has proven necessary. In response to these new requirements, ISO/IEC/ITU and ANSI X9 developed the X.509 Version 3 (v3) certificate format. The v3 format extends the v2 format by adding provision for additional extension fields. Particular extension field types may be specified in standards or may be defined and registered by any organization or community. In June 1996, standardization of the basic v3 format was completed. ISO/IEC/ITU and ANSI X9 have also developed standard extensions for use in the v3 extensions field. These extensions can convey such data as additional subject identification information, key attribute information, policy information, and certification path constraints. However, the ISO/IEC/ITU and ANSI X9 standard extensions are broad in their applicability and in order to develop interoperable implementations of X.509 v3 systems for Internet use it is necessary to specify a profile tailored for the Internet.

It is one goal of various RFCs to specify a profile for Internet WWW, electronic mail, and IPsec applications. Environments with additional requirements may build on this profile or may replace it. For example, the goal of the RFC 2459 specification is to develop a profile to facilitate the use of X.509 certificates within Internet applications for those communities wishing to make use of X.509 technology. Such applications may include WWW, electronic mail, user authentication, and IPsec. In order to relieve some of the obstacles to using X.509 certificates, one can define a profile to promote the development of certificate management systems, development of application tools, and interoperability determined by policy.

The primary goal of the TLS Protocol is to provide privacy and data integrity between two communicating applications [89]. The protocol is composed of two layers: the TLS Record Protocol and the TLS Handshake Protocol. At the lowest level, layered on top of some reliable transport protocol (e.g., TCP), is the TLS Record Protocol. The TLS Record Protocol provides connection security that has two basic properties:

1. The connection is private. Symmetric cryptography is used for data encryption (e.g., DES, RC4, etc.) The keys for this symmetric encryption are generated uniquely for each connection and are based on a secret negotiated by another protocol (such as the TLS Handshake Protocol). The TLS Record Protocol can also be used without encryption.

2. The connection is reliable. Message transport includes a message integrity check using a keyed MAC (message authentication code). Secure hash functions (SHA, MD5, etc.) are used for MAC computations. The TLS Record

Protocol can operate without a MAC, but is generally only used in this mode while another protocol is using the TLS Record Protocol as a transport for negotiating security parameters.

The TLS Record Protocol is used for encapsulation of various higher-level protocols. One such encapsulated protocol, the TLS Handshake Protocol, allows the server and client to authenticate each other and to negotiate an encryption algorithm and cryptographic keys before the application protocol transmits or receives its first byte of data. The TLS Handshake Protocol provides connection security that has three basic properties:

1. The peer's identity can be authenticated using asymmetric, or public key, cryptography (e.g., RSA, DSS, etc.). This authentication can be made optional, but is generally required for at least one of the peers.
2. The negotiation of a shared secret is secure: the negotiated secret is unavailable to eavesdroppers, and for any authenticated connection the secret cannot be obtained, even by an attacker who can place himself in the middle of the connection.
3. The negotiation is reliable: no attacker can modify the negotiation communication without being detected by the parties to the communication. One advantage of TLS is that it is application protocol independent. Higher-level protocols can layer on top of the TLS Protocol transparently. The TLS standard, however, does not specify how protocols add security with TLS. The decisions on how to initiate TLS handshaking and how to interpret the authentication certificates exchanged are left up to the judgment of the designers and implementors of protocols that run on top of TLS.

The Generic Security Service Application Program Interface (GSS-API), Version 2, as defined in RFC 2743 (which supersedes RFC 2078), provides security services to callers in a generic fashion, supportable with a range of underlying mechanisms and technologies, and, hence, allowing source-level portability of applications to different environments. This specification defines GSS-API services and primitives at a level independent of underlying mechanism and programming language environment.

RFC 2743 is to be complemented by other, related specifications: (i) documents defining specific parameter bindings for particular language environments, and (ii) documents defining token formats, protocols, and procedures to be implemented in order to realize GSS-API services atop particular security mechanisms.

A typical GSS-API caller is itself a communications protocol, calling on GSS-API in order to protect its communications with authentication, integrity, and/or confidentiality security services. A GSS-API caller accepts tokens provided to it by its local GSS-API implementation and transfers the tokens to a peer on a remote system. That peer passes the received tokens to its local GSS-API implementation for processing. The security services available through GSS-API are implementable (and have been implemented) over a range of underlying mechanisms based on secret-key and public-key cryptographic technologies. The GSS-API separates the operations of initializing a security context between peers, achieving peer entity authentication from the operations of providing per-message data origin authentication and data integrity protection for messages subsequently transferred in conjunction with that context [70]. Refer to RFC 2743 for additional information on GSS-API.

The primary motivations behind the GSI are [129]:

- The need for secure (authenticated and, perhaps, confidential) communication between elements of a computational grid
- The need to support security across organizational boundaries, thus making a centrally managed security system impractical
- The need to support "single sign-on" for users of the grid, including delegation of credentials for computations that involve multiple resources and/or sites

GSI and its public-key-based mechanism provides *single sign-on* authentication, communication protection, and some initial support for restricted *delegation* [114, 125]:

- *Single sign-on* allows a user to authenticate just once, and thus create a proxy credential that a program can use to authenticate with any remote service on the user's behalf. Once a user is authenticated, a proxy certificate is created and used when performing actions within the grid.
- *Delegation* allows for the creation and communication to a remote service of delegated proxy credentials that the remote service can use to act on the user's behalf, possibly with various restrictions. This capability is important for nested operations. For example, in a specific implementation one may use the GSI sign-in to grant access to the portal, or one may have one's own security for the portal. The portal, in turn, is responsible for signing in to the grid, either using the user's credentials or using a generic set of credentials for authorized users.

A remote delegation protocol of X.509 Proxy Certificates is layered onto TLS. GGF drafts define the delegation protocol for remote creation of an X.509 Proxy Certificate and GSS-API extensions that allow this API to be used effectively for Grid Computing environments (TLS and GSS-API extensions were discussed above). Restricted delegation, a capability of the X.509 Proxy Certificate Profile, is important in that it allows one entity to delegate a subset of its total privileges to another entity; these kinds of restrictions are needed to reduce the adverse effects of either intentional or accidental misuse of the delegated credential [114].

A central concept in GSI authentication is the certificate. Every user and service on the grid is identified via a certificate that contains information vital to identifying and authenticating the user or service. A GSI certificate includes four primary pieces of information [65]:

1. A subject name, which identifies the person or object that the certificate represents
2. The public key belonging to the subject
3. The identity of a CA (certificate authority) that has signed the certificate to certify that the public key and the identity both belong to the subject
4. The digital signature of the named CA

Note that a third party (a CA) is used to certify the link between the public key and the subject in the certificate. In order to trust the certificate and its contents, the CA's certificate must be trusted. The link between the CA and its certificate must be established via some noncryptographic means, or else the system is not trustworthy. GSI certificates are encoded in the X.509 certificate format, a standard data format for certificates established by the IETF. These certificates can be shared with other public key-based software, including commercial web browsers (e.g., Internet Explorer).

GRAM. The GRAM protocol and client API allow programs to be started and managed on remote resources, even in the face of resource heterogeneity. GRAM provides for the reliable, secure, remote creation and management of arbitrary computations, what are known as *transient service instances*. GRAM employs GSI mechanisms for authentication, authorization, and credential delegation to remote computations. RSL is used to communicate requirements.[5] A layered architecture allows application-specific resource brokers and coallocators to be defined in terms of GRAM services (integrated with Condor, MPICH-G2, and so on). Figure 6.3 depicts the resource management architecture [44]. A two-phase-commit protocol is used for reliable invocation, based on techniques used in the Condor system. Service creation is handled by a small, trusted "gatekeeper" process (a *factory*), while a GRAM reporter monitors and publishes information about the identity and state of local computations (*registry*) [114].

MDS-2 provides a framework (known as a *discovery* interface) for discovering and accessing system configuration and status information. Status information includes, but is not limited to, server configuration, network status, and the locations of replicated datasets. MDS-2 utilizes a soft-state protocol, the Grid Notification Protocol, for lifetime management of published information [114].

GRIP. The Grid Information Protocol (GRIP) supports discovery and enquiry. A user, program, or directory use GRIP to obtain information from an information

---

[5]The Globus Resource Specification Language (RSL) provides a common interchange language to describe resources. The various components of the Globus Resource Management architecture manipulate RSL strings to perform their management functions in cooperation with the other components in the system. The RSL provides the skeletal syntax used to compose complicated resource descriptions, and the various resource management components introduce specific <*attribute, value*> pairings into this common structure. Each attribute in a resource description serves as a parameter to control the behavior of one or more components in the resource management system (http://www.globus.org/gram/rsl_spec1.html).

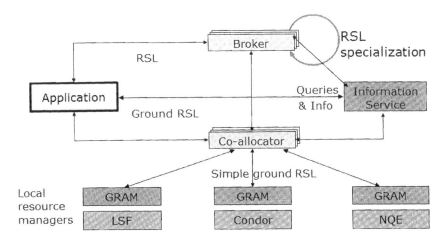

Figure 6.3   Resource management architecture. Copyright © 2002 University of Chicago and The University of Southern California. All Rights Reserved.

provider about the resources on which the provider has information.[6] Discovery is supported via a search capability. Given a set of "discovered resources," enquiry can then be used to refine the set of resources upon which a broker may schedule. Enquiry corresponds to a direct lookup of information: the enquiry supplies the resource name and the provider returns the resource description. LDAP is used in the GRIP context [14].

GridFTP. Data access and transfer is accomplished with GridFTP. GridFTP provides reliable, recoverable data transfers. GridFTP is an extended version of FTP, developed specifically for grid data access and transfer. The protocol is secure, efficient, reliable, flexible, extensible, parallel, and concurrent. It supports third-party data transfers and partial file transfers. It handles parallelism and striping. (We provide some additional details on GridFTP later in the chapter.)

6.1.3   Globus Toolkit Version 3

Globus Toolkit Version 3 is the reference implementation for OGSA (although the programming model from Version 2.2 was expected to change as part of the transi-

---

[6]An information provider is defined as a service that supports GRIP and GRid Registration Protocol (GRRP), to access information about entities and to notify aggregate directory services of the availability of this information, respectively. An information provider can be in possession of information on more than one entity. An aggregate directory is a service that uses GRRP and GRIP to obtain information (from a set of information providers) about a set of entities, and then replies to queries concerning those entities. These two protocols are the fundamental building blocks for the Globus Toolkit environment: for an entity to be known to VO participants, it must either speak these protocols directly (hence, being its own information provider) or interact with some other entity that acts as an information provider on its behalf [14].

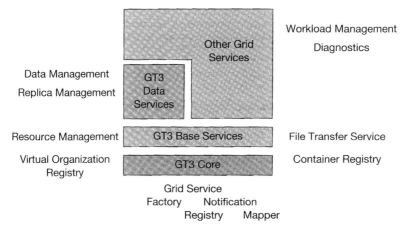

Figure 6.4 GT3 architecture.

tion, most of the actual APIs that are available with Globus Toolkit V2.2 will remain the same); see Figure 6.4 [32]. Globus Toolkit Version 3 provides a common and open standards-based set of ways to access various grid services using standards such as SOAP, WSDL, WSIL, UDDI, and XML, and a standard way to find, identify, and utilize new grid services as they become available [125]. Globus Toolkit 3.0 implements OGSI, and, in addition, provides some OGSI-compliant grid services that correspond to GT2 behaviors. It provides a set of grid services for security, resource management, information access, and data management. As noted, all major capabilities offered in GT2 (data transfer, replica management, job submission and control, secure communication, and access to system information) are offered in GT3; this observation is of interest to people who have already made major investments in GT2. GT3 implements standards that are being adopted by the e-science and e-business communities that are key to support intergrid interoperability.

There is an understanding in the grid computing industry that successful deployment of large-scale interoperable OGSA implementations would benefit from the definition of a small number of standard protocol bindings for grid service discovery and invocation [114]. Just as the ubiquitous deployment of IP allows any two entities to communicate, so ubiquitous deployment of such "intergrid" protocols will allow any two services to communicate. Hence, clients (grid users) can be relatively simple, since they need to know about only one set of protocols. Whether or not such intergrid protocols can be defined and gain widespread acceptance remains to be seen.

The Web services framework can be instantiated on a variety of different protocol bindings.[7] One example is SOAP (using HTTP) along with TLS for security;

[7]The rest of this section is based on [114]

other bindings can and have been defined. In selecting network protocol bindings[8] within an OGSA context, there are four primary requirements:

1. *Reliable transport.* The grid services abstraction typically requires support for reliable service invocation. One way to address this requirement is to incorporate appropriate support within the network protocol binding.
2. *Authentication and delegation.* The grid services abstraction typically requires support for communication of proxy credentials to remote sites. One way to address this requirement is to incorporate appropriate support within the network protocol binding, as, for example, in TLS extended with proxy credential support.
3. *Ubiquity.* The grid goal of enabling the dynamic formation of VOs from distributed resources means that, in principle, it must be possible for any arbitrary pair of services to interact.
4. *Grid service reference (GSR) format.* Grid service reference can take a binding-specific format. One possible GSR format that appears headed for commercial acceptance is a WSDL document.

The abstractions and services made available through OGSI/OGSA provide building blocks that can be used to implement a variety of higher-level grid services. Stakeholders, including the GGF, are working closely with the community to define and implement a wide variety of such services that will, collectively, address the diverse requirements of e-business and e-science applications. These higher-level services may include the following:

- *Distributed data management services,* supporting access to and manipulation of distributed data, whether in data bases or files. Services of interest include data base access, data translation, replica management, replica location, and transactions.
- *Workflow services,* supporting the coordinated execution of multiple application tasks on multiple distributed grid resources.
- *Auditing services,* supporting the recording of usage data, secure storage of that data, analysis of that data for purposes of fraud and intrusion detection, and so forth.
- *Instrumentation and monitoring services,* supporting the discovery of "sensors" in a distributed environment, the collection and analysis of information from these sensors, the generation of alerts when unusual conditions are detected, and so forth.
- *Problem determination services for distributed computing,* including dump, trace, and log mechanisms with event tagging and correlation capabilities.
- *Security protocol mapping services,* enabling distributed security protocols to

---

[8]A binding is the process of associating protocol (or data format) information with an abstract entity such as a message, an operation, or a portType (also see Glossary).

be transparently mapped onto native platform security services for participation by platform resource managers not implemented to support the distributed security authentication and access control mechanism.

It appears straightforward to reengineer the resource management, data transfer, and information service protocols used within the current Globus Toolkit to build GT3, based on these common mechanisms, as depicted in Figure 6.5.

6.1.4 Applications

The Globus Toolkit has become central to hundreds of science and engineering projects and the toolkit has been adopted for commercial offerings by major information technology companies. The combination of open source and open standards afforded by Globus Toolkit is driving increased adoption by users seeking to share resources seamlessly across distributed organizations [124]. A number of individuals, organizations, and projects have developed higher-level services, application frameworks, and scientific/engineering applications using the Globus Toolkit. For example, the Condor-G software provides a framework for high-throughput computing (e.g., parameter studies) using the Globus Toolkit for inter-organizational resource management, data movement, and security [33]. Globus Toolkit users fall into three principal classes: application framework developers, application developers, and grid builders [129]:

- *Application framework developers* are using Globus services to build software frameworks that facilitate the development and execution of specific types of applications. Examples include the CAVERNsoft framework for tele-immersive applications (University of Illinois at Chicago Electronic Visualization Laboratory), Condor-G for high-throughput computations and parameter studies (University of Wisconsin), the HotPage Grid portal framework (San Diego Supercomputer Center), the Linear Systems Analyzer (Indiana University), the MPICH-G implementation of the MPI Message Passing Interface (Northern Illinois University and Argonne National Laboratory), Nimrod-G for parameter studies (Monash University), the Parallel Application WorkSpace (PAWS: Los Alamos National Lab), and WebFlow (Syracuse University).

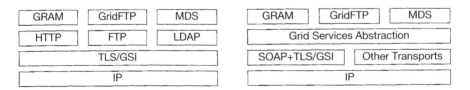

Figure 6.5 Globus Toolkit migration path. Left, some current Globus Toolkit protocols. Right, a potential refactoring to exploit OGSA mechanisms.

- *Application developers* use Globus services to construct innovative grid-based applications, either directly or via grid-enabled tools. Application classes include remote supercomputing (e.g., astrophysics at Max Planck Institute, Washington University), tele-immersion (e.g., NICE at EVL/University of Illinois at Chicago), distributed supercomputing (e.g., OVERFLOW at NASA Ames, SF-Express at Caltech), and supercomputer-enhanced scientific instruments (e.g., Advanced Photon Source, Argonne National Laboratory).

- *Grid builders* are using Globus services to create production grid computing environments. Major grid construction projects include NASA's Information Power Grid, two NSF grid projects (NCSA Alliance's Virtual Machine Room and NPACI), the European DataGrid Project, and the ASCI Distributed Resource Management Project.

6.2 GRID COMPUTING ENVIRONMENTS

6.2.1 Introduction

A Grid computing environment (GCE) is a set of tools and technologies that allow users "easy" access to grid resources and applications. Often, a GCE appears to the user as a Web portal that provides the user interface to a multitier grid application development stack. A GCE may also be a grid shell that allows a user access to and control over grid resources in the same way a conventional shell allows the user access to the file system and process space of a regular operating system. GCEs roughly describe the "user side" of a computing system; this is illustrated in Figure 6.6, where there is an overlapping division between GCE's and what is called a "basic" grid in the figure. The latter would include access to the resources and management of and interaction between them, security, and other such capabilities. The OGSA describes these "basic" capabilities and the Globus Project is the best-known "basic" software project. GCEs fulfill (at least) two functions [31]: (i) programming the user side of the grid, and (ii) controlling user interaction—rendering any output and allowing user input in some (Web) page. This includes aggregation of multiple data sources in a single portal page.

As we saw earlier, the Globus Toolkit is the most widely used grid middleware system, but it does not by itself provide "rich" support for building grid computing environments. To support basic capabilities, Java, CORBA, Python, and Perl commodity grid interfaces to the Globus Toolkit have been developed. Commodity grid toolkits (CoG kits), alluded to in Section 6.1.1, provide a mapping between computing languages, frameworks, environments, and grid services and fabric; a CoG Kit defines and implements a set of general components that map grid functionality into a commodity environment/framework [132]. The Grid Portal Development Toolkit (GPDK) is a suite of JavaBeans suitable for Java-based environments. Combined, the COG Kits and GPDK constitute the most widely used frameworks, at present, for building GCEs that use the Globus Toolkit environment for basic grid services. In turn, taken as an aggregated "basket" of capabilities, grid services, languages, frameworks, and GCEs create an effective development tool for developing grid-enhanced applications.

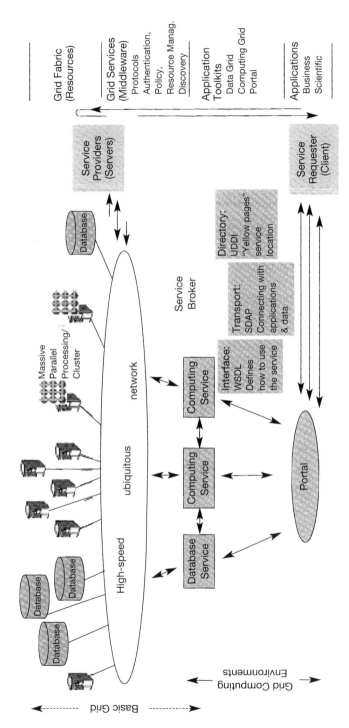

Figure 6.6 Grid computing environments.

6.2.2 Portal Services

Portal services control and render the user interface/interaction.[9] Figure 6.7 shows a key architectural idea emerging in this area. The assumption is made that all material presented to the user originates from a Web service that is called here a content provider. This content could come from a simulation, data repository, or stream from an instrument. Each such Web service has resource or service facing ports (RFIO in Figure 6.7), which are those used to communicate with other services. Here, one is more concerned with the user-facing ports that produce content for the user and accept input from the client devices. These user-facing ports use an extension of WSDL that is being standardized by the OASIS organization. This is called Web Services for Remote Portals (WSRP). It implements the so-called portlet interface that is being standardized in Java as part of a JCP (Java community process).

Most user interfaces need information from more than one content provider. For example, a computing portal could feature separate panels for job submittal, job status, visualization, and other services. One could integrate this in a custom application-specific Web service but it is attractive to provide a generic aggregation service. This allows the user and/or administrator to choose which content providers to display and what portion of the display "real estate" they will occupy. In this model, each content provider defines its own "user-facing document fragment" that is integrated by a portal. Such aggregating portals are provided by the major computer vendors. Portlets represent a component model for user interfaces in the same way that Web services represent a middleware component model. Using this approach has obvious advantages of reusability and modularity. One then has an elegant view with workflow-integrating components (Web services representing nuggets) in the middle tier and aggregating portals integrating them for the user interface.

OGSA/OGSI support GCE Portals. As noted earlier in the book, OGSA is a framework that addresses architectural issues related to the requirements and interrelationships of Grid services. OGSA consists of a set of basic grid Web services defined in terms of the OGSI specification (OGSI addresses detailed specifications of the interfaces that a service must implement in order to fit into the OGSA framework). As noted in Chapter 4, an OGSI-compliant grid service is a subclass of Web services whose ports all inherit from a standard grid service port (so a grid service is a Web service that conforms to a set of conventions that provide for controlled, fault-resilient, and secure management of stateful services). Using this port, there are standard ways that a remote portal can interrogate the service to discover such things as the other port types the service implements, what operations can be made on those ports, and the public internal state of the service. OGSI services can also implement a simple event subscription and notification mechanism in a standardized manner. OGSI also provides a mechanism for services to be group together into service collections.

The simple and standard nature of OGSI makes it possible for us to build on-the-fly compilers to generate portal portlet interfaces to any OGSI-compliant grid ser-

[9]This section is based in its entirety on [31].

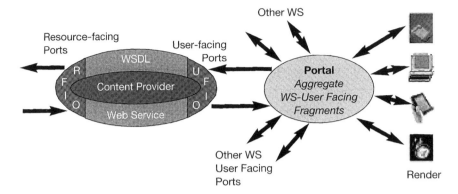

Figure 6.7 Portal providing aggregation service for document fragments produced by user-facing ports of a content-providing Web service.

vice. The basic services defined by OGSA include registries, directories and namespace binding, security, resource descriptions and resource services, reservation and scheduling, messaging and queuing, logging, accounting, data services (caches and replica managers), transaction services, policy management services, and work flow management and administration services. Each of these core services is rendered as a grid Web service. Applications that are designed for an OGSA-compliant grid can assume that these services are available and, with the proper authorization, that they can be used.

6.2.3 Database Access

This section opened by noting the need for a set of tools and technologies that allow users "easy" access to grid resources and applications. The same could be said with regard to data, specifically data bases: the goal is to ease application development through the provision of comparable components (services) [33]. Research and development activities relating to the grid have generally focused on applications in which data is stored in files; however, in many scientific and commercial domains, data base management systems have a central role in data storage, access, organization, authorization, and so on, for a large number of applications. The Data Access and Integration Services (DAIS-WG) working group of the GGF is studying the issue; the group seeks to promote standards for the development of grid database services, focusing principally on providing consistent access to existing, autonomously managed data bases (this is work in progress that should be tracked by IT planners).

6.3 BASIC GRID DEPLOYMENT AND MANAGEMENT ISSUES

In this section, we look at some basic deployment and management issues. This section only scratches the surface of the topic at hand.

6.3.1 Product Categories

The earlier sections (e.g., Chapter 2 and also above) alluded to various products that are available or are becoming available to support e-science and/or e-business applications. Naturally, companies looking to purchase products should ascertain that these are open/standards compliant (as described in Chapters 4 and 5), are secure (as discussed above), and are also easy to deploy and utilize (also as discussed above). Grid products include the following categories:

- Grid middleware
- Grid performance monitoring and forecasting
- Grid portals
- Grid programming environments
- Grid schedulers
- Grid systems
- Grid testbeds and development systems
- P2P systems

Capabilities that need to be included in the products span some or all of these:

- Service management
- Service communication
- Policy management
- Middleware selection and deployment
- Applications retuning
- Service control
- SLAs
- Fault/accounting/performance management
- Security

The reality is that, currently, industrial-strength products are still being developed. Hence, the pragmatic recommendation at this time in the development of the grid field is that organizations should begin by deploying small, limited grids first, focusing on nonmission-critical applications. This allows the organization to learn about grid behavior, peculiarities, challenges, installation idiosyncrasies, and management, before having to deal with more complicated and more extensive grid. Clearly, a "local grid" will be less complex than an "intragrid," which, in turn, is less complex than an "intergrid." Some of the deployment-related issues are covered in the sections that follow, at a very general level.

6.3.2 Business Grid Types

Figure 6.8 illustrates three kinds of business-oriented grid arrangements. The top portion depicts a situation in which the organization has deployed a local grid and

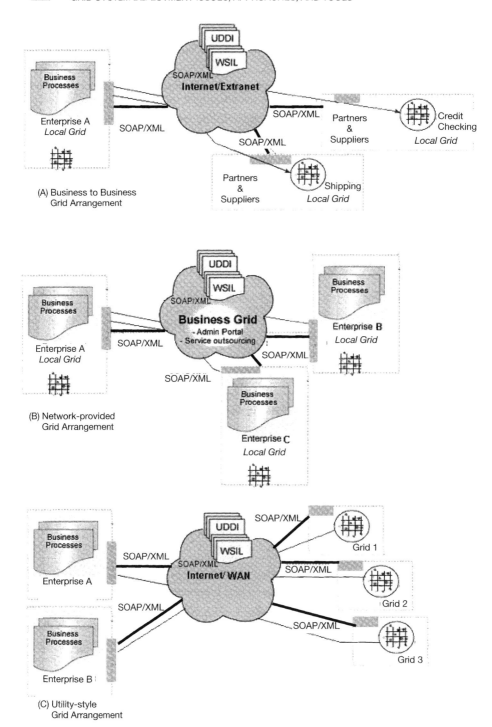

Figure 6.8   Three possible business-oriented grid applications/environments.

also has lined up partners and suppliers that either provide a business function or a grid-based service (e.g., number crunching, storage, etc.). This kind of arrangement can also be used in service outsourcing applications. The software needed by the outsourcing customer to support the connectivity portion of this arrangement is a connection adaptor handling the SOAP protocol (the number-crunching function will of course require its own application-specific software).

The middle portion of Figure 6.8 shows a collaborative logical grid solution for business process integration. In effect, the network-based grid is a business (similar to the concept of an ASP) that can line up other businesses to provide certain services to an enterprise (here, Enterprise A). One can refer to this logical grid as a "business grid" in which every service can be deployed as a grid service; hence, each such service can be accessed and used by other applications utilizing a standard OGSA/OGSI protocol. In a business grid environment, the enterprise customer does not need to install complex software; instead, the enterprise customer can use a Web browser to register with the business grid and subscribe to a specific business service (this is similar to the situation in which an organization deals with an ASP). The business grid connects to all the services "hosted" by the grid itself, or to service providers registered with the business grid. Application clients can utilize XML, SOAP over HTTP, SMTP, FTP, MQ, and so on, to interact with the business grid. The business grid also communicates with external legacy applications, Web services providers, and so on. It takes care of interoperability for connecting to multiple parties using different transports, data formats, and business protocols by offering a membership management service, data format translation service, business protocol translation service, advanced discovery service, and business flow management service [71].

The bottom portion of Figure 6.8 depicts a pure utility environment in which various enterprises can access grid resources (e.g., computing machine cycles) on a pure commodity basis. In turn, the utility provider can use any number of "horizontally based" grids, each of which may specialize to a given function.

As Figure 6.8 shows, grid service mechanisms can be employed to integrate distributed resources within internal commercial IT infrastructures as well as across virtual multiorganizational boundaries. In either case, a collection of grid services registered with appropriate discovery services can support functional capabilities delivering QoS-based services spanning distributed resource pools. Applications and middleware can utilize these services for distributed resource management across heterogeneous platforms with complete local and remote transparency [114].

6.3.3 Deploying a Basic Computing Grid

An "entry-level" grid may be developed and deployed by a few programmers in relatively little time. A certain fraction of applications can run with little or no modification, simply by linking with a grid-enabled version of an appropriate programming library. Even when appropriate high-level tools are not available, an important feature of the Globus Toolkit is that grid-like capabilities can often be incorporated into an existing application *incrementally,* producing a series of increasingly "grid-enabled" versions of the application [129]. More generally, applications may need to be "retuned" and/or ported and/or modified to some degree. Also, the scalability should be tested, as discussed earlier in this chapter. As the grid environment grows and gets larger, and as users become more dependent on it for mission-critical work, a more cohesive approach is needed.

To create a grid infrastructure one can, for example, download the Globus Toolkit and follow the instructions in the Globus Toolkit System Administrator's Guide. The documentation takes the planner through the process of building the Globus Toolkit software, setting up a grid information service, setting up a CA or using someone else's CA, installing the Globus resource management tools on the company's (or department's) servers, and installing Globus client tools and libraries for users. Even after installing the Globus Toolkit's grid services, each site within a grid retains control over access to its resources. When Globus Toolkit services are installed on a resource, the site administrator creates a Grid "mapfile" that contains mappings from grid credentials to local account names. The only people who can submit jobs to the organization's resources are those whose grid credentials are mapped to a valid local account. All job submissions are logged via syslog and an optional gatekeeper log, so one can easily determine who has attempted to use one's system. To use resources on the grid but residing at other institutions, the organization will need to have their grid credential added to the grid mapfile(s) at that site and/or institution. The administrators at those sites will map the organization's grid credential to a local account so that the organization can use their resources. Acquiring permission to use another site and requesting an entry in the site's mapfile is the responsibility of the organization [129].

Using the GT3 toolkit, or some generalized OGSA development toolkit, and/or, some grid computing environment development utility, the deployment of a grid service application involves the following main steps [71]:

1. Create a WSDL for the grid service definition
2. Generate Java proxies for the WSDL definition
3. Write server-side implementation code
4. Write client-side implementation code
5. Deploy and test grid service using the OGSA service browser

6.3.4   Deploying More Complex Computing Grids

The use of a grid is driven from a need for increased IT resources of some type and/or to reduce run-the-engine costs by achieving increased utilization. Another department, company, industry group, or even computing utility may have excess capacity in the particular resource. This requires the establishment of a nontrivial grid arrangement. As noted, complex grids will require a fair degree of effort. Tasks include, but are not limited to the following (also see Table 6.3):

• Identifying/designating resources to be included
• Selecting middleware

Table 6.3   Key grid administrator tasks

Planning	Solution architects and grid administrators need to understand the organization's functional and nonfunctional requirements along with business and systems needs, in order to be in a position to choose the appropriate grid technology.
Technology selection	A grid system suited to the needs at hand must be selected and procured. This may possibly entail the release of an RFI/RFP.
Installation	The selected grid system must be installed on a designated and appropriately configured set of processors and/or resources.
Network planning	Resources need to be interconnected using local, regional, or global networks with sufficient bandwidth and appropriate latency.
Business continuity planning	Solution architects and grid administrators need to understand the failover scenarios for the grid system, to ensure that the grid can continue operating even if some of the resources fail in some way. This includes processors, data bases, and authentication information.
Configuration	Grid software installed on donor machines or other resources may need to be customized. This software may be provided to potential donors on an FTP or equivalent server or be made available on physical media.
Administrative access	Grid software may need to be configured; hence, grid administrator will require "root" access to these grid resources. In some grid systems, the administrator will also need "root" access to the donor machines to install the software on these devices.
Managing enrollment of donors and users (ongoing)	Grid administrator is required to manage the members of the grid (resources and users). Administrator is responsible for controlling the rights of the users in the grid.
Managing security	Identity of users and resources must be established and entered in the CA. The user and the user's certificate credentials must be added to the subscriber list. The CA (external or internal) is a critical element of the grid. CA responsibilities include: positively identify entities requesting certificates; issuing, removing, and archiving certificates; protecting the CA server; and logging activity.
Resource management	Administrator is expected to manage the resources of the grid; e.g., setting permissions for users, tracking resource usage, implementing a corresponding accounting or billing system. Job schedulers typically have provisions for enforcing priorities and policies. It is the responsibility of the administrator to configure the schedulers to meet the goals of the overall organization.

- Selecting middleware management tools/procedures
- Undertaking applications retuning, if needed
- Installing, testing, tuning, and production-line-releasing the grid software and the application(s)
- Establishing service control
- Establishing security control tools/procedures
- Establishing service management and SLA management tools/procedures
- Establishing fault/accounting/performance management techniques
- Establishing appropriate networking infrastructure

One of the first considerations is the hardware needed and/or available. An organization will need to decide if the grid is a "local grid," an "intragrid," or an "Intergrid," more or less as implied in Figure 6.8 and Figures 1.1, 1.2, and 1.3 in Chapter 1. For business applications, the "local grid" and the "intragrid" are probably more likely, but a utility-based arrangement has a flavor of an "intergrid." "Local grids" are generally connected via a LAN or SAN, whereas "intragrids" are generally connected with a WAN intranet. ("Intergrids" often use the Internet or some other institutional network arrangement.) Note that the LANs just mentioned are typically in the data center and dedicated to the cluster-support function; they are not user facing or for carrying user traffic.

Next, an organization may want to add additional hardware to augment the capabilities of the grid. It is important to understand the applications to be used on the grid because their characteristics can affect the decisions of how to best choose and configure the hardware and its data connectivity [147]. Security needs to be properly planned for, as we have already stressed in this chapter and discuss further Section 6.4.

In regard to installing grid software, we have already noted that the administrator first installs the middleware, libraries, and so on, on the target resources (servers). At this stage, users install the provided grid software on their own machines (the user may optionally enroll his or her machine as a donor on the grid, and/or the times that the machine is usable by the grid, and other policy-related constraints). The grid user software may be automatically preconfigured by the grid management system to know the communication address of the management nodes in the grid and user or machine identification information; in less automated grid installations, the user may be asked to identify the grid's management node and possibly other configuration information [147]. After the installation procedure is completed, users are "enrolled" as legitimate grid users via the security apparatus (authentication via a CA).

6.3.5   Grid Networking Infrastucture Required for Deployment

In the bullet list provided in the previous subsection, we noted that the organization also needs to establish an appropriate networking infrastructure in order to deploy a grid. Networking can be perceived at two levels: lower layers, namely, the actual networks, links, transmission facilities, and so on; and at the upper layers, namely, the communications services such as file transfer, directory, messaging, and so on. Although the lower layers typically entail actual harware systems (e.g., routers, switches, LANs, SANs), the upper layers are typically implemented via some sort of end-system software/client. The lower layers of a grid networking system will be addressed in Chapters 8, 9, and 10. In the suctions that follow, we look at a particular upper layer of interest, namely, file transfer (GridFTP), as well as some other related work.

Communication-Related Standardization Work Underway. This subsection looks at some of the press-time initiatives related to communications and networking. This entire subsection is based on GGF sources [33].

GridFTP (GridFTP-WG). The GridFTP protocol is discussed in more detail in the section that follows. Here we examine some extension work under way. At press time, the GridFTP-WG was focusing on improvements to the FTP and GridFTP v1.0 protocols with the goal of producing bulk file transfer protocol suitable for grid applications. The new protocol should be backward compatible with RFC 959 FTP as much as possible, with new features added as (negotiable) extensions. Some desired extensions that are to be included in the future GridFTP v2.0 include:

- Parallel transfers
- GSI authentication
- Striped transfers

Additional information on GridFTP is provided below.

IPv6 (IPv6-WG). IPv6 is now emerging as a significant factor in operational networks, and continued scaling up of the Internet (and thus of grids) will require the additional address space and management features of IPv6. It is therefore important that all GGF specifications work as well (or better) with IPv6 as with IPv4. The goal of the IPv6-WG working group is to identify any GGF specifications that do not meet this requirement, to provide appropriate guidelines for future specifications, and to communicate any issues discovered with IPv6 to the IETF, the Java community, and so on. The scope is IP version dependencies in the output of all working groups in all GGF areas. Any specification that involves the handling of network I/O or IP addresses, or the processing or display of URLs, is likely to be affected. Some of the possible deliverables are:

1. IP version dependencies in GGF specifications. Identification of each GGF specification (approved or public draft) that contains dependencies on IPv4 (principally address format and length). It is intended to be used as a checklist for planning the necessary document revisions by the WGs concerned.
2. Issues in IPv6 specifications or support. If the work on the above two deliverables identifies any issues in the IETF specifications for IPv6, or in IPv6 sup-

port environments such as Java, additional deliverables describing these issues will be created as informational liaison documents to be sent to the IETF, the Java community, and so on.

Data Transport (DataTransport—Research Group). The goal of this group is to provide a forum where parties interested in the secure, robust, high-speed, wide-area transport of data and related technologies can discuss and coordinate issues and develop standards to ensure interoperability of implementations.

Grid High-Performance Networking (GHPN—Research Group). The Grid High-Performance Networking Research Group focuses on the relationship between network research and grid application and infrastructure development. The objective of GHPN-RG is to bridge the gap between the networking and grid research communities. It accomplishes its goal by serving as a forum for information exchange on advances and requirements in both fields, as well as by providing a focal point for liaison activities between the GGF and the various networking standards bodies.

GridFTP. This subsection discusses the GridFTP protocol [2]. This protocol builds on RFC 959 (the FTP RFC), RFC 2228 "FTP Security Extensions," RFC 2389 "Feature Negotiation Mechanism for the File Transfer Protocol," and the IETF draft draft-ietf-ftpext-mlst-16 "FTP Extensions," that are incorporated in GridFTP by reference.

GridFTP is an application-layer protocol in the parlance of the Open System Interconnection Reference Model. Chapters 8–10 will focus on the lowest three layers of the protocol stack: the physical layer, the data link layer, and the network layer.

In grid environments, access to distributed data is typically as important as access to distributed computational resources. Distributed scientific and engineering applications require:

- *Transfers* of large amounts of data (terabytes or petabytes) between storage systems
- *Access* to large amounts of data (gigabytes or terabytes) by many geographically distributed applications and users for analysis, visualization, and so on.

Unfortunately, the lack of standard protocols for transfer and access of data in the grid has led to a fragmented grid storage community. Users who wish to access different storage systems are forced to use multiple protocols and/or APIs, and it is difficult to efficiently transfer data between these different storage systems.

To addrss these issues, a common data transfer and access protocol called GridFTP that provides secure, efficient data movement in grid environments has been developed. This protocol, which extends the standard FTP protocol, provides a superset of the features offered by the various grid storage systems currently in use. GGF chose the FTP protocol because it is the most commonly used protocol for data transfer on the Internet, and of the existing candidates from which to start it comes closest to meeting the grid's needs. The GridFTP protocol includes the following features:

- GSI and Kerberos support
- Third-party control of data transfer
- Parallel data transfer (multiple TCP steams between two network endpoints)
- Striped data transfer (one or more TCP streams between m network endpoints on the sending side and n network endpoints on the receiving side)
- Partial file transfer
- Manual/automatic control of TCP buffer/window sizes
- Support for reliable and restartable data transfer
- Integrated instrumentation

Motivation. There are already a number of storage systems in use by the grid community. These storage systems have been created in response to specific needs for storing and accessing large data sets. They each focus on a distinct set of requirements and provide distinct services to their clients. For example, some storage systems (DPSS and HPSS) focus on high-performance access to data and utilize parallel data transfer streams and/or striping across multiple servers to improve performance. Other systems (DFS) focus on supporting high-volume usage and utilize data set replication and local caching to divide and balance server load. The SRB system connects heterogeneous data collections and provides a uniform client interface to these repositories, and also provides metadata for use in identifying and locating data within the storage system. Still other systems (HDF5) focus on the structure of the data, and provide client support for accessing structured data from a variety of underlying storage systems.

Unfortunately, most of these storage systems utilize incompatible and often unpublished protocols for accessing data, and therefore require the use of their own client libraries to access data. The use of multiple incompatible protocols and client libraries for accessing storage effectively partitions the data sets available on the grid. Applications that require access to data stored in different storage systems must either choose to only use a subset of storage systems, or must use multiple methods to retrieve data from the various storage systems.

One approach to breaking down partitions created by these mutually incompatible storage system protocols is to build a layered client or gateway that can present the user with one interface, but that translates requests into the various storage system protocols and/or client library calls. This approach is attractive to existing storage system providers because it does not require them to adopt support for a new protocol. But it also has significant disadvantages, including:

- Performance. Costly translations are often required between the layered client and storage-system-specific client libraries and protocols. In addition, it can be challenging to efficiently transfer a data set from one storage system to another.

- Complexity. Building and maintaining a client or gateway that supports numerous storage systems entails considerable work. In addition, staying up to date as each storage system independently evolves is very difficult. This is further exacerbated by the need to provide support for multiple client languages, such as C/C++, Java, Perl, Python, shells, and so on.

It would be mutually advantageous to both storage providers and users to have a common level of interoperability between all of these disparate systems: a common, but extensible, underlying data transfer protocol. Storage providers would gain a broader user base, because their data would be available to any client. Storage users would gain access to a broader range of storage systems and data. In addition, these benefits can be gained without the performance and complexity problems of the layered client or gateway approach.

Terminology. Terms used in connection with GridFTP include:

- Parallel transfer. A data transfer between two network endpoints that uses multiple TCP streams.
- Striped transfer. A data transfer between m network endpoints on the sending side and n network endpoints on the receiving side. This could involve multi-homed hosts or multiple hosts (a cluster).
- Data node. In a striped data transfer, a data node is one of the network endpoints returned in the SPAS command, or one of the network endpoints sent in the SPOR command.
- DTP. The data transfer process (DTP) establishes and manages the data connection. The DTP can be passive or active.
- PI (protocol interpreter). The user and server sides of the protocol have distinct roles implemented in a user PI and a server PI.
- Features. A response from a server indicating it supports a set of specified functionality. This is in accordance with RFC 2389.
- Options. A command to a server defining alternative behavior. This is in accordance with RFC 2389.

6.3.6 Grid Operation—Basic Steps

In this section, we briefly look at grid operations (this only scratches the surface.)

To use the grid, the user typically needs to log on to a system utilizing an ID that is enrolled in the grid. As discussed earlier in the chapter, the single log-on makes this convenient for grid users by making the grid look more like one large virtual computer rather than a collection of individual machines (e.g., the proxy log-in model of Globus discussed above). After a validated log-in process, the user can query the grid and submit jobs. The user may be able to perform queries (e.g., via a GUI) to determine how busy the grid is, to monitor how his or her submitted jobs are progressing, and to identify resources on the grid.

Job submission usually consists of three phases (even if there is only one command required) [147]:

1. "Staging the input data." Some input data and possibly the executable program or execution script file are sent to the machine to execute the job (alternatively, the data and program files may be preinstalled on the grid machines or accessible via a mountable networked file system). Note that when the grid consists of heterogeneous machines, there may be multiple executable program files, each compiled for the different machine platforms on the grid. Some grid systems require that the program and input data be preprocessed in some way by the grid system, to add protective execution controls around the application.

2. "Execution." The job is executed on the grid machine. The grid software running on the donating machine executes the program in a process on the user's behalf. Some grid systems implement a protective "sandbox" around the program so that it cannot cause any disruption to the donating machine if it encounters a problem during execution. Rights to access files and other resources on the grid machine may be restricted. Note that the data accessed by the grid jobs may be moved among resource elements in the grid system. However, depending on its size and the number of jobs, this can potentially become a network burden. Also, the latency involved may adversely impact the performance of the application. Optimal deployment of data (if it requires caching and/or duplication) is needed to minimize data movement on the grid. There are many considerations in efficiently planning the distribution and sharing of data on a grid: this analysis is a true "must" for large, mission-critical jobs.

3. "Output." The results of the job are sent back to the submitter (in some implementations, intermediate results can be viewed by the user who submitted the job).

Monitoring and recovery mechanisms are important, particularly for commercial and/or mission-critical applications (such as in the financial services industry). It is to be expected that the user should be able to query the grid system to determine how the application (and possibly the constituent microjobs) are progressing toward task completion. A grid system, particularly the job scheduler, needs to provide automatic monitoring and recovery of microjobs that happen to fail; failures may be due to programming errors (including infinite loops) or hardware failures of various kinds in the grid resources and/or networking system. The scheduler will attempt to determine the reason for a job's failure. Whenever possible, schedulers automatically resubmit jobs. However, this is not always achievable and for this category of problems, the user is informed about the failure and he/she must decide whether to rerun the failed jobs.

6.3.7 Deployment Challenges and Approaches

Currently there are a number of practical challenges to deploying a robust mission-critical grid. Grid systems have the goal of allowing applications to share, in an efficient manner, data and computing resources that belong to the organization, and, when appropriate, to enable access to data and computing resources across multiple organizations, also, of course, in an efficient manner. Grid services need to be deployed on different hosting environments, even on different operating systems. The Globus services discussed above that have been developed to date help organizations overcome some of the barriers to grid computing; however, many challenging problems remain to be overcome before we can say that we have a fully functional grid environment [129]. We identified a number of such challenges in Chapter 2. Some additional information is presented herewith (security, one of the major challenges, is covered in the next section).

In particular, to facilitate the deployment of such distributed solutions, the IT professional needs to be able to rely on standards. As covered in previous chapters and alluded to earlier in this chapter, there is activity under way in this regard, but more work in this space is required. It is imperative that a stable level of standardization be achieved in this industry (and its products) because we currently are where data networks were in the late 1980s before standardization (and ensuing expansion) was achieved by the Internet and Web protocols. For example, current software-based grid solutions typically are installed behind an organization's firewall. These "local" grid solutions are not only platform dependent, but they also typically utilize incompatible communication protocols. In addition, these "local" grids offer limited integration mechanisms for communicating with other "local" grids. Hence, due to proprietary interfaces, it is difficult to add new applications from different vendors to an existing "local" grid. Consistent interfaces to the various services and grid resources are required [71]. The interfaces of grid services need to address discovery, dynamic service instance creation, lifetime management, notification, and manageability; the conventions of grid services need to address naming and upgrading issues. A related challenge is to find ways to quickly and easily integrate with external business processes and services provided by different "local" grids using a secure, low-cost, and manageable solution [71].

Obviously, administrators need capabilities to control access to grid resources and enable users or applications to access grid resources. There is a need for a grid administration capability to handle *resource provisioning.* Resource provisioning allows the administrator to make visible computing resources to a grid solution developer, but do so in a secure manner. There is also a need to register (publish) a new grid service. For intragrids, a private UDDI registry is typically used for this registration process.

Related challenges to be faced in deployment include, but are not limited to the following:

- Activity/performance monitoring
- Administration of the grid
- Affordable networking bandwidth availability
- Application integration

- Application portability
- Comprehensive administration
- Data sharing and access
- Decision making with regard to, for example, virtualization and classical out-sourcing solutions (are these alternatives easier/cheaper than going down the path of grid enablement?), or open-source outsourcing based on grid/Web services
- People/talent
- Policy-based grid management mechanisms
- Resource provisioning
- Right-sizing the technology for companies other than Fortune 500

Among the major research challenges that the industry is addressing in current work are [129]:

- End-to-end resource management and adaptation techniques able to provide application-level performance guarantees despite dynamic resource properties
- Automated techniques for negotiation of resource usage, policy, and accounting in large-scale grid environments
- High-performance communication methods and protocols
- Infrastructure and tool support for data-intensive applications and new problem-solving environment techniques

A key desideratum is to find scalable grid systems to integrate data and other computing resources with applications in a distributed environment. We already alluded in earlier chapters to the need for scalability, particularly in a linear fashion. Assume that the typical number of transactions that a planner needs to run is n_t. This means that the IT planner, upon testing a grid system on a small batch of transactions, say on $n_1 = 0.01 \cdot n_t$ and finding that it requires R_1 resources, and, further, that upon testing the grid system $n_2 = 0.05 \cdot n_t$ and finding that it requires R_2 resources, and, further, that upon testing the grid system $n_3 = 0.25 \cdot n_t$ and finding that it requires R_3 resources, that the following is true to a large extent:

$$0.01 \cdot n_t/R_1 = 0.05 \cdot n_t/R_2 = 0.25 \cdot n_t/R_3$$

so that the prediction

$$R_t = R_1/0.01 \qquad \text{(that is, } R_1 = R_t \cdot 0.01\text{)}$$

holds true, or something very similar to it.

6.4   GRID SECURITY DETAILS—DEPLOYMENT PEACE OF MIND

6.4.1   Basic Approach and Mechanisms

Because of its critically, in this section we revisit the issue of security in some detail. Computational grid infrastructure software (such as the Globus Toolkit) enables a user to identify and use the available machine irrespective of location and ownership, and can facilitate execution of a single computation across multiple machines. However, without adequate security both the grid user and the system administrator who contributes resources to a grid can be at risk [72].

The security topic is extensive and could be the subject of a separate book. As we noted in the previous chapter, OGSA's security model must address authentication, confidentiality, message integrity, policy expression and exchange, authorization, delegation, single log-on, credential life span and renewal, privacy, secure logging, assurance, manageability, firewall traversal, and security at the OGSI layer. The OGSA specification defines requirements for interfaces, behaviors, and models used to structure and achieve interactions among grid services and their clients. *While it is assumed that such interactions must be secured, the details of security are out of scope of the OGSA specification.* Security is addressed in related specifications that define how abstract interactions are bound to specific communication protocols, how service behaviors are specialized via policy-management interfaces, and how security features are delivered in specific programming environments. One can expect that existing and evolving standards will be adopted or recognized in the Grid Security Model [69].

Below, we briefly revisit the basic mechanism used by the Globus Toolkit that we introduced in Section 6.1 because many (if not all) of these techniques carry forward (security mechanisms for grid computing are based on public key cryptography). Then we look at some security guidelines based on the document "Security Implications of Typical Grid Computing Usage Scenarios, GFD-I.12" [72].

Mutual Authentication.[10] If two parties have certificates, and if both parties trust the CAs that signed each other's certificates, then the two parties can prove to each other that they are who they say they are. This is known as *mutual authentication.* For example, the GT3 GSI uses SSL for its mutual authentication protocol (as noted earlier, SSL is also known by a new IETF standard name: Transport Layer Security, or TLS). Before mutual authentication can occur, the parties involved must first trust the CAs that signed each other's certificates. In practice, this means that they must have copies of the CAs' certificates, which contain the CAs' public keys, and that they must trust that these certificates really belong to the CAs.

To mutually authenticate, the first person (A) establishes a connection to the second person (B). To start the authentication process, A gives B his certificate. The certificate tells B who A is claiming to be (the identity), what A's public key is, and what CA is being used to certify the certificate. B will first make sure that the cer-

---

[10]The information provided in the next few paragraphs is based directly on [65].

tificate is valid by checking the CA's digital signature to make sure that the CA actually signed the certificate and that the certificate has not been tampered with. (This is where B must trust the CA that signed A's certificate.)

Once B has checked out A's certificate, B must make sure that A really is the person identified in the certificate. B generates a random message and sends it to A, asking A to encrypt it. A encrypts the message using his private key, and sends it back to B. B decrypts the message using A's public key. If this results in the original random message, then B knows that A is who he says he is. Now that B trusts A's identity, the same operation must happen in reverse. B sends A her certificate, A validates the certificate and sends a challenge message to be encrypted. B encrypts the message and sends it back to A, and A decrypts it and compares it with the original. If it matches, then A knows that B is who she says she is.

At this point, A and B have established a connection to each other and are certain that they know each others' identities.

Confidential Communication. As an illustrative example, GT3 GSI does not automatically establish *confidential* (encrypted) communication between parties. Once mutual authentication is performed, the GSI gets out of the way so that communication can occur without the overhead of constant encryption and decryption. The GSI can easily be used to establish a shared key for encryption if confidential communication is desired. Recently relaxed United States export laws now allow us to include encrypted communication as a standard optional feature of the GSI.

A related security feature is *communication integrity.* Integrity means that an eavesdropper may be able to read a communication between two parties but is not able to modify the communication in any way. For example, the GT3 GSI provides communication integrity by default (it can be turned off if desired). Communication integrity introduces some overhead in communication, but it is not as large an overhead as encryption.

Securing Private Keys. As an illustrative example, the core GSI software provided by the Globus Toolkit expects the user's private key to be stored in a file in the local computer's storage. To prevent other users of the computer from stealing the private key, the file that contains the key is encrypted via a password (also known as a *pass phrase*). To use the GSI, the user must enter the pass phrase required to decrypt the file containing their private key. The Globus Project has also prototyped the use of cryptographic smartcards in conjunction with the GSI. This allows users to store their private key on a smartcard rather than in a filesystem, making it still more difficult for others to gain access to the key.

Delegation and Single Sign-On. As an illustrative example, the GT3 GSI provides a *delegation* capability. Delegation is an extension of the standard SSL/TLS protocol that reduces the number of times the user must enter his pass phrase. If a grid computation requires that several grid resources be used (each requiring mutual authentication), or if there is a need to have agents (local or remote) requesting services on behalf of a user, the need to reenter the user's pass phrase can be avoid-

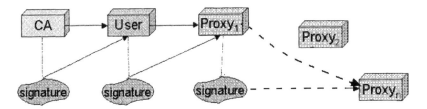

Figure 6.9 Proxies.

ed by creating a *proxy*. A proxy consists of a new certificate (with a new public key in it) and a new private key. The new certificate contains the owner's identity, modified slightly to indicate that it is a proxy. The new certificate is signed by the owner rather than a CA (see Figure 6.9). The certificate also includes a time notation after which the proxy should no longer be accepted by others. Proxies have limited lifetimes.

The proxy's private key must be kept secure, but because the proxy has a limited lifetime, this key does not necessarily have to be kept as secure as the owner's private key. Hence, it is possible to store the proxy's private key in a local storage system without being encrypted, as long as the permissions on the file prevent anyone else from looking at them easily. Once a proxy is created and stored, the user can use the proxy certificate and private key for mutual authentication without entering a password.

When proxies are used, the mutual authentication process differs slightly. The remote party receives not only the proxy's certificate (signed by the owner), but also the owner's certificate. During mutual authentication, the owner's public key (obtained from her certificate) is used to validate the signature on the proxy certificate. The CA's public key is then used to validate the signature on the owner's certificate. This establishes a chain of trust from the CA to the proxy through the owner.

Note that that at press time, the GSI (and software based on it—the Globus Toolkit, GSI-SSH, and GridFTP) was the only software that supported the delegation extensions to TLS. The Globus Project is working with the GGF and the IETF to establish proxies as a standard extension to TLS so that GSI proxies may be used with other TLS software.

6.4.2 Additional Perspectives

This section[11] has the goal of providing some explicit perpectives related to the kind of infrastructure that is required to support secure interactions between grid users and services. This is accomplished through a number of use scenarios.

This discussion starts with some nomenclature. The term *grid user* or *user* refers to the person who is attempting to access a resource. *Principal* is used to mean any

---

[11]The entire security section that follows is based directly on [72].

entity, either human or process, that has an identity associated with it and wants to make use of or to provide resources. *Stakeholders* are people or organizations who set the use policy for a resource. A *grid gateway* is a process that accepts remote requests to use resources. A *grid resource gateway* is the process that actually controls the use of the resource (this may be legacy code). A *grid administrator* is a grid-aware person with responsibility for the overall functioning of the grid (note that there will probably exist multiple grid administrators with nonoverlapping realms of responsibility in a single grid). *Site administrators* are responsible for the functioning of a single site. The user's *home organization* is the administrative domain to which the user belongs that may have trust relationships or service agreements with some of the resource providers.

Next, some assumptions are documented. A fundamental assumption that is made is that each user and principal will have a grid-wide identity that all the other grid principals can verify. Another assumption is that some local resource managers will require legacy local user IDs for users of their resources, so there must be a way to map from Grid IDs to local user IDs. Access control will be enforced both by local resource managers, often using legacy access control mechanisms and by grid-aware services that may want to use grid-centered access policies. In either case, there must be simple ways for users to request access rights and allocations and the stakeholders to grant them. The use of short-term proxy certificates in place of the long-term grid ID is a desirable feature of a distributed system, since it limits the exposure of long-term private keys. The delegation of rights and/or identity from a user to the servers operating on his behalf is also required. Finally, a user may want a way to specify security options for a session.

Most computing resources that are shared among a group of users require that a user have established a unique ID and allocation before he can use the resource. When using a grid, a person should have an ID that is unique within the grid and recognized by all the local resource managers. As noted, Globus uses X.509 identity certificates that can either be issued by a grid-centric CA or by individual corporate CAs for grid identities. Legion has a similar capability to issue an unforgeable "token" that a user can present to identify himself or herself as a Legion user. A CA can require different levels of verification that the person requesting the certificate (usually via e-mail) is in fact the person named in the certificate (such as showing up in person, presenting various documents such as a driver's license, passport, and/or undergoing a background check). Currently, neither Globus nor Legion require or support certification beyond a rudimentary level. Both Globus and Legion also support Grid IDs established through external means such as the enterprise-wide Kerberos [155] installations that exist in certain national laboratories. A grid is likely to have multiple CAs in place. If so, a crucial step is for the each CA to carefully establish a policy for its signed certificates, and for the CAs to cross-certify each other if possible. That is, a CA must clearly identify what, if anything, another CA's certificates mean to it.

Once the grid ID is established, a user may request allocations on each machine on which she might potentially execute a task. Some sites (such as supercomputer centers) may require that each individual have a local user ID and allocation; other sites may allow group allocations or simply require that a user be permitted to use the resource possibly in a constrained manner (e.g., only on weekends or late nights). Establishing permissions and allocations on a resource depends on the resource owner's policy and may require sending e-mail to the system administrator of the resource in question. For each resource that requires local user IDs, the grid system administrator on the resource must determine the mapping of grid ID (such as brown@xyz.org) to local user ID (such as *stebrown*) into a grid-specific mechanism such as the /etc/gridmap file in Globus.

The need for short-term proxy certificates arises from the fact that the client tends to interact with grid services through a sequence of requests over the network to various entities. During each of these transactions, the client must present his grid ID and use his private key to verify his identity. In order to minimize the use of long-term credentials, it is desirable to allow the user to set up a short-term (typically 12 hours) proxy credential to represent the user in interactions with the grid. This short-lived keypair can be kept unencrypted by the local grid software and can be used for signing, verifying, and (optionally) encryption. A second certificate is generated, signed by the long-lived keypair, stating that for the next 12 hours the public key of the user is the public key of the short-lived pair. The Globus security infrastructure software supports this through *grid-proxy-init*.

The last steps a user may take before interacting with services or resources on the grid is to establish parameters that may be in place for the life of the session. For example, a user may specify his or her specific role that he or she wants to assume, such as system administrator for a particular resource or ordinary user. There are different rights and responsibilities for each role. In addition, a user may also set security-related parameters, such as the hosts that are trusted, the level of message integrity and confidentiality that is required, and a limit to the amount of allocation that can be used.

6.4.2.1 *Use Scenarios.* These scenarios are ordered by increasing complexity, where complexity is loosely defined as the number of decisions that must be made by the grid software, or the scope of those decisions. Six categories are described: immediate job execution, job execution that requires advance scheduling, job control, accessing grid information services, setting or querying security parameters, and auditing use of grid resources. The security implications of each scenario is discussed in turn.

6.4.2.1.1 Immediate Job Execution

6.4.2.1.1.1 RUN A JOB ON A SPECIFIED GRID COMPUTER; LOCAL I/O REQUIRED. In a basic approach to remote job submission, the user specifies the execution host to be used and submits a job for which either the code already exists on the target machine or else is uploaded as part of the request. The job uses only remote computation cycles and possibly temporary file storage. The input data is uploaded as part of the job submission, and the output is returned through the connection that was established at job submission. The security requirements and implications in this grid usage scenario are:

1. Mutual authentication of user and grid gateway on specified host
2. Grid gateway on specified host must map grid ID to local ID
3. Grid gateway must submit the request to resource gateway in a manner so that the job will run as the authorized local user

In this scenario, authorization to use the target machine is performed by the grid gateway. Although in general the resource gateway may be a separate entity from the grid gateway, in this case it is likely that the resource gateway is the queuing system on the target machine and does not need to perform a specific authorization step. In this case, the target job does not have to be given a copy of the credential of the user presented to the grid gateway as part of the mutual authentication step.

6.4.2.1.1.2   RUN A JOB ON A SPECIFIED GRID COMPUTER; NON-LOCAL FILE I/O RE-QUIRED.   As an extension of scenario 6.4.2.1.1.1, the remote job must access nonlocal files, either by copying them from the remote site to the local site or by obtaining the nonlocal data as needed during the execution of the job. In addition to the security requirements and implications of Scenario 6.4.2.1.1.1,

1. If the model of execution is such that the grid gateway (or some other grid infrastructure process) performs the file transfer before execution, then this process must be given authorization to obtain these files on behalf of the user.
2. If (1) is not true, then the job obtains the data directly from the files and therefore must be given the appropriate credentials upon startup to obtain the data.
3. If the remote job writes output to the local file server, and the file server is AFS or DFS, then the remote job needs the user's Kerberos ticket (which may or may not be the same as the credentials used to authenticate to the grid gateway).
4. If the output of the job is in the form of local files, and these files are required to be copied back to the machine from which the user submitted the job, then either the job itself or the grid process that does the copying must be given the necessary credentials to be authorized with the grid gateway on the local machine.

In contrast to Scenario 6.4.2.1.1.1, in this example *delegation* is necessary to access services or information. In step (1), the grid gateway does not automatically have permission to copy arbitrary files from the grid environment, so it must be given the explicit capability. Similarly, in step (4), the grid process needs the user's credentials to be granted access to copy the local files back to the machine on which the job was submitted. In each of the delegation requirements, note that the specific right to be delegated is different. Delegation is an extremely difficult problem that is the subject of current research, so the current approach is to solve these require-ments with unlimited delegation.

6.4.2.1.1.3   RUN A JOB ON "BEST" GRID COMPUTER. Instead of mandating the spe-cific grid computer in the request, a user may wish to run a job on the "best" com-puter, defined to be the host that is "quickest" or "cheapest" according to some met-ric. The choice is made by a third-party service, such as one of the emerging "super schedulers" as exemplified by the default scheduler in Legion. The user may speci-fy a specific group from which to choose, or the user may leave it to the super scheduler to locate the set from which to choose. The additional security ramifica-tions posed by this scenario are:

1. If the set of candidate hosts has not been identified by the user, the super scheduler will need to interact with the information services component(s) of the grid to identify the set of possible hosts. Since access to such information is restricted, the super scheduler may need a delegated credential to act on be-half of the user.
2. The super scheduler must determine if the target user is allowed to execute on each of the target grid machines, and, if so, the remaining allocations of the user in the particular role in which the user is acting. This information is deter-mined by asking information services or querying each grid machine directly. Again, the super scheduler may need a delegated credential from the user.
3. The super scheduler needs to query the grid hosts to determine available cy-cles so that it can make an informed scheduling decision. Account informa-tion is largely regarded as being important to keep secret, so it will probably be the case that an arbitrary entity will not be allowed to ask information ser-vices where a particular user can execute or how much allotment she has left. Therefore, either the super scheduler, as a principal, must be granted broad access to such information and trusted not to leak such information to any one except the affected user, or the super scheduler must be explicitly granted the right to ask on behalf of the user. In this scenario, the super scheduler needs the rights to read information about a user's access and allocations on specific machines form the grid information service and possibly from the grid hosts themselves.

6.4.2.1.1.4   RUN A JOB THAT REQUIRES RESOURCES AT MULTIPLE SITES. In a sce-nario that many see as the defining contribution of grid computing technology, a user may wish to combine resources from multiple sites into a single, coordinated job. For example, a user could generate a large amount of data from a major shared instrument (e.g., an accelerator or microscope study). This data needs to be up-loaded to a large data store that in turn can be accessed by a powerful computing engine. Once preliminary data analysis has taken place, intermediate data may need to be saved and also passed on to a different computing engine for further analysis such as visualization procedures. This scenario requires a remote job to start other remote jobs or access files on behalf of the original user. The significant new security requirement of this scenario is that a controlling agent or each remote job in a sequence has to be able to request resources on behalf of the user, perhaps through subsequent calls to a super scheduler.

The controlling agent or each remote job must be given a delegated credential from the user to further submit jobs on behalf of the user. If the agent has been granted a credential with unlimited delegation, it can impersonate the user for any purpose, such as a malicious activity. Thus, it is more desirable for the agent to have a delegated credential that allows it to perform only limited tasks for a limited period of time. An issue that might arise here is the exact time during which the delegated credential is valid. If it expires too soon, the agent cannot perform its intended action, but if the timeout is too long, it could be a potential point of compromise. Another issue that can arise is how many levels of credential delegation should be allowed. If the user is using application code that he did not write, he may not know how many levels of servers and objects are going to be instantiated. Allowing unlimited levels of delegation may be the only practical approach, but that exposes the user's credential to more possible misuse.

6.4.2.1.2   Job Execution that Requires Advance Scheduling

6.4.2.1.2.1   MAKE A RESERVATION FOR SIMULTANEOUS RESOURCES. If Scenario 6.4.2.1.1.4 involved the use of an instrument that produces a large data flow that must be processed in real time, it will require the advance reservation of data storage, network bandwidth, and, possibly, computing cycles. Advance reservations require:

1. Delegation of the user's rights to a super scheduler and bandwidth broker to make the reservations on behalf of the user.
2. Assurance that if a user has been granted a reservation for the future, she will have access at the time the reservation is claimed.
3. Bandwidth reservations usually require service agreements for priority bandwidth between ISPs and computing sites. This implies that a bandwidth broker needs to know at reservation time that user's connection will come from an authorized site. If the model of execution is such that the bandwidth broker returns a *claim ticket* to the super scheduler, the transfer of the claim ticket from the super scheduler to the user must be protected, and the claim ticket itself must be nonforgeable.

6.4.2.1.2.2   CLAIM THE RESERVATIONS. The execution of a job on reserved resources can require multiple concurrent claiming procedures. In this model, a user directly interacts with the individual resource gateways to claim the reservation. In general, reservation claiming requires:

1. The user must be able to identify himself or herself as the entity that made the reservation. The reservation may have been made on behalf of a group of users, in which case the user has to prove himself or herself to be a member

of the group. Another way of handling the situation where one person makes a reservation and a different person wants to claim it is to allow the claim tickets to be transferred. In this case, the resource gateway must be able to verify that the claim has been legitimately transferred by the person who made the original reservation to the current claimant.

2. The user should still have access to all the resources that he has reserved, except in extreme cases, such as when the user is no longer associated with the organization that is going to pay for the resource use, or the organization has failed to pay its bills.

3. In the case of a user losing access to a resource, a check should be made of advance reservations in his name, and the appropriate parties should be notified of the change.

This scenario contains two important requirements in grid computing: *group membership* and *nonrepudiation.* Group membership is nontrivial because, whereas individual users should be able to define groups, it is not clear how exactly to do this. Nonrepudiation in this context refers to the requirement that the resource gateway should not be able to arbitrarily deny that it granted a reservation.

6.4.2.1.3   Job Control

6.4.2.1.3.1   ALLOW A USER TO MONITOR OR ATTACH TO A RUNNING JOB. A standard requirement of people who start long-running remote jobs is to be able to disconnect from a job and then at a later time and possibly from a different location reattach to it. The user may just want to monitor the progress of a job, or may want to enter some steering information at specific points in the run. Monitoring a job's progress may be as simple as knowing where logging files are being written and having the access to read them. Steering implies that the user has defined "entry points" into the computation and has some way of controlling who may connect to them. In the collaborative environment facilitated by the grid, a different user may want to use the monitoring or attachment points as well. In this case,

1. The resource that is being protected is access to a running job created by a user, who will set the access policy and later be granted access by that policy. This can perhaps be most easily accomplished if the policy and code to enforce access is part of the job.

2. The point of entry is probably directly to the computation itself as opposed to through the grid gateway or the resource gateway, so the potential collaborator must be able to authenticate to the computation itself. The user in this scenario would probably rely on predefined libraries generated by security developers rather than creating an individual security solution. Utilizing well-accepted libraries facilitates interoperability.

6.4.2.1.3.2   SYSTEM ADMINISTRATOR(S) TERMINATES OUT-OF-CONTROL JOB. Certainly, situations will occur in which a grid user has submitted a job and does not realize that the job is behaving abnormally and perhaps consuming more resources that expected or allowed. Whereas the grid administrator(s) might be aware of an out-of-control job, only the system administrator of one or more physical resources will have the rights to terminate the job. This scenario requires:

1. A system administrator must detect the out-of-control process and trace its origin to a particular grid user. Alternatively, grid-monitoring software might detect the out-of-control process and notify the system administrator.
2. The system administrator can optionally inform the grid administrators that the process is about to be terminated. The grid administrators need this information to coordinate the termination of this job across multiple grid sites.
3. The grid administrators either attempt to terminate the individual components of the job by directly interacting with the job or by asking the system administrators to terminate those processes of the job that are on their respective machines.
4. The job owner must be notified by the grid administrator that his job has been terminated.

Although the steps described here are a function of the unique implementation of the grid design, this scenario presents a number of interesting requirements. First, resources of a grid are used both by "local" users and grid users, so it is not necessarily obvious from where an out-of-control process originated. Therefore, grid software must keep audit records, or at least provide a means by which this determination can be made. Second, there will generally not be a single person who has the power to kill a single "grid computation," because it will span multiple resources of multiple administrative domains. As such, ideally, a coordinated effort must be made if a single job is to be prematurely terminated (note that this is unlikely, at least in the near term). Third, the user must be told at the very least that her job has been prematurely terminated, as opposed to the computation just disappearing. The exact mechanism for doing this is not clear, nor are the security implications.

6.4.2.1.4 Accessing Grid Information Services

6.4.2.1.4.1 READ/QUERY INFORMATION FROM ONE INFORMATION SERVER. The ability to locate services and to determine the status and availability of those services will be crucial in a well-functioning computational grid. In most computational grid architectures, there exist information services whose purpose is to be a centralized repository for information about the many services in the grid. To avoid being a single point of failure, information services can be replicated, organized hierarchically, or organized geographically. However, high availability does not mean easily and uniformly accessible; many services require carefully controlled access to information regarding the services they provide, their current status, and who can use them. In general, when a grid user queries a single information server,

1. Mutual authentication should take place between the user and the information services.
2. The information services should implement the access control policy as desired by the service.

Although the information services require the user to authenticate, it is not strictly necessary for information services to authenticate to the user, for example, if the user subsequently authenticates to the service itself. The extra cost of mutual authentication in general can be weighed against the potential effects of malicious information. With regard to the information services providing the actual information requested, it could be the case that the individual services are allowing the information services to determine an "appropriate" access policy. However, a more general scenario is to allow each publisher to set the policy. In this case, the publisher and the information services must agree on a policy language. Subsequently, the publisher must trust that the information services accurately implements the policy.

6.4.2.1.4.2   PUBLISH/UPLOAD INFORMATION TO AN INFORMATION SERVICE. Scenario 6.4.2.1.4.1 partially describes the implications of publishing information into a centralized repository from the perspective of the publisher. In order for a service to upload the information that the information services is providing to others,

1. Mutual authentication must take place between the publisher and the information services.
2. Confidentiality or message integrity on the communication from the publisher to the information services could be required by the publisher.

If there are no constraints on the information being provided by the publisher, then neither (1) nor (2) are necessary. However, in most cases, the publisher cares about who sees the information, making mutual authentication and confidentiality and/or message integrity important.

6.4.2.1.4.3   QUERY ACROSS MULTIPLE INFORMATION SERVICES. The case in which a user wishes to interact with multiple information services, receives noncontradictory information, and combines the information himself is a relatively straightforward extension to Scenario 6.4.2.1.4.1. However, when the information services return information that is not consistent with each other, if there is no obvious reason to believe one piece of information over another, the determination of which information to believe must be based on a trust relationship established with one of the information services.

Anytime an information service is accessed, the user must trust the information being returned to a certain extent. It is important to note that authentication does not directly address trust, as authentication merely ascertains that a particular entity is who it claims to be, not that the entity is "doing the right thing." In the case where one information service contradicts another, the user needs to have his own policy for establishing trust that could be based on who wrote each service, who deployed each service, where each service is executing, how useful the information has been in the past from each of the information services, and so on. This policy is set by the user and cannot be mandated by the grid administrator(s).

6.4.2.1.5 Setting or Querying Security Parameters

6.4.2.1.5.1 USER SETS MESSAGE INTEGRITY AND CONFIDENTIALITY PARAMETERS. An individual grid user should have the capability to constrain the manner in which she interacts with the collective grid services. One way in which to personalize a user's interaction with a grid is for the user to define message integrity and confidentiality parameters. For example, a user can state that all communication between grid services as a direct or indirect result of the user be encrypted stronger than a selected minimum amount (e.g., encryption algorithm and key size). The implications of this requirement are:

1. Message integrity implies supporting MAC algorithms
2. Confidentiality requires a key agreement protocol to be supported
3. Services must recognize the rationale for per-user security configuration and be designed accordingly
4. There must exist an easy mechanism for users to specify such constraints
5. There must be a secure and efficient mechanism to propagate or otherwise convey a particular user's integrity and confidentiality parameters from the user to the services

This scenario exemplifies one of the key challenges in constructing a grid, namely, that there is a tension between support for heterogeneity and a requirement that services implement some subset of shared functionality. Many users will implement and deploy services for a grid, each with a different API and different functionality. However, their utility will be significantly impeded if they mandate how users are to interact with them, as opposed to how the users would like to interact with them. Requirements for message integrity or confidentiality are an example of requirements that may be imposed across a class of applications from their perspective users.

6.4.2.1.5.2 RESOURCE PROVIDER SETS MESSAGE INTEGRITY AND CONFIDENTIALITY PARAMETERS. Resource providers can also set security requirements on a per-resource or per-site basis. An example requirement is that all traffic into and out of a site be encrypted with a particular algorithm. The implications of this requirement are:

1. Services must be aware of and adhere to the message integrity and confidentiality rules of the resource provider on which they execute.
2. Users must be made aware of and adhere to the message integrity and confidentiality rules of the resource provider on which the services that they invoke execute.

Adhering to resource providers' requirements in general is difficult, if only because resource providers generally will not publish security requirements. Many resource providers believe in "security through obscurity" and thus will not publish such information. Other resource providers do not have a single clear document that details such security requirements, but rather the policy has been formed ad hoc in response to individual events.

6.4.2.1.5.3   *QUERYING INDIVIDUAL'S ACCESS TO A RESOURCE.* There are several scenarios in which one principal wants to know what his own or another's access rights are with respect to a resource:

1. A user may want to determine his access to a resource before attempting to use or schedule use of the resource.
2. A stakeholder may want to see what access a user has.
3. A super scheduler may need to see what machines and resources a user has access to.

In each case, either the resource gateway or an independent policy analyzer must be able to determine a user's access given the grid ID of the user and decide if the principal asking the question has the right to see the answer.

6.4.2.1.5.4   *STAKEHOLDERS SET AUTHORIZATION POLICY.* A stakeholder for a resource on a remote machine may want to set or modify the policy for the use of a resource. He or she may want to see what the existing policy is before modifying it. In this scenario, it is assumed that there is an authentication policy interpreter separate from the resource itself. To determine if a request is authorized, the resource gives the identity of the authenticated user and the authorization policy to the separate policy interpreter and is returned a yes/no/maybe answer. The implications of this scenario are:

1. For policy information stored on the resource gateway, the stakeholder must be able to securely connect to the gateway machine (and subsequently hand-edit a policy file) or authenticate himself to a server on the gateway machine that can modify the policy information.
2. In the case in which the server maintains the authorization policy, the server must be able to check that the stakeholder is authorized to change the information.
3. In the case in which the server maintains the authorization policy, the server can require message integrity or confidentiality when it reads the policy.
4. If the policy information can be stored locally to the stakeholder, the authorization policy must be kept securely.
5. Policy information may need a validity period or a priority assigned to it if the policy is intended to be temporary.

A challenge in supporting this scenario is that that may be multiple stakeholders that have jurisdiction over different usage rights of a single resource. Therefore, the server that maintains the policy must carefully enforce the policy regarding each stakeholder's ability to change the access policy.

6.4.2.1.5.5 STAKEHOLDER WANTS TO REVOKE ACCESS IN A TIMELY FASHION. A stakeholder may want to deny access to a user or set of users and have the ban take effect promptly. Although the semantics of "timely" or "promptly" vary from case to case, the general scenario is that an entity that was previously allowed access should not be denied access. The implications of this requirement are:

1. Any caching of access rights must be short-lived and/or provide a way of being flushed.
2. If policy information is stored in distributed places or multiple copies are kept, it must be linked together or indexed in some way so that all the copies can be deleted.
3. If capabilities are used, they must be very short-lived or else kept in known places from which they can be removed.

There are similar issues that a CA must address when a certificate must be revoked.

6.4.2.1.5.6 USER REQUIRES CONFIDENTIALITY ON STORED DATA. In an extension to Scenario 6.4.2.1.5.1, a user may want to specify that certain files be encrypted or all the data at a given site be encrypted. The user may also wish to specifically mandate that a server that acts on her behalf store all data related to her encrypted. This scenario implies:

1. There may exist a need to share an encryption key with the program or server that is writing the file to storage.
2. The user and/or server will need long-term storage and escrow of encryption keys.
3. A secure system is required to associate keys with particular files.

In general, proper key management is a requirement for many of the scenarios. For example, certain administrative domains within a grid may require *smart cards* for key management, as opposed to a password-based authentication scheme. The requirements for key management must be properly conveyed to the users by the grid administrators. Managing keys and understanding each of the individual key management requirements will be a challenge for the user, as a grid may cross multiple administrative domains.

6.4.2.1.5.7 USER, SERVICE PROVIDER, OR ADMINISTRATOR SPECIFIES TRUSTED GRID HOSTS. As part of a session-specific configuration or in a directed scheduling request, a user may want to specify what hosts she is willing to use. If a job is going to use several hosts, this information has to be passed along to the scheduler or the job controller. Similarly, a service provider may mandate that requests for service must arrive from a particular subset of hosts, perhaps because the other hosts are not trusted or because of billing requirements (if the service is not free). Lastly, a grid administrator may specify that no user or service is allowed to interact with users or services from another administrative domain. For example, if NASA trusts DoD, but DoD does not trust NASA, then the DoD grid administrator(s) might require that DoD users cannot use NASA machines in DoD-related computations. To support the specification of trusted grid hosts or trusted grid domains,

1. Grid hosts must be able to authenticate and possibly prove membership in a particular grid domain. This can be done through host TLS/SSL credentials or secure DNS and IPSec.
2. Servers in this category require a protocol in which both the identity and location/domain from which the request originated are authenticated. Clients must be ready to provide this information.
3. Grid administrators must be able to enforce these requirements.

Implementation of these requirements can be problematic with regard to all entities that could specify a set of trusted grid hosts. For example, if the computation scenario is such that there is a *chaining* of services (e.g., user asks server 1, server 1 asks server 2, server 2 asks server 3, . . . , server n returns information back to the user), then the entire chain might be required to be authenticated before server n performs the requested action and subsequently returns the information to the user. Similarly, server $n - 1$ must realize the user's restricted set of hosts before contacting server n (if server n is not contained in the list of trusted hosts, then server $n - 1$ should not attempt to use server n). This is also not easy for the grid administrator to enforce.

6.4.2.1.6   Auditing Use of grid Resources

6.4.2.1.6.1   SYSTEM ADMINISTRATOR WANTS TO CHECK A LIST OF PAST REQUESTS. Ether a site system administrator or a grid administrator may need to monitor all accesses to a resource. This information may be used for accounting purposes, for a routine security review, or for a real-time intrusion-detection procedure. The system administrator may wish to check both the accesses allowed and the accesses rejected. This scenario implies:

1. The resource gateway server must keep an unforgeable log of all access, by unique user identification and time of access.
2. The format of the entries to this log must be negotiated between the system administrator and the resource gateway.
3. Access to this log should be carefully restricted.

4. The system administrator must ask the resource gateway server to signal especially troublesome resource access requests via a mechanism separate from this log.

The system administrator and the resource gateway server must agree upon both the content and format of logged entries. Arbitrary servers running on a grid host can also be required to do similar logging. How useful such a log file is will depend on how trusted the server is. Presumably, it is trusted to some extent or else the local system administrator would not allow it to run at all, but, by definition, it is not a standard part of the grid services.

6.4.2.1.6.2 STAKEHOLDER WANTS TO CHECK WHO HAS BEEN ACCESSING RESOURCE. Whereas in Scenario 6.4.2.1.6.1 the system administrator can require access to all access logs for services on the system administrator's machines, a stakeholder should be able to only access the logs for which he has authorization. A stakeholder may want the same information as in Scenario 6.4.2.1.6.1 but only for resources for which he is the stakeholder. To meet this requirement,

1. Stakeholders should have limited access to the access logs.
2. There is a need to identify a stakeholder with a resource. A stakeholder may wish to review the resource logs to determine the overall usefulness of the service or to determine whether the service is oversubscribed.

6.4.3 Conclusion

This section provided some explicit scenarios to highlight to the kind of infrastructure that is required to support secure interactions between grid users and services. These scenarios are indicative of actual dynamics and provide input to people who are developing security mechanisms for grid applications. The section should provide organizations a certain comfort level by knowing that a nontrivial amount of thought and effort are being expended in this direction.

CHAPTER 7

Grid System Economics

This chapter looks at some macroeconomic issues related to grid computing. It focuses on mechanisms to support chargebacks in a virtualized data center/grid environment that might be supported by a computing utility. Chargeback mechanisms are fundamental to a commercial grid environment that aims at supporting, in a financially viable and sustainable manner, coordinated resource sharing and problem solving in dynamic, multiinstitutional VOs. These techniques can also be employed to support "open outsourcing."

We will not, out of pragmatic necessity, provide financial modeling that IT planners can utilize to calculate various total cost of ownership calculations. We have already provided anecdotal information throughout the text that various savings in run-the-engine costs can be achieved. For example, in Chapter 1 we quoted results from [111] that reported that

> Grid middleware vendors indicate that cluster computing yields reductions in information technology costs and costs of operations that are expected to reach 15% by 2005 and 30% by 2007–8 in most early adopter sectors. Use of Enterprise Grids is expected to result in a 15% savings in IT costs by the year 2007–8, growing to a 30% savings by 2010 to 2012.

See Figure 7.1.

In the text we also cited IBM studies that show that mainframes are generally idle 40% of the time, that in many instances Unix servers are actually "serving" something less than 10% of the time, and that most PCs are lightly used for 95% of a typical day. This is an inefficient situation for organizations [43]. We noted that (some) firms that have implemented grid architectures have observed measurable changes: processor utilization rates have grown to 80%, while costs have dropped in some cases by as much as 90% [130]. Firms have found that Intel-based Linux servers, often used in grid deployments, can be between 1 to 10% of the total cost of "heavy iron" machines like mainframes or high-end UNIX servers based on proprietary operating systems. For example, IBM reportedly worked with Hewitt Associates, the global outsourcing and HR consulting firm, to build a grid, Linux, and WebSphere-based solution for the company's pension modeling application. On the grid, Hewitt reduced its transaction costs by some 90%, without rewriting their ap-

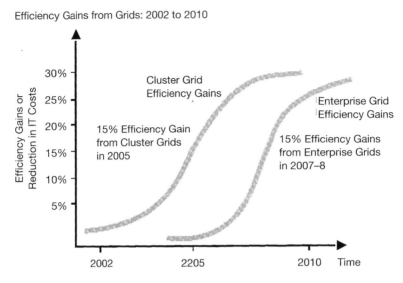

Figure 7.1 Efficiency gains from grids: 2002 to 2010. Reprinted from [111] by courtesy of the publisher.

plications. And in [118] it was noted that "IBM's ultimate vision for grid is a utility model over the Internet . . . with more than 60% of IT budgets dedicated to maintenance and integration . . . the need to reduce complexity and management demands is a pressing one." All of this leads to anecdotal inferences to potential savings with grid computing.

At a higher level, grid researchers make the case that [105]:

- Grids exploit synergies that result from cooperation of autonomous entities and a grid-based economy provides incentive needed for sustained cooperation.
- Grid technologies support the emergence and operation of virtual enterprises (and/or virtual data centers).
- Grid service brokers allow users to dynamically lease grid services at run time based on their quality, cost, availability, and users QoS/SLA requirements.
- Grids are enabling the emergence of a new service-oriented computing industry.
- Grid computing standards can become the foundation for "open-source outsourcing."

The rest of the chapter focuses on chargeback mechanisms in grid environment that can be used by the IT organization to pass costs back to the ultimate corporate users, or by a computing utility. Chargeback models for enterprise IT computing were developed starting in the late 1960s: enterprise IT chargeback models have been discussed at the technical level in this arena in the past decade. Many of these classical chargeback concepts carry over to the grid computing space.

7.1 INTRODUCTION

Intergrids (and also enterprise or intragrids) aim at exploiting synergies that result from cooperation of autonomous distributed entities. As we noted in the opening pages of this book, the term "synergistic" implies "working together so that the total effect is greater than the sum of the individual constituent elements." Synergies include, to list just a few, resource sharing (processors, data storage, software, sensors, etc.), "on-demand" virtual enterprises creation, and on-demand aggregation of resources. For e-science projects, the majority (if not all) of the funding is from government sources of all types. For this cooperation to be sustainable, however, especially in commercial settings, participants need to have an economic incentive. It follows that, "incentive" mechanisms need be considered as one of the design/development facets of grid computing.

We had noted in Chapter 2 that some of the challenges related to grid computing dealt with the following predicament [105]: users, resources, and owners are geographically distributed. A challenge relates to the fact that resources, users, and applications are heterogeneous. Another challenge deals with the fact that resource availability and capabilities vary with time. Furthermore, policies and strategies are heterogeneous and decentralized. Quality of service (service-level agreements) are also heterogeneous. Finally, cost and price vary based on resources, users, time, and demand.

On this topic Buyya [10] argues that the heterogeneity and decentralization that is present in a grid is very similar to that present in "standard human economies," where market-based mechanisms have been used successfully to manage them. Therefore, "market-pricing" models for managing grid resources may be applicable by treating their services as information and communication technologies (ICT)[1] commodities/utilities. This approach provides a workable paradigm for managing self-interested and self-regulating entities (resource owners and consumers). It helps in regulating the supply and demand for resources: services can be priced in such a way that equilibrium is maintained. The "market pricing" is understood by the user (and is user-centric) as being a utility-oriented model. The approach is also scalable since there is no need for a central coordinator (during negotiation) so resources (sellers) and also users (buyers) can make their own decisions and try to maximize utility and profit. It facilitates the offering of different QoS (SLAs) to different applications depending on the value users place on them. Market (commodity) pricing improves the utilization of resources, as one can see from the operation of any stock or commodity market (for tangible as well as intangible assets).

Grid consumers are interested in executing jobs for solving problems of varying size and complexity. The strategy is to minimize expenses. Grid users (service con-

---

[1]The U.S. reader is more familiar with the term IT: information technologies. Information and communication technologies (ICT) covers those industries that facilitate processing, transmission and display of information by electronic means. ICT is a term used in the rest of the world (ROW), for example, the Eurostat Task Force on Information Society Statistics. It includes data operations/IT services, consultancy services, software producers, producers/suppliers of data, radio and television equipment, telecommunications, multimedia and the Internet, and system integrators.

sumers) benefit by selecting and aggregating resources optimally. There is typically a trade-off between response time (time frame) and cost. Grid providers contribute ("idle") resources for executing consumer jobs. The strategy is to maximize return on investment. Hence, grid service providers benefit by maximizing resource utilization. There is, typically, a trade-off between local utilization of the resources (to meet local requirements, if any) and the market opportunity. Therefore, resource owners deal with questions such as [10, 105]:

- How do I decide prices? (economic models?)
- How do I specify them?
- How do I translate price to resource allocation?
- How do I enforce them?
- How do I advertise and attract consumers?
- How do I do accounting and handle payments?

And resource consumers (users) deal with questions such as [10, 105]:

- How do I decide expenses?
- How do I express QoS requirements?
- How I trade between time frame and cost?

A number of models have been advanced to handle such questions. Table 7.1 summarizes, for illustrative purposes, proposals by Buyya [10].

Table 7.1 Buyya's economic models for trading grid services [10]

Pricing—based on supply, demand, value, and wealth of economic system:
- Commodity Market Model
- Posted Price Model
- Bargaining Model
- Tendering (Contract Net) Model
- Auction Models
  - English auction (first-price sealed-bid or open bid),
  - Vickrey auction (second-price sealed-bid)
  - Dutch (consumer: low, high, rate; producer: high, low, rate),
  - Continuous double auction
- Proportional Resource-Sharing Model

Mechanism (new components) of market-based grid systems:
- An information and market directory for publicizing grid services
- Models for establishing the value of resources
- Resource-pricing schemes and publishing mechanisms
- Economic models and negotiation protocols
- Mediators to act as a regulatory agency for establishing resource value, currency standards, and crisis handling
- Accounting, billing, and payment mechanisms
- Users' QoS requirements driving brokering/scheduling systems

The Global Grid Forum has focused on the importance of these issues. Although OGSA provides an infrastructure for virtualizing resources of many types (computing, storage, software, networking, etc.) into grid services, it is unlikely that any sustainable commercial infrastructure will be provided by any nonresearch organization without financial compensation. Keep in mind, in this regard, the concept of the commercialization of the Internet in the early 1990s. For grid services to be provided on demand (that is, to deploy the utility infrastructure that has been one of the goals of grid technology), organizations will want to be paid for contributing these resources to the grid. Because of these considerations, a Grid Economic Services Architecture (GESA) is being developed within the GESA Working Group (GESA-WG) to define the additional service data and ports needed to describe the economic grid services—the enabling infrastructure [30]. The next section will describe GESA. GESA does not aim to describe the economic models themselves that will be built on such an infrastructure, but rather the service data and ports to support contributors' financial compensation.

7.2   GRID ECONOMIC SERVICES ARCHITECTURE

7.2.1   Introduction[2]

The ability to virtualize any resource as a service through a standard framework, such the OGSA, will enable many different forms of interaction between these diverse service offerings. However, the provisioning of these services is currently dependent on "best efforts" from the academic and research community. For grid services compatible with the OGSA to be provided reliably to the users in a community, the users must expect to fund these services in some manner. By integrating the ability to charge for grid services within the core OGSA infrastructure, one expects to enable new models of service provisioning such as utility computing.

One such effort to develop such an architecture is taking place in the United Kingdom through the successful funding of the UK e-Science Core Programme Project—A Market for Computational Services. As part of this activity, the GESA Working Group of the GGF was developing at press time an infrastructure to enable the trading of grid services as defined through the OGSA. GESA defines extensions to the standard grid services that will enable the construction of such a marketplace. By definition, any such marketplace must support interoperable protocols. GESA interacts with other activities within the GGF, notably the Resource Usage Service (RUS), the Grid Resource Allocation Agreement Protocol (GRAAP), OGSA, OGSI, and the Usage Record (UR) Working Group. The mechanisms needed to trade services are well established from initiatives in traditional economic areas; therefore, GESA deliberately excludes the detailed mechanisms as how to price these services (any such discussion is for illustrative purposes only). However, GESA does focus on the static and dynamic metadata that needs to be generated and maintained within the service data elements (SDEs) exposed through the rele-

---

[2]This section is based in its entirety on [30].

vant ports defined by the OGSI. It should be understood that this is work in progress; specifically, there are a number of unresolved issues.

7.2.2 Overview

7.2.2.1 Architecture. A straw-man architecture of the Grid Economic Services Architecture infrastructure is illustrated in Figure 7.2, showing how the grid service that is to be sold as a Chargeable Grid Service (CGS) interacts with the grid payment system (GPS) and the resource usage service.

Early discussions identified one key requirement: that the underlying OGSA service interface should not be changed, only extended, by the wrapping of a grid service as a CGS. This would allow existing clients to interact with a CGS even if the client interface had been generated for the underlying grid service. The service data elements and service interface for the CGS is described in Section 7.2.3 and the GPS in Section 7.2.4. A variant of the GPS, the GPSHold Service, which allows "reservations" to be made on a user's money is defined in Section 7.2.5. Section 7.2.6 describes a PortType for Currency Exchange. Section 7.2.7 contains a simple example as to how a system using these protocols might work.

This basic architecture exploits the transient nature of a grid service to encapsulate the cost of using the service within its SDEs. All changes in state of the grid service (from the initial advertisement, establishing the cost of its use, to the acceptance of this cost, through to its eventual use) are encapsulated through the creation of new services (see Figure 7.3). This sequence shows how the user finds a service and requests a price through the RequestPricing operation. The pricing is encapsulated in a short-lived service (30 seconds in this example) that is not acceptable to the user and the chosen economic model supports a second-round pricing request

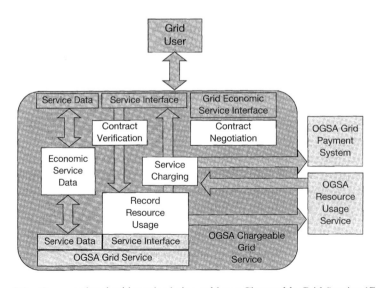

Figure 7.2   Computational grid service being sold as a Chargeable Grid Service (CGS).

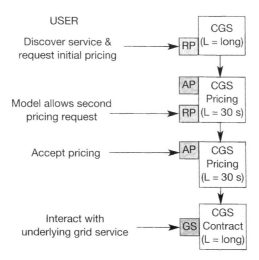

Figure 7.3   Pricing model.

that is triggered by the second call to the RequestPricing operation to produce a second short-lived service. The user has only two choices: to reject the price and let the service destroy itself after 30 seconds or to accept the pricing (acceptPricing operation), which produces a long-lived service specifically created for the user. The pricing of this service may have two stages (as in this example), a single stage, or many stages. The detailed protocols need to support this form of interaction is described in the GESA document.

7.2.2.2 *Definitions and Notational Conventions.* Throughout GESA, the term "user" is used as a generic term for a client to a CGS that may be an interactive user client, a broker acting on the user's behalf, or any other such entity. This specification uses namespace prefixes throughout; they are listed in Table 7.2. Note that the choice of any namespace prefix is arbitrary and not semantically significant.

Table 7.2   Prefixes and namespaces used in GESA

Prefix	Namespace
ogsi	http://www.gridforum.org/namespaces/2003/03/OGSI
gwsdl	http://www.gridforum.org/namespaces/2003/03/gridWSDLExtensions"
sd	"http://www.gridforum.org/namespaces/2003/03/serviceData"
wsdl	"http://schemas.xmlsoap.org/wsdl/"
http	"http://www.w3.org/2002/06/wsdl/http"
xsd	"http://www.w3.org/2001/XMLSchema"
xsi	"http://www.w3.org/2001/XMLSchema-instance"
ur	"http://www.gridforum.org/namespaces/2003/??/UR" ???
rus	"http://www.gridforum.org/namespaces/2003/??/RUS" ???
gesa	"http://www.gridforum.org/namespaces/2003/??/GESA" ???

7.2.2.3 Scope. The remainder of GESA defines the structure of the CGS and GPS, and how they interact with other services such as RUS. For each of these services, one needs to define the:

- Service Data Elements: The additional SDEs that are needed within the grid service to express the service's economic meta-data.
- Service Interface Definition: The operations needed to support interaction with the grid service.
- Implementation: Observations on implementing the operations.
- Other Issues: The goal is to build on the OGSI and related standards and specifications. In some areas, these may need clarification or further development.

GESA also defines subsidiary services that are needed to support these primary services.

7.2.3 The Chargeable Grid Service (CGS)

The CGS represents the abstraction of a grid service that has been enabled to support economic interaction.

7.2.3.1 Service Data Elements. The SDEs provided by the CGS (see Table 7.3) are in addition to those defined within the OGSI Specification. These contain static and dynamic metadata relating to the economic use of the service. This list of SDEs is not exhaustive and should be expanded and adapted as the requirements from the economic models develop. For instance, instead of using real currency within the refund or compensation SDEs, a service may choose to give credit. This could be represented as a currency exchangeable only with the services run by a specific service provider. The SDEs constitute a service-specific advertising element and some of these SDEs may only be relevant at different stages of the CGS lifetime.

Table 7.3   SDEs provided by the CGS

SDE	Occurence	Provided by	Comment
pricing	1+	Service administrator	Supported pricing mechanisms
usage	0/1 Static	Service administrator	A GSH to a trusted RUS
price	1	Service administrator	Price generated through the pricing mechanism
liability	0/1 Static	Service administrator	Route to "human compensation"
testimonial	0+ Dynamic	Third party service	A digitally signed declaration from a grid entity as to the reliability of the CGS

7.2.3.1.1 Pricing. The primary purpose of a CGS is to let other grid entities know how to obtain pricing information relating to the use of the service:

```
<serviceData name="gesa:pricing" pricingName="FixedPriceCharge" >
. . .
</serviceData>
```

A CGS must support a pricing SDE for each pricing mechanism supported by the service and these are differentiated by their different names and the service data elements that they encompass. There may not be any difference between a pricing mechanism named "DutchAuction" and one named "FixedPrice." The service provider may add attributes to the pricing service data to describe the mechanisms used within the pricing mechanism. It is therefore possible to have pricing systems named "FixedPrice" and "FixedPriceWithCompensation" that have different pricing strategies as they offer different levels of compensation—some and none. The presence of a pricing element implies that there is a pricing capability accessible through the CGS::requestPricing operation that will, if invoked, produce a new service instance containing pricing information within a price SDE. Within the pricing element, we define the default and maximum lifetime of any quotation (i.e. the underlying service) provided by the CGS.

A pricing SDE can contain a number of important subelements that are listed in Table 7.4. These elements are now described in turn.

Table 7.4 Subelements of a pricing SDE

Element name	Occurence	Provided by	Comment
pricingType	1	Service administrator	Basic pricing mechanism that will be used
duration	0/1	Service administrator	Limits of/expected service duration
currency	1 + Static	Service administrator	A declaration of the currencies provided by a GSH that are acceptable
paymentMechanism	1 + Static	Service administrator	Acceptable payment mechanisms
resources	0+	Service administrator	The consumed resources that will incur cost to the user
compensation	0 + Static	Service administrator	Pays a proportion of the survice cost to the user if the service fails to deliver
refund	0 + Static	Service administrator	Refunds a proportion of any paid charges to the user if the servicefails to deliver
product	0+	Service administrator	

7.2.3.1.1.1 *PRICING TYPE.* Within each pricing element the service provider *must* provide precisely one element that characterizes the pricing method:

```
<serviceData name="gesa:pricing" pricingName="FixedPriceCharge" >
<gesa:pricingType>
<gesa:FixedPrice />
</gesa:pricingType>
</serviceData>
```

The above example provides an example of how to describe a CGS's pricing mechanism by using a fixed price, that is, one that is fixed after an AcceptPricing. The elements used to classify the pricing mechanism (i.e., a very simple lightweight ontology) are given in Table 7.5. Further elements (obviously) need to be defined.

7.2.3.1.1.2 *DURATION.* As part of the pricing SDE, a service *may* specify information about how long the service will last. Any combination of minimum, maximum, and default time (in seconds) may be defined:

```
<serviceData name="gesa:pricing" pricingName="FixedPriceCharge" >
. . .
<gesa:Duration default="3600" maximum="3600" />
</serviceData>
```

7.2.3.1.1.3 *CURRENCY.* All transactions within a CGS will incur some "cost" on a GPS for the resources that are consumed. This cost may be charged in real money (through some later off-line reconciliation) or through some form of site/organization-specific service tokens:

```
<serviceData name="gesa:pricing" pricingName="FixedPriceCharge" >
. . .
<gesa:currency currencyName="HeyPounds" email="cash@hey.ac.uk" />
</serviceData>
```

The currencies that are usable within each CGS are declared through one or more of the above elements. In order to complete a transaction, the service must "know of" a GPS that supports this currency type (and the desired payment method) and the grid entity requesting to use the service. A GPS that supports this currency

Table 7.5 Elements used to classify pricing mechanisms

Element	Definition
FixedPrice	The price for the service is set in a single non-negotiable stage.
Auction	Indicates that the price is set through a multi-stage action.
EnglishAuction	Indicates a particular approach to setting a service price.

may be found by searching the service registry and its interface and SDEs are defined later.

7.2.3.1.1.4   PAYMENT METHOD. Each transaction within a CGS will be resolved using a specific underlying payment method. This element describes a particular payment method:

```
<serviceData name="gesa:pricing" pricingName="FixedPriceCharge" >
. . .
<gesa:paymentMethod paymentMethodName="CreditCard" />
<gesa:paymentMethod paymentMethodName="Invoice/Purchase Order" />
</serviceData>
```

As with currency above, the user must know of a GPS that supports this payment method.

7.2.3.1.1.5   RESOURCES. Any service invocation will consume a plethora of resources. However, a service provider may only be interested in a relatively small subset of these resources for the purposes of deciding a price to charge for a service. The Usage Records Working Group within the GGF have defined an initial subset of base properties, such as:

- Network
- Disc
- Memory
- Wall clock time
- Processor Time
- Node count
- Processors

These could be used as part of the service pricing policy. The resources that the CGS will charge for are specified in the gesa:pricing element. This element also specifies if an estimate of this resource is required by the service to provide any pricing for service use. The default value for the "estimateRequried" attribute is false:

```
<serviceData name="gesa:pricing" pricingName="FixedPriceCharge" >
. . .
<gesa:chargedResources >
<ur:memory />
</gesa:chargedResources>
<gesa:chargedResources estimateRequired="true" >
<ur:cpuTime />
<ur:processors />
```

```
</gesa:chargedResources>
</serviceData>
```

7.2.3.1.1.6 COMPENSATION. A statement of compensation is required for any organization offering a service for monetary reward. The level and complexity of compensation may vary from one organization to another and for different pricing methods offered by the same organization or even the same service. For instance, "GoldStarFixedPrice" might provide some specified compensation. Consider a simple case:

```
<serviceData name="gesa:pricing" pricingName="FixedPriceCharge" >
. . .
<gesa:compensation percentage="0" />
</serviceData>
```

This option allows the client is to refund to the client the agreed cost of invoking the service, even if the server defaults on the delivery of the service before it is invoked. This element will, by default, have the amount contained within the "percentage" attribute refunded to the client on failure:

- percentage="0" means that the client will receive no compensation for any service failure.
- percentage="100" means the client will receive in compensation the agreed-upon cost of using the service even if the client has not yet paid for any part of the service.
- percentage="200" means the client will receive in compensation twice the agreed-upon cost of using the service.

By default, the percentage value is set to zero, meaning the client will receive no compensation for any service failure. Any nonzero positive value of this variable will result in the service provider paying out to the client, with no income if the service fails. (There is also a need to handle staged payments: 10% on reservation, 60% on job startup, 30% on completion of a job. This option is a work In progress.)

7.2.3.1.1.7 REFUND. A refund could be set up as

```
<serviceData name="gesa:pricing" pricingName="FixedPriceCharge" >
. . .
<gesa:compensationRefund percentage="100" />
</serviceData>
```

This option allows the client to obtain a full refund of any money paid to the service provider if it does not deliver the service. This element will have the amount contained within the "percentage" attribute refunded to the client on failure:

- percentage="0" means that the client will receive no refund for any service failure.
- percentage="100" means the client will receive a full refund for any service failure; that is, if money is deducted from the client's account, it will be refunded.
- percentage="200" means the client will receive a full refund in addition to an equal amount of compensation for the money deducted.

By default, the percentage value is set to 100, meaning that if the service fails to deliver, then, from a financial perspective, all cost transactions are rolled back.

7.2.3.1.1.8 PRODUCT. Although it is possible to encapsulate the "product" that is being sold within the grid service itself, this does not always fully capture the behavior of the service being sold. For instance, the grid service may sell access to a mechanism to download a product, for example, an operation that downloads an MP3 track or retrieves an electronic book. One alternative to this approach is to encapsulate each product and pricing mechanism within a separate grid service:

```
<serviceData name="gesa:pricing" pricingName="FixedPriceCharge" >
<gesa:Duration default="3600" maximum="3600" />
<gesa:chargedResources>
<ur:invocation/>
</gesa:chargedResources>
<gesa:product element=http://softwareprovider.com/schema.xml>
<sp:availablePlatforms name="redhat-8.0" />
<sp:product name="SicLib" version="1.2"/>
<sp:duration time="24h" />
</gesa:product />
<gesa:product element=http://softwareprovider.com/schema.xml>
<sp:availablePlatforms name="solaris-2.8" />
<sp:product name="SicLib" version="1.2"/>
<sp:duration time="24h" />
</gesa:product />
<gesa:currency currencyName="SciPounds"
email=cash@softwareprovider.com />
<gesa:paymentMethod paymentMethodName="CreditCard" />
<gesa:paymentMethod paymentMethodName="Invoice/Purchase Order" />
. . .
</serviceData>
```

7.2.3.1.2 Usage. All invocations within a service are recorded in a resource usage service instance. This service is used to collect the resources consumed by a service for the purposes of calculating a charge to the user for service use:

```
<serviceData name="gesa:resourceUsage" GSH="GSH for the RUS" />
```

7.2.3.1.3 Price. Once the price has been generated by invoking the CGS::requestPricing operation, it needs to be displayed through a SDE within the new service instance. This allows the user to hold several service instances that may be used by the user and to examine the economic state of each:

```
<serviceData name="gesa:price" currencyName="HeyPounds">
<gesa:sum>
<gesa:mult>
<gesa:const value=1.0 />
<gesa:totalUse name="ur:WallDuration" units="hour"/>
<gesa:maxUse name="ur:processors" />
</gesa:mult>
<gesa:mult>
<gesa:const value=10.0 />
<gesa:maxUse name="ur:memory" units="GB"/>
</gesa:mult>
</gesa:sum>
</serviceData>
```

The price may be set for a particular resource by different measures. The consumption of resources may be charged for at a rate (e.g., Mbps), total consumed resources (e.g., number of processors), maximum value (e.g., temporary disk space), and so on. The above example indicates that the cost for using the CPU will be charged at 1 HeyPounds for each processor for each hour of the job's duration. There is an additional charge for memory at 10 HeyPounds per GB of memory, based on the maximum memory usage.

7.2.3.1.4 Liability. Liability defines the organization responsible for providing the service. This is an area in need of further exploration. Effectively, this is an informational SDE element that a user agent may (or may not) search for and provides the required information if needed by the user:

```
<serviceData name="gesa:liability"
organisationName="London e-Science Centre"
email="lesc-admin@doc.ic.ac.uk" >
Complaints Department
London e-Science Centre
180 Queen's Gate
London, SW7 2AZ, UK
</serviceData>
```

Note: There may be an additional element with a certificate and digital signature allowing this statement of liability to be authenticated in an automatic manner.

7.2.3.1.5 Testimonial. Many usage scenarios include some mechanism to "rate" the "quality" of a service. This is from a technical context a fairly ill-defined problem that could be resolved in several ways:

- Broker. The broker collects services instances, testing them and adding rating information into the metadata before repackaging the service as one that they can provide.
- Dynamic SDE. An alternative approach is to provide clients with the opportunity to update the SDE with their views as to the service's performance. This could include text, a numerical rating, and a digital signature to provide credibility.
- Testimonial Server. Independent third-party service that maintains a list of services and user-supplied comments (positive/negative).

Testimonial elements have not currently been defined.

7.2.3.1.6 Unresolved Issues. Items still to be resolved:

- Difference between compensation and refund is unclear.
- Need more work on pricing.
- The compensation and refund elements are activated on the "failure" of the service to deliver on something (e.g., SLA). How is this failure detected? Through resource monitoring provided by RUS?
- Resource-specific compensation mechanisms? Can resources be refunded?
- Is an insurance or warranty action needed beyond the compensation mechanism? Is this an extension of the testimonial action?
- How is this linked into SLAs?

7.2.3.2 Service Interface Definition. It is proposed that the Grid Economic Service Interface (GESI) should support a number of operations to facilitate the GESA. The first of these is a factory operation to allow the creation of new instances of this particular CGS. It is envisaged that many of these CGSs will have a multistage process to define the final cost of the service to the user, for example, negotiation, auctioning, and so on. To enable each mode of interaction, the initial act of any CGS on being contacted by a user will be to create a new service instance to deal with the requested interaction method.

7.2.3.2.1 CGS::requestPricing. The requestPricing is a service operation provided by the GESI that will create a new service instance containing information relating to the price charged for using the service. This operation extends the Factory::createService operation. A service containing this operation has the ability to provide a quotation for the use of the service. This quotation is encapsulated within a new service instance created by this operation. The following is a list of inputs:

- *TerminationTime* (optional). The earliest initial termination time that is acceptable to the client. This is effectively the length of time that the client wishes to retain the right to use the service. If not specified, then it defaults to the duration in seconds specified by the gesa:defaultDuration element. After this time, the service instance and, therefore, the right to use the service, will be destroyed through the lifetime management provided by the container.

- *EconomicParameters* (optional). This factory-specific element contains data used by the GESA Factory element to help instantiate a new service instance and set the cost for its use. The EconomicParameters element must contain the following child elements:

  ○ *PricingMechanism.* This element specifies the pricing mechanism that is to be used.

  ○ *Product* (Optional/Required). If multiple product elements are specified within the economic SDE, then one of these must be specified within the pricing request.

  ○ *AllowedUser.* This element specifies the distinguished name of the users (or other user agents, e.g., brokers) that are allowed to access the created service. Multiple elements are allowed.

  ○ *Currency.* Specifies the currency that will be used to record payment. If only one currency element is specified by the CGS, this element becomes optional. If multiple currency elements are specified in the SDE, this becomes required.

  ○ *PaymentMethod.* Specifies the mechanism by which currency will be transferred. This needs to be specified here so that the CGS knows which GPS to use. If only one mechanism is specified by the CGS, this element becomes optional. If multiple mechanisms are specified in the SDE, this becomes required.

  ○ *ConsumedResources.* Elements specifying the estimated consumed resources that the service invocation will use. This may be used by the CGS to adjust the returned costing.

  ○ *ServiceTerminationTime.* The time beyond which the user will no longer require the service if they decide to make use of the service offering. This element may need to include "goodUntil" and "goodFrom" to define an advanced reservation.

The following addresses outputs and faults:

- The pricing for the service use is encapsulated in the new service instance within the price element. The only mechanism for changing the price of a service is through the creation of a new service instance within the requestPricing element. Therefore, at any point in this process a GSH with a price element within the SDE encapsulates the cost of using this service instance, whereas a pricing element represents the presence of a requestPricing operation and a means of creating a new price. This approach allows the "root" service instance to declare how prices are to be set but to make the final price a result of a multistage negotiation through a series of service instances.

- Any use of the underlying CGS interface must result in failure until the price for using the service is explicitly accepted by invoking CGS::acceptPricing.

7.2.3.2.2   CGS::acceptPricing. This operation may only appear if a price element is contained within the SDE. By invoking this operation, which extends the Factory::createService operation, a new service instance is created, embedding the terms and conditions in the new service. The existing service is destroyed as the quotation it refers to is no longer valid because it has been accepted by a user. The new service provides access to the underlying grid service to the specified user community and will record the service invocations in RUS and calculate the resulting cost for recording in the GPS. This instance of the CGS, from a user perspective, is identical to the grid service it encapsulates. The overall lifetime of this new service is set from the parameters used by the proceeding CGS::requestPricing call. Within those constraints, it must support standard OGSI lifetime management interfaces to manage the service lifetime.

There are no inputs. For output one has:

- *Contract.* Returns a signed XML document consisting of the economic SDEs signed by the service provider (i.e., the hosting environment or host certificate). This provides the user with a document (for offline storage) stating the terms and conditions for using the grid service that cannot be denied by the service provider at a later date.

Faults are as defined in the OGSI specification document.

7.2.4   The Grid Payment System

The GPS provides a service to a payment infrastructure that is itself defined outside the GESA document. The purpose of this section is to define the interaction between the GPS and other entities within the GESA (e.g., the CGS, a user). No implementation details are specified within the GESA. However, the GPS could be implemented by any infrastructure with an account-based abstraction. This could include systems based around electronic cash, credit cards, accountancy packages with periodic reconciliation, prepaid accounts, service tokens, and so on. The "currency" used in these transactions need not be recognized or supported by a large community. A currency could relate to service tokens allocated within a specific service center or virtual organization. If the CGS is willing to accept more than one currency to pay for service usage, then this may be specified within its economic SDEs.

7.2.4.1   Service Data Elements. (See Table 7.6)

Table 7.6 Service data elements

SDE	Occurence	Provided by	Comment
Currency	1+	Service administrator	Supported GESA currencies
Bakcer	1+	Service administrator	Backers for currencies
Payment/Method	1+	Service administrator	Supported payment mechanisms
TrustedUser	0+	Service administrator	Declare trusted users
PrivilegedUser	0+	Service administrator	Declare priviliged users

7.2.4.1.1   Currency (1+). The currency is the greatest differentiator between different instances of a GPS. The "gesa::currency" element declared earlier should be used to indicate the currency supported by this GPS. It is possible that multiple currencies can be handled by a single GPS:

    <serviceData name="gesa:currency" currencyName="HeyPounds"
    email="cash@hey.ac.uk" />

7.2.4.1.2   Backer (1+). Each currency SDE should have a backer SDE associated with it, linked by the currencyName attribute:

    <serviceData name="gesa:backer" currencyName="HeyPounds"
    organisationName="London e-Science Centre"
    email=lesc-admin@doc.ic.ac.uk >
    Banking Department
    London e-Science Centre
    180 Queen's Gate
    London, SW7 2AZ, UK
    </serviceData>

Note: There should probably be an additional element with a certificate and digital signature allowing this statement of liability to be authenticated in an automatic manner.

7.2.4.1.3   PaymentMethod (1+). The payment method is the second greatest differentiator between different instances of a GPS. It is possible that a single GPS will be able to handle multiple payment methods. The "gesa::paymentMethod" element declared earlier should be used to indicate the currency supported by this GPS:

    <serviceData name="gesa:paymentMethod"
    paymentMethodName="CreditCard" />
    <serviceData name="gesa:paymentMethod"
    paymentMethodName="Invoice/Purchase Order" />

7.2.4.1.4   TrustedUser (0+). This is an element that places the specified user into the role of a trusted user:

<serviceData name="gesa:TrustedUser">
/C=UK/O=eScience/OU=Imperial/L=LeSC/CN=steven newhouse
</serviceData>

7.2.4.1.5   PrivilegedUser (0+). This is an element that places the specified user into the role of a privileged user:

<serviceData name="gesa:PrivilegedUser">
/C=UK/O=eScience/OU=Imperial/L=LeSC/CN=steven newhouse
</serviceData>

7.2.4.2   Interface Definition. The GPS supports the following operations. As the current authorization model for GridServices has yet to be defined, one uses the following classification for these operations:

- Unprivileged. A normal GSI authenticated client connection is sufficient.
- Trusted. A GSI authenticated client whose DN is registered as an account holder in the GPS or is contained in the TrustedUser SDE (defined earlier).
- Privileged. A GSI authenticated client whose DN is contained in the PrivilegedUser SDE (defined earlier.)

In the above and the following, DN is defined as the Distinguished Name of the X.509 certificate.

7.2.4.2.1   GPS::isDNAccountHolder. This operation determines if the specified DN has an account with this GPS instance. This is used by CGS and other entities to validate that the user exists. The input is:

- *DN.* String parameter containing the DN of the user. The presence of a user's account in a GPS should not be public information.

The output is:

- *Result.* Returns are true (the account exists) or false (the account does not exist).

The faults are:

- *Fault.* Any fault that occurred.

This is operation is available to trusted clients; for example, a client of a bank is allowed to query the existence of other clients.

7.2.4.2.2   GPS::creditCheck. Before agreeing to provide a service to a client, it may be necessary to check that the client has the funds available to support the proposed cost of the service invocation. The inputs are:

- *DN.* String parameter containing the DN of the user.
- *Amount.* The amount of the currency that the client is asking to be available.

The output is:

- *FundsAvailable.* Returns true if the funds are available and false if they are not.

The faults are:

- *NoDNFault.* The user's DN does not exist in the GPS.
- *Fault.* Any other fault.

This operation is only available to trusted clients.

7.2.4.2.3   GPS::getLastTransactions. This allows users to retrieve their recent transactions or privileged users to view another user's recent transactions. The inputs are:

- *DN.* String parameter containing the DN of the user.
- *NumberOfTransactions.* Specifies the number of transactions that should be returned.

The output is:

- *Statement.* An XML document containing the details of each transaction.

The faults are:

- *NoDNFault.* The user's DN does not exist in the GPS.
- *Fault.* Any other fault.

7.2.4.2.4   GPS::getTransactionsByDate. This allows users to retrieve their recent transactions or privileged users to view another user's recent transactions by specifying a date range. The inputs are:

- *DN.* String parameter containing the DN of the user.
- *StartDate.* Specifies the date from which transactions should be viewed.
- *EndDate* (optional). Specifies the date before which transactions should be viewed. If not specified this, is taken to be the current date.

The output is:

- *Statement.* An XML document containing the details of each transaction.

The faults are:

- *NoDNFault.* The user's DN does not exist in the GPS.
- *Fault.* Any other fault.

7.2.4.2.5 *GPS::transferOut.* This operation transfers money from one account to another within the GPS. This is initiated by the user clients effectively "putting" money from their accounts into someone else's account. The inputs are:

- *DN.* String parameter containing the DN of the user account the money is to be transferred to.
- *Amount.* The amount of the currency that the client is asking to be transferred.

The faults are:

- *NoDestinationDNFault.* The DN the money is to be transferred to does not exist in the GPS.
- *NoSourceDNFault.* The DN the money is to be transferred from does not exist in the GPS.
- *NoMoneyFault.* Insufficient funds exist in the user's account to complete the transfer.
- *Fault.* Any other fault.

Only the holders of the accounts can initiate this operation. They are identified by the GSI authenticated connection. This operation must either be originated from the user (inconvenient) or allow a delegated proxy to perform the transaction (insecure?).

7.2.4.2.6 *GPS::transferOutExternal.* This operation transfers money from one account to another account in another GPS. This is initiated by the user clients effectively "putting" money from their accounts into someone else's. The inputs are:

- *externalGPS.* Handle to the external GPS.
- *DN.* String parameter containing the DN of the user account the money is to be transferred to.
- *Amount.* The amount of the currency that the client is asking to be transferred.

The faults are:

- *NoExternalGPSFault.* The external GPS handle was not valid.
- *NoDestinationDNFault.* The DN the money is to be transferred to does not exist in the GPS.

- *NoSourceDNFault.* The DN the money is to be transferred from does not exist in the GPS.
- *NoMoneyFault.* Insufficient funds exist in the user's account to complete the transfer.
- *Fault.* Any other fault.

Only the holders of the accounts can initiate this operation. They are identified by the GSI authenticated connection. This operation must either be originated from the user (inconvenient) or allow a delegated proxy to perform the transaction.

7.2.4.2.7 *GPS::transferIn.*

This operation transfers money from one account to another within the GPS. The recipients are effectively "getting" money from someone's account and placing it in their own. The inputs are:

- *DN.* String parameter containing the DN of the user account the money is to be transferred from.
- *Amount.* The amount of the currency that the client is asking to be transferred.

The faults are:

- *NoDestinationDNFault.* The DN the money is to be transferred to does not exist in the GPS.
- *NoSourceDNFault.* The DN the money is to be transferred from does not exist in the GPS.
- *NoMoneyFault.* Insufficient funds exist in the user's account to complete the transfer.
- *Fault.* Any other fault.

This operation is essential for third-party transactions but will expose the client (and the GPS) to potential abuse unless a secure mechanism for the users approving a transaction out of their accounts is achieved. The recipient of the money initiates this operation.

7.2.4.2.8 *GPS::transferInExternal.*

This operation transfers money from an external account in another GPS to the user's account in "this" GPS. The recipients are effectively "getting" money from someone's account and placing it in their own. The inputs are:

- *externalGPS.* Handle to the external GPS.
- *DN.* String parameter containing the DN of the user account the money is to be transferred from.
- *Amount.* The amount of currency that the client is asking to be transferred.

The faults are:

- *NoExternalGPSFault.* The external GPS handle was not valid.
- *NoDestinationDNFault.* The DN the money is to be transferred to does not exist in the GPS.
- *NoSourceDNFault.* The DN the money is to be transferred from does not exist in the GPS.
- *NoMoneyFault.* Insufficient funds exist in the user's account to complete the transfer.
- *Fault.* Any other fault.

This operation is essential for third-party transactions but will expose the client (and the GPS) to potential abuse unless a secure mechanism for the users approving a transaction out of their accounts is achieved. The recipient of the money initiates this operation.

7.2.4.2.9 GPS::createAccount. This operation creates an account for the specified user with the stated amount. The inputs are:

- *DN.* String parameter containing the DN of the user that the account is to be created for.
- *Amount.* The initial amount of the currency that will be in the account.

The fault is:

- *Fault.* Any fault.

This operation is restricted to privileged users only.

7.2.4.2.10 GPS::deleteAccount. This operation removes the specified account. The input is:

- *DN.* String parameter containing the DN of the user whose account is to be deleted.

The faults are:

- *NoAccountFoundFault.* The specified DN does not have an account in the GPS.
- *Fault.* Any other fault.

This operation is restricted to privileged users. In reality, the account should probably be disabled as opposed to purged from the GPS.

7.2.4.2.11 GPS::createHold. During a long-running execution, it may be desirable to put a "hold" on a certain proportion of a user's account to ensure that there is money left to pay for the consumed resources on completion. This operation creates a GPSHold instance that encapsulates the "reservation" on the specified user's account. This operation extends the Factory::createService operation. The inputs are:

- *TerminationTime.* The earliest initial termination time that is acceptable to the client. This is effectively the length of time that the client wishes to retain a "hold" on the money. Must be specified.
- *HoldParameters.* This factory-specific element contains data used by the GPS Factory element to help instantiate a new service instance and set the cost for its use. The HoldParameters element *must* contain the following child elements:
  - ○ *User.* This element specifies the DN of the account from which the specified amount of currency will be held.
  - ○ *Amount.* Specifies the amount of money that is to be held.

The ouputs and faults are as defined in the OGSI specification document.

7.2.5   GPSHold Service

This service instance encapsulates the duration and amount of money being held on behalf of the client. OGSI standard lifetime management tools can be used to extend the length of this service (perhaps to a defined maximum from the GPS SDE). The hold on the currency ends when this service instance expires or is terminated by the client. The service can only be terminated by the declared owner—the entity requesting the hold.

7.2.5.1 Service Data Elements. The current OGSI specification does not deal with any form of access control within SDEs. Therefore, how much of the state encapsulated by the hold should be made public? If information were to be made public, the SDE would be of the form:

```
<serviceData name="gesa:GPSHold">
<gesa:account>
/C=UK/O=eScience/OU=Imperial/L=LeSC/CN=steven newhouse</gesa:
    account>
<gesa:amount>500</gesa:amount>
<gesa:currency name="HeyPounds" email="cash@hey.ac.uk" />
<gesa:owner>
/C=UK/O=eScience/OU=Imperial/L=LeSC/CN=LeSC Broker</gesa:owner>
</serviceData>
```

In this example, the account is the Grid ID of the account holder, the owner of the hold is the entity requesting (and therefore controlling) the hold, and the amount is the amount of money (and in which currency) from that account that is to be held.

7.2.5.2 Interface Definition. This service's primary function is to encapsulate the "reservation" of money from a specified account. The amount encapsulated by the reservation is immutable. Therefore, the only operations that may be performed on the service instance are effectively related to the lifetime management of the reservation (extended or terminated), and this may be restricted by the GPS during service creation.

7.2.6 The Grid CurrencyExchange Service

The Grid CurrencyExchange Service (GCES) provides a service to a currency exchange infrastructure that is itself defined outside this document. Our purpose within this section is to define the interaction between the GCES and other entities within the GESA (e.g., the CGS, a user). No implementation details are specified within this document; however, the GCES could be implemented by any infrastructure with a currency-based abstraction. This could include systems based around electronic cash, credit cards, service tokens, and so on.

The "currency" used in these transactions need not be recognized or supported by a large community. A currency could relate to service tokens allocated within a specific service center or virtual organization.

7.2.6.1 Service Data Elements. (See Table 7.7).

7.2.6.1.1 exchangeRate (1+). The exchangeRate SDE contains the exchange rates between the currencies supported by the GCES:

```
<serviceData name="gesa:exchangeRate">
<from>
<gesa:currency name="HeyPounds" email="cash@hey.ac.uk"/>
</from>
<to>
<gesa:currency name="HeyDollars" email="cash@hey.edu"/>
```

Table 7.7 Service data elements

SDE	Occurence	Provided by	Comment
exchangeRate	1+	Service administrator	Exchange rates between different currencies
exchangeCommission	0/1	Service administrator	A fixed charge when currency exchanges take place

```
</to>
<rate>1.4</rate>
</serviceData>
<serviceData name="gesa:exchangeRate">
<from>
<gesa:currency name="HeyDollars" email="cash@hey.edu"/>
</from>
<to>
<gesa:currency name="HeyPounds" email="cash@hey.ac.uk"/>
</to>
<rate>0.7</rate>
</serviceData>
<serviceData name="gesa:exchangeRate">
<from>
<gesa:currency name="HeyPounds" email="cash@hey.ac.uk"/>
</from>
<to>
<gesa:currency name="HeyDollars" email="cash@hey.edu"/>
</to>
<rate>1.4</rate>
</serviceData>
```

7.2.6.1.2 exchangeCommission (0/1). Whenever a currency exchange takes place, the GCES may charge a commission. If the exchangeCommission SDE does not exist, then the GCES will not charge any commission:

```
<serviceData name="gesa:exchangeCommission">
<currency currencyName="HeyPounds" email="cash@hey.ac.uk"/>
<amount>3</amount>
</serviceData>
```

7.2.6.2 Interface

7.2.6.2.1 GCES:exchangeCurrency. Through this operation, a consumer may convert one currency to another. The exchange rate to be used for the conversion is advertised through the exchangeRate SDE. The GCES may charge commission for the exchange rate. The amount to be charged is advertised through the ex-changeCommission SDE. The inputs are:

- *Amount.* The amount to be converted.
- *Currency.* The currency to which the amount is to be converted.

The outputs are:

- *Amount.* The amount returned after the conversion.
- *CommissionCharged.* The currency and amount charged for commission.

The faults are:

- *ConversionNotSupported.* Either the given currency or the requested currency is not supported.
- *Fault.* Any other fault.

7.2.7 An Example

To illustrate the previous service definitions, we consider the simple economic use of a service to purchase the use of a service by a client, to record the use of the service within RUS, and to have the appropriate amount of money deducted from an account within a grid payment system. The relationship between the client and the service instances is defined below (see Figure 7.4). The protocol communication passing between these entities is identified in the figure and expanded upon in the following sections.

7.2.7.1 Economically Enabled Counter Service. The economically enabled counter service has the same basic interface as the counter service, with the addition of an operation to support the pricing of the service. This is described through the economic SDEs:

```
<serviceData name="gesa:pricing" pricingName="FixedPriceCharge" >
<gesa:Duration default="30" maximum="60" />
<gesa:PricingType>
<FixedPrice />
</gesa:PricingType>
<gesa:chargedResources>
```

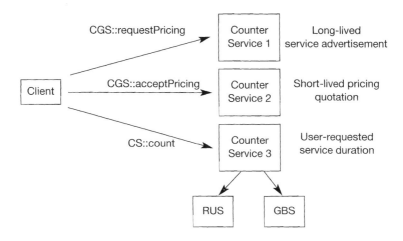

Figure 7.4   Example.

```
<ur:invocation/>
</gesa:chargedResources>
<gesa:currency
currencyName="HeyPounds" email="cash@hey.ac.uk" />
. . .
</serviceData>
```

These SDEs (expiry information has been ignored) describe the economic state of the counter service. The service is being offered for a fixed price per invocation, where the price in "HeyPounds" will by default be held open for 30 seconds before expiry.

7.2.7.2 *Requesting a Price.* From this information, the clients are able to compose a response describing the service invocation they require, and this is passed to the requestPricing operation:

```
<gesa:requestPricing>
<gesa:pricingMechanism name="FixedPriceCharge">
<gesa:allowedUser>/C=UK/O=eScience/OU=Imperial/L=LeSC/CN=steven
Newhouse</gesa:allowedUser>
<gesa:allowedUser>/C=UK/O=eScience/OU=Imperial/L=LeSC/CN=anthony
mayer </gesa:allowedUser>
<gesa:currency name="HeyPounds" email="cash@hey.ac.uk" />
<gesa:consumedResources>
<gesa:resourceUsage name="ur:invocation">5</gesa:resourceUsage>
<gesa:consumedResources>
<gesa:terminationTime duration="180" />
</gesa:requestPricing>
```

This is a request to set up a service instance that is accessible by two users who estimate that they will use the service five times and are willing to pay in "HeyPounds." If this request is successful, then a new instance of the service is created and a GSH is returned to the user. The service is expected to be used for 180 seconds.

7.2.7.3 *New Service Instance.* The new service instance created by the previous activity results in a service instance with the following SDEs:

```
<serviceData name="gesa:price" currencyName="HeyPounds">
<gesa:mult>
<gesa:const value=2.0 />
<gesa:totalUse name="ur:invocation" />
</gesa:mult>
</serviceData>
```

The standard SDE structure will provide information relating to the service's expiry. The SDEs provides all the information relating to the service use specifying that 2 "HeyPounds" will be charged for each service invocation, which is only accessible to the two specified users.

In creating the service instance, there is an opportunity to perform several checks. A GPS can be found for the specified currency and the proposed users checked to ensure that they possess accounts and have sufficient funds to support the proposed charge for the estimated number of invocations.

7.2.7.4   Accept Pricing.
The user is able to browse the SDE and view the offered service contract. This offer *will* be limited by the overall lifetime of the service as defined in the SDE and as requested by the user. If the users wish to commit to the pricing, they should invoke the CGS::acceptPricing operation. This will create a new service instance for the users to interact with.

7.2.7.5   Grid Service Instance.
This grid service instance is dedicated to the use of the stated user community at the previously quoted price. The user interacts with this service instance in the same manner as a non-GESA-enabled service using the provided GSH.

7.2.7.6   Service Use.
This uses the GSH to invoke operations in the grid service encapsulated by the CGS, in this case the GS:count operation. On each invocation, the user access is checked against the allowed users.

7.2.7.7   Resource Use.
The consumed resources are recorded in an instance of the RUS. The consumed resources for this service would have the form:

```
<ur:UsageRecords>
<ur:UsageRecord>
<UserIdentity>
<ds:KeyInfo xmlns:ds="http://www.w3.org/2000/09/xmldsig#">
<X509Data>
<X509SubjectName>
/C=UK/O=eScience/OU=Imperial/L=LeSC/CN=anthony mayer
</X509SubjectName>
</X509Data>
</ds:KeyInfo>
</UserIdentity>
<MachineName>eric.mvc.mcc.ac.uk</MachineName>
<SubmitHost>eric.mvc.mcc.ac.uk</SubmitHost>
<ur:GlobalJobID>GSH of the invoking service</ur:GlobalJobID>
<ur:Resource name="UseCount">1</ur:Resource>
<ur:UsageRecord>
<ur:UsageRecords>
```

This would be passed to the RUS::insertUsageRecords operation within the RUS service instance defined within the gesa:resourceUsage element of the CGS SDE.

7.2.7.8 Service Charging. Following service invocation (or on service destruction or periodically), the service use has to be charged for. All records relating to this service instance are retrieved from the RUS and the total cost for using the service calculated from the total resource usage and the stated charging policy. This cost is passed onto the CGS and any holds on the user's accounts released.

Usage information would be extracted from the RUS through the RUS::extractUsageByGlobalJobID operation specifying the GSH of the service. The usage record(s) returned by this search need to be aggregated and combined with the charging information to define a cost for the service use. This cost is used within the GPS::transferIn operation (invoked by the service provider) and specifying the DN of the entity that created this instance of the CGS as the entity that is to be charged for the use of the service.

7.2.8 Security Considerations

GESA assumes the availability of the security provisions from the OGSI. There is a need to be able to specify access to services on a per-user basis.

Communication Systems for Local Grids

The remainder of the book develops more explicitly our focus on networking. We have mentioned a number of times the importance and role of the network throughout the previous chapters. In the last three chapters, we describe key networking technologies to support and enable the deployment of grid computing services at the local (Chapter 8), intragrid (Chapter 9), and intergrid (Chapter 10) level. These technologies apply at the SAN, LAN, MAN, WAN, and GAN levels. Motivations for looking at grids from a networking perspective include the fact that grid systems have network bandwidth dependencies and network latency dependencies, and also the fact that grid systems suffer from synchronization protocol inefficiencies. Network security is also an issue (although, host security is even more critical). Hence, an appropriate networking infrastructure is required to make grid computing a reality.

Our treatment is by necessity terse, but rather up-to-date. The interested reader can consult a number of textbooks on telecommunications, including, but not limited to [77]. In 1987, in internal Bellcore/Telcordia Special Reports, in a section called "Network for a Computing Utility," we stated that networks were critical for utility computing [6]:

> The proposed service provides the entire apparatus to make the concept of the Computing Utility possible. . . . This service is basically feasible once a transport and switching network with strong security and accounting [chargeback] capabilities is deployed. A high degree of intelligence in the network is required . . . a physical network is required . . . security and accounting software is needed . . . protocols and standards will be needed to connect servers and users, as well as for accounting and billing. These protocols will have to be developed before the service can be established. . . .

8.1 INTRODUCTION AND POSITIONING

As we have seen, grid computing can be considered as a network of computation. It supports the concept of "utility computing," with which users can get "on-demand" "machine cycles off a grid" without having to own the physical assets. Also it sup-

A Networking Approach to Grid Computing. By Daniel Minoli
ISBN 0-471-68756-1 © 2005 John Wiley & Sons, Inc.

ports the concept of the enterprise grid (intragrid), with which organizations make more synergistic use of often underutilized assets they already own. Grid computing is supported by tools and protocols for coordinated resource sharing and problem solving among pooled assets [91]. With grid computing, specialized equipment, data stores, and computers that are remotely deployed on the network can be managed as virtualized assets, thereby reducing the necessity of the organization to purchase multiple devices, in much the same way today that people in the same office share Internet access or a printing resources across a LAN [130]. Ultimately, organizations will be able to obtain computing services over a network from a remote computing-service provider. Broadband networks play a key enabling role in making grid computing possible. Transmission of content and control within the grid are important for sending jobs and their required data to remote points within the grid. Some jobs require a large amount of data to be processed that may not always reside on the processor running the job.

As we have also seen, OGSA is a standardized model for grid computing that employs distributed resources over the interconnecting network [119]. OGSA facilitates the overall management of distributed computing resources available over a local or wide-area network that appear to an end user or application as one large virtual computing system. The networking mechanism is the most fundamental resource for the grid because, clearly, without networking grid computing would not be possible. The growth in communication capacity in the recent past makes grid computing practical, compared to the limited bandwidth available when distributed computing was first emerging in the 1970s, 1980s, and 1990s.

Figure 3.7 in Chapter 3 provided an example of a protocol stack and network-enabled services in a grid environment. Figure 8.1 amplifies this protocol stack by identifying key networking technologies of interest at the "lower layers" of the protocol model, namely, the physical, data link, and network layers. The bandwidth available for the subtending communication links can often be a critical resource that can limit utilization of the grid. LANs and SANs support local clusters; high-capacity, high-quality intranet support intragrids and long-haul global connectivity (including Internet-provided capacity) make intergrids possible. LAN connectivity now is in the 1–10 Gbps range, and practically affordable WAN/intranet connectivity for companies is in the 45–155 Mbps range. As previously noted, speeds in the 2.4–10 Gbps range are commonplace within the inner workings of carriers, but these kinds of speeds are not generally affordable to Fortune 500 companies. A single fiber can now carry in the range of 0.5 Tbps using high-density DWDM, but these speeds begin to be out of reach for all but the largest carriers, at least as of 2004. Internet-based virtual private networks (VPNs) provide bandwidths in the 1–10 Mbps range, but QoS continues to be an issue, particularly given (i) the slowdown in deployment of telecom assets in the early 2000s, and (ii) the growth in the early 2000s (as much as 50% a quarter) of bandwidth-consuming e-mail spam.

Local grids rely on LANs and SANs. The availability of powerful workstations, processors, servers, and "blade technology," along with high-speed networks [such as Gigabit Ethernet (GbE)], as commodity components has led to the emergence of

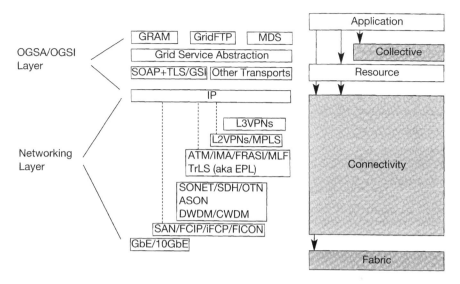

Figure 8.1 Key high-speed networking technologies usable in grid environments.

local clusters for high-performance computing. The availability of such clusters within many organizations has fostered a growing interest in aggregating distributed resources to solve large-scale problems of multiinstitutional interest. As we saw in Chapter 2, key networking, storage, and platform vendors are already working together to enable enhanced grid services for SAN-based environments. The physical-layer network technologies that are used (or have been used) for this function include Fibre Channel (FC) and Fast and Gigabit Ethernet. Network protocols that are used at higher layers of the protocol stack include Fibre Channel Protocol for SCSI (SCSI FCP), TCP/IP, VI, CIFS, and NFS [140].

Intragrids rely on WANs. Supported by innovations in optics, the theoretical performance of WANs has increased significantly in recent years. The "affordable" bandwidth has also grown in the past 5–10 years. Furthermore, the integration of intelligent services into the network helps simplify data access across the grid and resource sharing and management [35] (this topic is treated in Chapter 9).

Intergrids often rely on the Internet. As Web services proliferate, concerns include the overall demands on network bandwidth and, for any particular service, the effect on performance as demands for that service rise. Security becomes even more critical in the Internet environment. A number of new products have emerged that enable software developers to create or modify existing applications that can be "published" as Web services [15] (this topic is covered in Chapter 10).

In the sections (and chapters) that follow, we cover some (but not all) of the key technologies shown in Figure 8.1. We focus here on the lower layers. Services such as Web services (HTTP, SOAP, WSDL, UDDI, etc.), are not discussed further (refer to Chapters 3 and 4 for basic information on these protocols and capabilities).

8.2 SAN-RELATED TECHNOLOGY

SANs can be viewed as a subset of high-speed networks that support the networking (extension) of the *channel* of a processor. A channel is generally a higher-speed port than a communication port, although this distinction is becoming less of an issue of late. A channel is generally a parallel communication link that needs to be serialized before it can be transported at any distance greater that a few feet. Within the native channel environment, data is known as "block-level data." One can think of the channel as an extension of the processor's internal bus (although there is a subtle difference for the purists). We emphasize, once again, that channel-level communication is applicable to a number of instances (e.g., real-time mirroring), but the *marketing* focus has been on storage. Hence, for all intents and purposes, SANs and channel-based communication have lately been thought of as one and the same. (This author wrote what is believed to be the first textbook treatment of WAN-based channel extension technologies, with extensive materials on this topic in the 1991 book, *Telecommunications Technology Handbook,* now in its second edition [77].)

Channel-based communication has always been problematic, not only because it is intrinsically a parallel-originated stream and is high-speed oriented, but, more importantly, because the developers failed over the years to make use of available telecom standards such as SONET in the wide area, and Ethernet in the local area.

Channel-oriented streams can be carried generally in four modes:

- Native over a channel-specific communication fiber (typically dark fiber), as is the case of FICON and even FC (Fibre Channel)
- Mapped over a physical-layer service such as SONET, SDH, or the evolving OTN/ASON
- Mapped over a data-link layer service such as ATM
- Tunneled over a network layer (that, is IP) service (most recently)

A service or protocol is "mapped" when it is carried by a service/protocol that operates at the same protocol layer; it is "tunneled" when it is carried within the payload of a protocol that operates at a layer(s) higher than the protocol in question.

Although native or mapped service is typically best from a bandwidth and latency perspective, it may be expensive to support and may be inflexible (by requiring fiber or high-speed availability at various locations). Tunneled service makes the assumption that IP networking is ubiquitously available as an intranet- or Internet-delivered service, and, hence, channel communication based on it is more flexible. Tunneled service, however, is faced with bandwidth and latency issues. Mid-range applications (e.g., tape backup) can use tunneled services; higher-range applications (e.g., real-time disk mirroring and scientific-data-intensive applications), may need native or mapped services.

When storage is all within a data center, SANs can run the native FC medium; when storage is remote, a native, mapped, or tunneled service is required. Another approach is not to use channel-based communications to begin with, but to use the communication port, specifically GbE or 10 GbE, instead. As mentioned, in the past, the channel "port" of a processor, server, or storage system was faster that the "communication" port; these days, both ports can operate at the same speed, hence, the designer, in fact, has a choice.

8.2.1 Fibre Channel Technology—Native Mode

8.2.1.1 FC. This section looks at some basic concepts of Fibre Channel. "Fibre" is a generic term used to cover all physical media supported by the FC protocols, including optical fiber, twisted pair, and coaxial cable. Logically, FC is a bidirectional point-to-point serial data channel, designed for high-performance information transport. Physically, FC is an interconnection of one or more point-to-point links. Each link end terminates in a port that is specified in the Physical Interface (FC-PI) specification and in the Framing and Signaling (FC-FS) specification. FC is defined by a set of ANSI standards, including the following:

- Fibre Channel Framing and Signaling (FC-FS); ANSI INCITS*/Project 1331-D; Draft Standard Rev. 1.9, Apr. 9, 2003
- Fibre Channel Physical and Signaling Interface (FC-PH); ANSI INCITS 230-1994 (R1999), formerly ANSI X3.230-1994 (R1999), Nov. 14 1994
- ANSI X3.297:1997, Information Technology—Fibre Channel—Physical and Signalling Interface-2 (FC-PH-2)
- ANSI X3.303:1998, Fibre Channel—Physical and Signalling Interface-3 (FC-PH-3).
- ANSI X3.272:1996, Information Technology—Fibre Channel—Arbitrated Loop (FC-AL).
- ANSI NCITS 332-1999, Fibre Channel—Arbitrated Loop (FC-AL-2).
- Fibre Channel Switch Fabric—2 (FC-SW-2); ANSI INCITS 355-2001, 2001
- Fibre Channel Switch Fabric—3 (FC-SW-3); ANSI INCITS/Project 1508-D; Draft Standard Rev. 6.3, Feb. 19 2003
- Fibre Channel Methodologies for Interconnects (FC-MI-2); ANSI INCITS/Project 1599-DT; Draft Standard Rev 2.03, June 4 2003

FC is structured as a set of hierarchical and related functions, FC-0 through FC-3 (see Figure 8.2). Each of these functions is described as a level. FC does not restrict implementations to specific interfaces between these levels. The Physical Interface (FC-0) consists of transmission media, transmitters, receivers, and their interfaces. The Physical Interface specifies a variety of media, and associated drivers and receivers capable of operating at various speeds. The Transmission Protocol (FC-1), Signaling Protocol (FC-2), and Common Services Protocol (FC-3) are fully specified in FC-FS and FC-AL-2. Fibre Channel levels FC-1 through FC-3 specify the rules and provide mechanisms needed to transfer blocks of information end to end, traversing one or more links. An Upper Level Protocol mapping to FC-FS consti-

*INCITS: International Committee for Information Technology Standards (formerly NCITS).

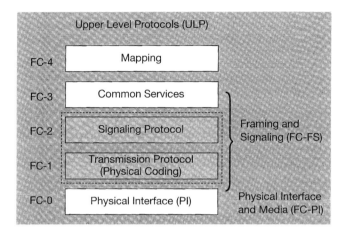

Figure 8.2 FC layers.

tutes an FC-4 that is the highest level in the FC structure. FC-2 defines a suite of functions and facilities available for use by an FC-4. A fibre channel node may support one or more N_Ports (the endpoints for fibre channel traffic) and one or more FC-4s. Each N_Port contains FC-0, FC-1, and FC-2 functions. FC-3 optionally provides the common services to multiple N_Ports and FC-4s.

An encapsulated description of the FC functionality follows [78].[2] Table 8.1 defines some key terms needed to describe the concepts.

Fibre channel is a frame-based, serial technology designed for peer-to-peer communication between devices at gigabit speeds and with low overhead and latency. Figure 8.3 depicts the basic fibre channel environment.

The Fibre Channel Network. The fundamental entity in FC is the fibre channel network. Unlike a layered network architecture, a fibre channel network is largely specified by functional elements and the interfaces between them. As shown in Figure 8.4, these consist, in part, of the following:

1. N_PORTs. The endpoints for fibre channel traffic. In the FC standards, N_PORT interfaces have several variants, depending on the topology of the fabric to which they are attached. As used in this specification, the term applies to any one of the variants.

2. FC Devices. The fibre channel devices to which the N_PORTs provide access.

3. Fabric Ports. The interfaces within a fibre channel network that provide attachment for an N_PORT. The types of fabric port depend on the fabric topology and are discussed later.

4. The network infrastructure for carrying frame traffic between N_PORTs.

---

[2]The rest of this section is based on [78].

Table 8.1 Key FC and iFCP terms

Address-translation mode	A mode of gateway operation in which the scope of N_PORT fabric addresses for locally attached devices are local to the iFCP gateway region in which the devices reside.
Address-transparent mode	A mode of gateway operation in which the scope of N_PORT fabric addresses for all fibre channel devices are unique to the bounded iFCP fabric to which the gateway belongs.
B_Port	A bridging function typically interfaces between FC switch fabrics that provide an extension port for traffic that must be routed to a remote destination, not part of the local switching domain. The bridging function identifies its connection endpoints as B_Ports, indicating its transparent behavior to the Fabric. A bridging function delivers all frames except link service frames received on one port transparently to its opposite and does not have responsibilities in routing frames [177].
Bounded iFCP fabric	The union of two or more gateway regions configured to interoperate together in address-transparent mode.
DOMAIN_ID	The value contained in the high-order byte of a 24-bit N_PORT fibre channel address.
E_Port	An expansion port on a switch used to link multiple switches together into a FC Fabric. A network built with two or more FC switches is used to establish a consistent addressing scheme, exchange of routing information and name service information for discovery and monitoring of devices.
F_Port	The interface used by an N_PORT to access fibre channel switched fabric functionality.
Fabric	The entity that interconnects N_PORTs attached to it and is capable of routing frames by using only the address information in the fibre channel frame.
Fabric port	The interface through which an N_PORT accesses a fibre channel fabric. The type of fabric port depends on the fibre channel fabric topology. In this specification, all fabric port interfaces are considered to be functionally equivalent.
FC-2	The fibre channel transport services layer described in [21].
FC-4	The fibre channel mapping of an upper layer protocol, such as [23], the fibre channel to SCSI mapping.
Fibre channel device	An entity implementing the functionality accessed through an FC-4 application protocol.
Fibre channel network	A native fibre channel fabric and all attached fibre channel nodes.
Fibre channel node	A collection of one or more N_PORTs controlled by a level above the FC-2 layer. A node is attached to a fibre channel fabric by means of the N_PORT interface described in [21].

(continued)

Table 8.1 *Continued*

G_Port	Generic switch port that can be either an F_port or an E_port. Port function is automatically determined during login [177]
Gateway region	The portion of an iFCP fabric accessed through an iFCP gateway by a remotely attached N_PORT. Fibre channel devices in the region consist of all those locally attached to the gateway.
iFCP	The Internet Fibre Channel Protocol (iFCP). A gateway-to-gateway protocol that supports FC Layer 4 FCP over TCP/IP.
iFCP frame	A fibre channel frame encapsulated in accordance with the FC Frame Encapsulation Specification and iFCP.
iFCP portal	An entity representing the point at which a logical or physical iFCP device is attached to the IP network. The network address of the iFCP portal consists of the IP address and TCP port number to which a request is sent when creating the TCP connection for an iFCP session.
iFCP session	An association comprised of a pair of N_PORTs and a TCP connection that carries traffic between them. An iFCP session may be created as the result of a PLOGI fibre channel login operation.
iSNS	The server functionality and IP protocol that provide storage name services in an iFCP network.
L_Port	An arbitrated loop port.
Locally attached device	With respect to a gateway, a fibre channel device accessed through the fibre channel fabric to which the gateway is attached.
Logical iFCP device	The abstraction representing a single fibre channel device as it appears in an iFCP network.
N_PORT	An iFCP or fibre channel entity representing the interface to fibre channel device functionality. This interface implements the fibre channel N_PORT semantics specified in [21]. Fibre channel defines several variants of this interface that depend on the fibre channel fabric topology. As used in this document, the term applies equally to all variants.
N_PORT alias	The N_PORT address assigned by a gateway to represent a remote N_PORT accessed via the iFCP protocol.
N_PORT fabric address	The address of an N_PORT within the fibre channel fabric.
N_PORT ID	The address of a locally attached N_PORT within a gateway region. N_PORT IDs are assigned in accordance with the fibre channel rules for address assignment specified in [21].
N_PORT network address	The address of an N_PORT in the iFCP fabric. This address consists of the IP address and TCP port number of the iFCP Portal and the N_PORT ID of the locally attached fibre channel device.
Port login (PLOGI)	The fibre channel Extended Link Service (ELS) that establishes an iFCP session through the exchange of identification and operation

Table 8.1   *Continued*

	parameters between an originating N_PORT and a responding N_PORT.
Remotely attached device	With respect to a gateway, a fibre channel device accessed from the gateway by means of the iFCP protocol.
Unbounded iFCP fabric	The union of two or more gateway regions configured to interoperate together in address-translation mode.

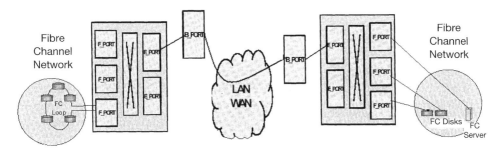

Figure 8.3   Fibre channel environment.

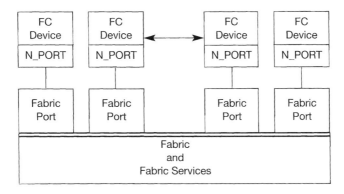

Figure 8.4   A fibre channel network.

5. Within a switched or mixed fabric (see below), a set of auxiliary servers, including a name server for device discovery and network address resolution. The types of service depend on the network topology.

The following subsections describe fibre channel network topologies and give an overview of the fibre channel communications model.

Fibre Channel Network Topologies. The principal fibre channel network topologies consist of the following:

1. Arbitrated Loop. A series of N_PORTs connected together in daisy-chain fashion. In [21], loop-connected N_PORTs are referred to as NL_PORTs. Data transmission between NL_PORTs requires arbitration for control of the loop in a manner similar to a token ring network.
2. Switched Fabric. A network consisting of switching elements, as described below.
3. Mixed Fabric. A network consisting of switches and "fabric-attached" loops. A description can be found in [20]. A loop-attached N_PORT (NL_PORT), is connected to the loop through an L_PORT and accesses the fabric by way of an FL_PORT.

Depending on the topology, the N_PORT and its means of network attachment may be one of the types listed in Table 8.2. The differences in each N_PORT variant and its corresponding fabric port are confined to the interactions between them. To an external N_PORT, all fabric ports are transparent and all remote N_PORTs are functionally identical.

SWITCHED FIBRE CHANNEL FABRICS. An example of a multiswitch fibre channel fabric is shown in Figure 8.5. The interface between switch elements is either a proprietary or the standards-compliant E_PORT interface described by the FC-SW2 specification [24].

MIXED FIBRE CHANNEL FABRIC. A mixed fabric contains one or more arbitrated loops connected to a switched fabric, as shown in Figure 8.6. As noted previously, the protocol for communications between peer N_PORTs is independent of the fabric topology, N_PORT variant, and type of fabric port to which an N_PORT is attached.

FIBRE CHANNEL LAYERS AND LINK SERVICES. As noted, FC consists of the following layers:

- FC-0. The interface to the physical media.

Table 8.2 N_PORT types

FC Network Topology	Network Interface	N_PORT Variant
Loop	L_PORT	NL_PORT
Switched	F_PORT	N_PORT
Mixed	FL_PORT via L_PORT	NL_PORT
	F_PORT	N_PORT

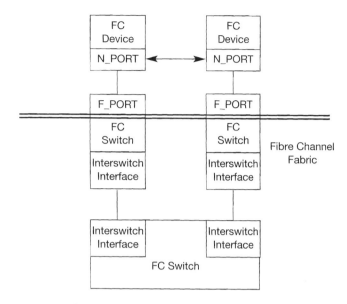

Figure 8.5   Multiswitch fibre channel fabric.

- FC-1. The encoding and decoding of data and out-of-band physical link control information for transmission over the physical media.
- FC-2. The transfer of frames, sequences, and exchanges comprising protocol information units.
- FC-3. Common services.

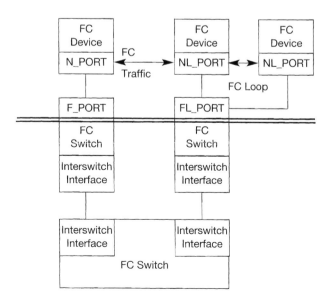

Figure 8.6   Mixed fibre channel fabric.

- FC-4. Application protocols such as the Fibre Channel Protocol for SCSI (FCP). FCP is the ANSI SCSI serialization standard to transmit SCSI commands, data, and status information between a SCSI initiator and SCSI target on a serial link, such as a FC network (FC-2).

In addition to the layers defined above, FC defines a set of auxiliary operations called link services, some of which are implemented within the transport layer fabric. These are required to manage the fibre channel environment, establish communications with other devices, retrieve error information, perform error recovery, and other similar services. Some link services are executed by the N_PORT. Others are implemented internally within the fabric. These internal services are described in the next subsection.

Fabric-Supplied Link Services. Servers internal to a switched fabric handle certain classes of Link Service requests and service-specific commands. The servers appear as N_PORTs located at the "well-known" N_PORT fabric addresses specified in [21]. Service requests use the standard fibre channel mechanisms for N_PORT-to-N_PORT communications.

All switched fabrics must provide the following services:

- Fabric F_PORT server. Services N_PORT requests to access the fabric for communications.
- Fabric Controller. Provides state change information to inform other FC devices when an N_PORT exits or enters the fabric (see below).
- Directory/Name Server. Allows N_PORTs to register information in a data base, retrieve information about other N_PORTs, and discover other devices as described below.

A switched fabric may also implement the following optional services:

- Broadcast Address/Server. Transmits single-frame, class 3 sequences to all N_PORTs.
- Time Server. Intended for the management of fabric-wide expiration timers or elapsed time values and not intended for precise time synchronization.
- Management Server. Collects and reports management information, such as link usage, error statistics, link quality, and similar items.
- Quality of Service Facilitator. Performs fabric-wide bandwidth and latency management.

FIBRE CHANNEL NODES. A fibre channel node has one or more fabric-attached N_PORTs. The node and its N_PORTs have the following associated identifiers:

1. A worldwide unique identifier for the node.
2. A worldwide unique identifier for each N_PORT associated with the node.
3. For each N_PORT attached to a fabric, a 24-bit fabric-unique address having

the properties defined below. The fabric address is the address to which frames are sent.

Each worldwide unique identifier is a 64-bit binary quantity having the format defined in [21].

FIBRE CHANNEL DEVICE DISCOVERY. In a switched or mixed fabric, fibre channel devices and changes in the device configuration may be discovered by means of services provided by the fibre channel name server and fabric controller.

The name server provides registration and query services that allow a fibre channel device to register its presence on the fabric and discover the existence of other devices. For example, one type of query obtains the fabric address of an N_PORT from its 64-bit worldwide unique name. The full set of supported fibre channel name server queries is specified in [22].

The fabric controller complements the static discovery capabilities provided by the name server through a service that dynamically alerts a fibre channel device whenever an N_PORT is added or removed from the configuration. A FC device receives these notifications by subscribing to the service as specified in [21].

FIBRE CHANNEL INFORMATION ELEMENTS. The fundamental element of information in fibre channel is the frame. A frame consists of a fixed header and up to 2112 bytes of payload having the structure described below. The maximum frame size that may be transmitted between a pair of fibre channel devices is negotiable up to the payload limit, based on the size of the frame buffers in each fibre channel device and the path maximum transmission unit (MTU) supported by the fabric.

Operations involving the transfer of information between N_PORT pairs are performed through "exchanges." In an exchange, information is transferred in one or more ordered series of frames referred to as sequences.

Within this framework, an upper layer protocol is defined in terms of transactions carried by exchanges. Each transaction, in turn, consists of protocol information units, each of which is carried by an individual sequence within an exchange.

FIBRE CHANNEL FRAME FORMAT. A fibre channel frame (see Figure 8.7) consists of a header, payload, and 32-bit CRC bracketed by SOF and EOF delimiters. The header contains the control information necessary to route frames between N_PORTs and manage exchanges and sequences. Figure 8.7 gives a schematic view of the frame.

The source and destination N_PORT fabric addresses embedded in the S_ID and D_ID fields represent the physical addresses of originating and receiving N_PORTs, respectively.

N_PORT Address Model. N_PORT fabric addresses (see Figure 8.8) are 24-bit values having the following format defined by the fibre channel specification [21].

A FC device acquires an address when it logs into the fabric. Such addresses are volatile and subject to change based on modifications in the fabric configuration. In a fibre channel fabric, each switch element has a unique Domain ID assigned by the

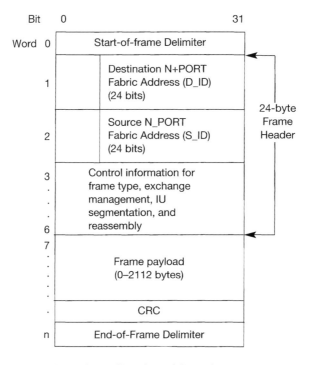

Figure 8.7 Fibre channel frame format.

principal switch. The value of the Domain ID ranges from 1 to 239 (0xEF). Each switch element, in turn, administers a block of addresses divided into area and port IDs. An N_PORT connected to a F_PORT receives a unique fabric address consisting of the switch's Domain ID concatenated with switch-assigned area and port IDs.

A loop-attached NL_PORT (see Figure 8.6) obtains the Port ID component of its address during the loop initialization process described in [19]. The area and domain IDs are supplied by the fabric when the fabric login (FLOGI) is executed.

FIBRE CHANNEL TRANSPORT SERVICES. N_PORTs communicate by means of the following classes of service specified in the FC standard [21]:

- Class 1. A dedicated physical circuit connecting two N_PORTs.
- Class 2. A frame-multiplexed connection with end-to-end flow control and delivery confirmation.

Figure 8.8 Fibre channel address format.

- Class 3. A frame-multiplexed connection with no provisions for end-to-end flow control or delivery confirmation.
- Class 4. A connection-oriented service, based on a virtual circuit model, providing confirmed delivery with bandwidth and latency guarantees.
- Class 6. A reliable multicast service derived from Class 1.

Class 2 and Class 3 are the predominant services supported by deployed FC storage and clustering systems.

Class 3 service is similar to User Datagram Protocol (UDP) or IP datagram service. FC storage devices using this class of service rely on the ULP implementation to detect and recover from transient device and transport errors.

For Class 2 and Class 3 service, the FC fabric is not required to provide in-order delivery of frames unless explicitly requested by the frame originator (and supported by the fabric). If ordered delivery is not in effect, it is the responsibility of the frame recipient to reconstruct the order in which frames were sent based on information in the frame header.

LOGIN PROCESSES. The login processes are FC-2 operations that allow an N_PORT to establish the operating environment necessary to communicate with the fabric, other N_PORTs, and ULP implementations accessed via the N_PORT. Three login operations are supported:

1. Fabric Login (FLOGI). An operation whereby the N_PORT registers its presence on the fabric, obtains fabric parameters such as classes of service supported, and receives its N_PORT address.
2. Port Login (PLOGI). An operation by which an N_PORT establishes communication with another N_PORT.
3. Process Login (PRLOGI). An operation that establishes the process-to-process communications associated with a specific FC-4 ULP, such as FCP-2, the fibre channel SCSI mapping.

Since N_PORT addresses are volatile, an N_PORT originating a login (PLOGI) operation executes a name server query to discover the fibre channel address of the remote device. A common query type involves use of the worldwide unique name of an N_PORT to obtain the 24-bit N_PORT fibre channel address to which the PLOGI request is sent.

8.2.1.2 10GFC. The 10 Gigabit Fibre Channel Standard[3] describes in detail extensions to fibre channel signaling and physical layer services introduced in FC-PH, to support data transport at a rate in excess of 10 gigabits per second. This standard was developed by Task Group T11 of Accredited Standards Committee INCITS during 2000–2001. 10GFC describes the signaling and physical interface services that may be utilized by an extended version of the FC-2 level to transport data at a

---

[3]10GbE Alliance Tutorial Materials.

rate in excess of 10 gigabits per second over a family of FC-0 physical variants. 10GFC additionally introduces port management functions at the FC-3 level.

FC-3 General Description. The FC-3 level of 10GFC extends the FC-3 levels of FC-FS and FC-AL-2 by adding a port management interface and register set and low-level signaling protocol. The port management interface and register set provide an interconnection between manageable devices within a port and port management entities.

The link signaling sublayer (LSS) is used to signal low-level link and cable plant management information during the Idle stream. The WAN interface sublayer (WIS) is an optional sublayer that may be used to create a physical layer that is data-rate and format compatible with the SONET STS-192c transmission format defined by ANSI, as well as the SDH VC-4-64c container specified by ITU. The purpose of the WIS is to support 10GFC data streams that may be mapped directly to STS-192c or VC-4-64c streams at the PHY level, without requiring higher-layer processing. The WIS specifies a subset of the logical frame formats in the SONET and SDH standards. In addition, the WIS constrains the effective data throughput at its service interface to the payload capacity of STS-192c/VC-4-64c, that is, 9.58464 Gbps. Multiplexed SONET/SDH formats are not supported.

FC-2 General Description. The FC-2 level of 10GFC extends the FC-2 levels of FC-FS and FC-AL-2 to transport data at a rate of 10.2 Gbps over a family of FC-0 physical variants. 10GFC provides the specification of optional physical interfaces applicable to the implementation of 10GFC Ports. These interfaces include the 10 Gigabit Media Independent Interface (XGMII) and the 10 Gigabit Attachment Unit Interface (XAUI). One or both of these interfaces may typically be present within a 10GFC port.

XGMII, the 10 Gigabit Media Independent Interface, provides a physical instantiation of a 10.2 Gbps parallel data and control transport within FC-2. Its implementation is typically an internal chip interconnect or chip-to-chip interconnect. The XGMII supports 10.2 Gbps data transport through its 32-bit-wide data and 4-bit-wide control transmit and receive paths.

XAUI, the 10 Gigabit Attachment Unit Interface, provides a physical instantiation of a 10.2 Gbps four-lane serial data and control transport within FC-2 or between FC-2 and lower levels, including FC-1 and FC-0. The XAUI is defined as an XGMII extender. Its implementation is typically a chip-to-chip interconnect including chips within transceiver modules. The XAUI supports 10.2 Gbps data transport through its four 8B/10B-based serial transmit and receive paths.

FC-1 General Description. The FC-1 level of 10GFC provides the ability to transport data at a rate of 10.2 Gbps over a family of FC-0 physical variants. 10GFC provides the following FC-1 functions and interfaces:

- Direct mapping of FC-1 signals to 10GFC ordered sets.
- 8B/10B transmission code that divides FC-2 data and ordered sets among four serial lanes.

- 64B/66B transmission code that supports FC-2 data and ordered sets over a single serial lane.
- An optional physical interface for use by single-lane serial FC-0 variants. This interface is known as the 10 Gigabit Sixteen Bit Interface (XSBI).

FC-1 signals convey FC-2 data as well as frame delimiters and control information to be encoded by FC-1 transmission code. The same conveyance exists in the reverse direction.

8B/10B transmission code is the same as that specified in FC-FS. It is intended for 10.2 Gbps data transport across printed circuit boards, through connectors, and over four separate transmitters and receivers. These four transmitters and receivers may be either optically multiplexed to and from a single fiber-optic cable or directly conveyed over four individual fibers.

64B/66B transmission code is intended for 10.2 Gbps data transport across a single fiber-optic cable. The primary reason for the development of this code is to provide minimal overhead above the 10.2 Gbps serial data rate to allow the use of optoelectronic components developed for other high-volume 10 Gbps communications applications such as SONET OC-192.

The 10 Gigabit Sixteen Bit Interface provides a physical instantiation of a 16-bit-wide data path that conveys 64B/66B encoded data to and from FC-0. The XSBI is intended to support serial FC-0 variants.

FC-0 General Description. The FC-0 level of 10GFC describes the FC link. The FC-0 level covers a variety of media and associated transmitters and receivers capable of transporting FC-1 data. The FC-0 level is designed for maximum flexibility and allows the use of a large number of technologies to meet the broadest range of fibre channel system cost and performance requirements.

The link-distance capabilities specified in 10GFC are based on ensuring interoperability across multiple vendors supplying the technologies (both transceivers and cable plants) under the tolerance limits specified in 10GFC. Greater link distances may be obtained by specifically engineering a link based on knowledge of the technology characteristics and the conditions under which the link is installed and operated. However, such link-distance extensions are outside the scope of 10GFC.

FC-PI describes the physical link, the lowest level, in the fibre channel system. It is designed for flexibility and allows the use of several physical interconnect technologies to meet a wide variety of system application requirements.

Optical variants. Multiple optical serial physical full-duplex variants are specified to support the transport of encoded FC-1 data transport over fiber-optic medium.

Copper physical variant. A four-lane electrical serial full duplex physical variant is specified to support the transport of encoded FC-1 data transport over copper medium.

Ports, links, and paths. Each fiber set is attached to a transmitter of a port at one link end and a receiver of another port at the other link end. When a fabric is present in the configuration, multiple links may be utilized to attach more than one N_Port to more than one F_Port. Patch panels or portions of the active fabric may function as repeaters, concentrators, or fiber converters. A path between two N_Ports may be made up of links of different technologies. For example, the path may have single fiber multimode fiber links or parallel copper or fiber multimode links attached to end ports but may have a single-fiber, single-mode fiber link in between.

FC-PI defines the optical signal characteristics at the interface connector receptacle. Each conforming optical FC attachment must be compatible with this optical interface to allow interoperability within an FC environment. FC links must not exceed the BER objective (10^{-12}) under any conditions. The parameters specified in FC-PI support meeting that requirement under all conditions, including the minimum input power level.

The following physical variants are included:

- 850 nm Parallel (four-lane) Optics. Specified in this standard.
- 850 nm Serial. Fully specified in IEEE P802.3ae Clause 52.
- 850 nm WDM (four-wavelength). Specified in this standard.
- 1310 nm Serial. Fully specified in IEEE P802.3ae Clause 52.
- 1310 nm WDM (four-wavelength). Fully specified in IEEE P802.3ae Clause 53.

The 850 nm Parallel (four-lane) variant supports MM short wavelength (SW) data links. The laser links operates at the 3.1875 GBd (gigabaud) rate. The specifications are intended to allow compliance to Class 1 laser safety. Reflection effects on the transmitter are assumed to be small but need to be bounded. A specification of maximum relative intensity noise (RIN) under worst-case reflection conditions is included to ensure that reflections do not impact system performance. The receiver must operate within a BER of 10^{-12} over the link's lifetime and temperature range.

The 850 nm WDM (four wavelength) variant has the following four wavelengths for the spectral specifications:

1. 771.5–784.5 nm
2. 793.5–806.5 nm
3. 818.5–831.5 nm
4. 843.5–856.5 nm

8.2.2 Fibre Channel Technology—Tunneled Modes

This section describes tunneled methods for handling of channel communication. The jargon "IP storage" is being used to describe this set of approaches. A number of transport protocol standards, specifically, Internet Small Computer Systems Interface (iSCSI), Fibre Channel over TCP/IP (FCIP), and The Internet Fibre Channel Protocol (iFCP), have emerged of late. This affords organizations additional choices for accessing data over IP networks. "IP storage products" are also appearing.

The benefits of the tunneled approach to channel communications ("IP storage networking" in the new jargon) relate to leveraging the large installed base of Ethernet and IP intranets, extranets, and the Internet. This facilitates storage to be accessed over LAN, MAN, or WAN environments, without needing to deal with native channel interfaces across the network itself. This is of interest in grid computing. Tunneling enables the rapid deployment of IP-based SANs linking to FC devices. It allows the organization to implement enterprise-level solutions based on existing applications that already communicate with the FCP layer, alluded to in the previous section. These protocols enable scalable implementations using existing FC storage products via TCP/IP networks of any distance, using standard GbE layer 2 switches and layer 3 routers. They can also be used to facilitate grid deployment.

To facilitate the IP-based movement of block-level data that is stored as either direct-attached storage (DAS) or on a FC SAN requires new tunneling (transport) protocols. The tunneling protocols enable organizations to create and manage heterogeneous data storage environments (e.g., for backup, disaster recovery, and grid computing) where DAS and FC SANs can be integrated over a common IP network backbone.

Even without the full power of grid computing, these developments in IP storage networking are being viewed by proponents as a storage virtualization that enables managers to create virtual storage pools among geographically dispersed DAS, NAS, and SAN data resources [46]. This is kind of an entry-level approach to a full-fledged data grid.

The tunneling protocols that have emerged of late are:

- The Internet Small Computer Systems Interface (iSCSI). Defines the mechanisms to transmit and receive block storage data over TCP/IP networks by encapsulating SCSI-3 commands into TCP and transporting them over the LAN/intranet/extranet/Internet via IP (that is, iSCSI SANs can be deployed within LAN, MAN, or WAN environments). TCP provides the required end-to-end reliability. The iSCSI protocol runs on the host initiator and the receiving target device. ISCSI is being developed by the IETF.

- Fibre Channel over TCP/IP (FCIP). Provides a mechanism to "tunnel" FC over IP-based networks. This enables the interconnection of SANs, with TCP/IP used as the underlying reliable wide-area transport. FCIP is a protocol for connecting FC SANs over IP networks. It provides the functionality and speed of fibre channel with the ability to manage the networks using the same tools used today [41]. FCIP is being developed by the IETF.

- The Internet Fibre Channel Protocol (iFCP). A gateway-to-gateway protocol that supports FC Layer 4 FCP over TCP/IP. TCP/IP routing and switching components complement, enhance, or replace the FC fabric. That is to say, the iFCP specification defines a protocol for the implementation of a FC fabric in which TCP/IP switching and routing elements replace FC components: the lower-layer FC transport is replaced with IP-based networks (along with TCP) and standard LANs, such as GbE. The protocol enables the attachment of existing FC storage products to an IP network by supporting the fabric services required by such devices. iFCP supports FCP. It replaces the transport layer (FC-2) with an IP network (i.e., Ethernet), but retains the upper-layer (FC-4) information, such as FCP. This is accomplished by mapping the existing FC transport services to IP [144].

These tunneling/IP storage networking protocols are different, but they all deal with transporting block-level data over an IP network. These protocols enable end users to [12]:

- Leverage existing storage devices (SCSI and FC) and networking infrastructures, such as GbE-based LANs and IP-based intranets.
- Optimize storage resources to be available to a larger pool of applications.
- Reduce the geographical limitations of DAS and SAN access.
- Extend the reach of existing storage applications (backup, disaster recovery, and mirroring), without "upper-layer" modification.
- Manage storage networks with traditional IP tools.
- Provide enablements: the ease of deployment and management, support association, scalability, and flexibility that come with IP networking. This in turn, is expected to provide impetus to grid-based solutions (as well as more traditional data center solutions).

iSCSI has been developed to enable access of the embedded base of DAS over IP networks. The protocol enables block-level storage to be accessed from FC SANs using IP-based networks. iSCSI defines the mechanisms to handle block storage applications over TCP/IP networks. At the physical layer, iSCSI supports a Gigabit Ethernet interface; this means that systems supporting iSCSI interfaces can be directly connected to standard switches and/or IP routers. The iSCSI protocol is positioned above the physical and data-link layers and interfaces to the operating system's standard SCSI Access Method command set. iSCSI can be supported over any physical media that supports TCP/IP, but the most typical implementations are on GbE. iSCSI can run in software over a standard GbE network interface card (NIC), or can be optimized in hardware for better performance on an iSCSI host bus adapter (HBA). iSCSI also enables the access of block-level storage that resides on fibre channel SANs over an IP network via iSCSI-to-fibre channel gateways such as storage routers and switches. Initial iSCSI deployments are targeted at small to medium-sized businesses and departments or branch offices of larger enterprises that have not deployed fibre channel SANs [46] (see Figure 8.9, top).

By allowing greater access to DAS devices, possibly in a grid computing setting, these storage resources can be optimally utilized. Applications such as remote backup, disaster recovery, storage virtualization, and grid computing can be supported. The recent standardization efforts in this arena, the iSCSI-compliant products that are becoming available, and the SNIA IP Storage Forum's multivendor interoperability validations enable users to rapidly deploy "plug-and-play" IP SAN environments.

FCIP encapsulates FC packets and transports them via IP. It is a tunneling protocol that uses TCP/IP as the transport, while keeping FC services transparent to the end-user devices. FCIP relies on IP-based network services (e.g., alternate routing, QoS, etc.), and on TCP for congestion control and end-to-end integrity (that is, data-error and data-loss recovery). FCIP is intended to support the installed base of FC SANs, as shown in Figure 8.9, middle, and the need to interconnect these SANs

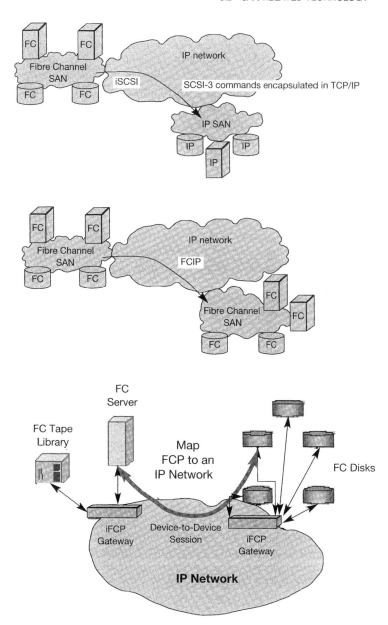

Figure 8.9   Tunneling arrangements. Top: iSCSI permits SCSI-3 commands to tunneled and delivered reliably over IP networks. Middle: FCIP permits multiple local FC SANs to be interconnected over an IP network backbone. Bottom: IFCP permits FC SANs to be interconnected via IP networkds of any length, using traditional IP network elements (e.g., routers).

over a geographic area to support mission-critical environments. This approach enables applications that have been developed to run over SANs to be supported over WANs. This same arrangement can be used to support data grids. SANs extensions (SANs interconnections to meet the needs for remote storage access) provide the high performance and reliability (diversified routing) required for business continuity and disaster recovery, including remote backup/archiving and remote mirroring. FCIP also provides centralized management. By combining IP networking, FCIP allows an organization to extend the interconnectivity of SANs across regional and national distances. FCIP provides the transport for traffic between SANs over LANs, MANs, and WANs. It also enables organizations to leverage their current IP infrastructure and management resources to interconnect and extend SANs.

iFCP is a TCP/IP-based protocol for interconnecting FC storage devices or FC SANs using an IP infrastructure in conjunction with, or, better yet, in place of, FC switching and routing elements. iFCP also has the capability to create fully autonomous regions by interconnecting islands of existing FC hub- and switch-based SANs, if present. iFCP represents an ideal customer solution for transitioning from FC to native IP storage networks [41]. Like FCIP, the primary drivers for iFCP are the large installed base of FC devices, combined with the broad availability of IP networking. Through the use of TCP, iFCP can accommodate deployment in environments in which the underlying IP network may not by itself be reliable. iFCP's primary advantage as a SAN gateway protocol is the mapping of FC transport services over TCP, allowing networked, rather than point-to-point, connections between and among SANs without requiring the use of FC fabric elements. Existing Fibre Channel Protocol (FCP)-based drivers and storage controllers can safely assume that iFCP, also being FC-based, provides the reliable transport of storage data between SAN domains via TCP/IP, without requiring any modification of those products. iFCP is designed to operate in environments that may experience a wide range of latencies [144]. iFCP maps FC transport services to an IP fabric as outlined in Figure 8.9, bottom. Gateways are used to connect existing FC devices to an IP network, and, as such, will include physical interfaces for both FC and IP. FC devices (e.g., disk arrays, switches, and HBAs) connect to an iFCP gateway or switch. Each FC session is terminated at the local gateway and converted to a TCP/IP session. The remote gateway or switch receives the iFCP session and initiates a FC session at the remote location. iFCP is a TCP/IP protocol that transports encapsulated FC-4 frame images between gateways. iFCP session endpoints are Fibre Channel N_Ports [144]. Table 8.3 compares iFCP and FC. Data center functionalities such as centralized backup, remote mirroring, storage management, and storage virtualization are supported within an iFCP environment due to the ability to create a scalable, peer-to-peer fibre channel/IP storage network [46].

8.3 LAN-RELATED TECHNOLOGY

In recent years, there has been major progress in developing high-speed LAN systems to support in-building corporate enterprise connectivity and data center con-

Table 8.3 Comparison of iFCP and FC

	iFCP	FC

General services	• IP based
	• Name services, security key distribution, time services, zoning
	  ○ iSNS, TLS, etc.
	• Time services
	  ○ /TBS/
		• Name services, security key distribution, time services, zone management, fabric configuration services, management services
		• Based on FC-GS2

Fabric services	• Class 2, Class 3	• Class 1, Class 2, Class 3 per FC-FS
Routing	• OSPF or any other IP routing protocol	• FSPF (Fabric Shortest Path First), a variant of OSPF

nectivity. In the late 1990s, Ethernet operating at 1 Gbps was standardized; we refer to these systems here as GbE. In the early 2000s, systems operating at 10 Gbps were standardized; we refer to these here as 10GbE. There is work underway to increase the speed to 40 and/or 100 Gbps.

This section provides some very basic information on LANs, GbE, and 10GbE. A full treatment of Ethernet requires a (large) book of its own; below, we provide just the most basic introduction to some topics of interest. The IEEE is the standardization body (in conjunction with ISO/IEC) that handles LAN standardization. Figure 8.10 depicts the basic IEEE LAN protocol model that is consistent across all LAN technologies.

8.3.1 Standards

Over the years, the IEEE has published a comprehensive set of International Standards for LANs employing Carrier Sense Multiple Access (CSMA/CD) as the ac-

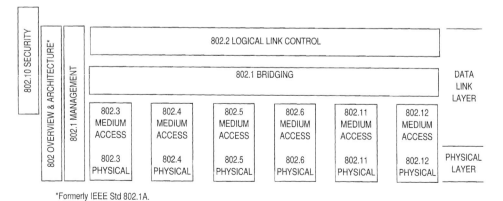

*Formerly IEEE Std 802.1A.

Figure 8.10 Basic IEEE LAN protocol model.

cess method. These standards are embodied in the IEEE 802.3 family. CSMA/CD is a low-complexity multiplexing scheme that allows multiple users to share the common medium (e.g., the four-wire bus that comprises the LAN); this is accomplished using channel-contention schemes. More recently, switched configurations that eliminate altogether the need for multiplexing and, hence, contention, have emerged. A majority of installations now use these switched configurations operating at 10- or 100-Mbps speeds. LANs were invented in the 1970s. Digital Equipment Corporation, Intel, and Xerox (DIX) brought out the first "standard" version of a LAN, calling this version Ethernet. The formal standardization that followed in the early 1980s by the IEEE was based on DIX-advanced technology.

IEEE 802.3 is intended to encompass several media types and techniques for signal rates from 1 Mbps to 10000 Mbps. The latest edition of the IEEE 802.3 standard provides the necessary specifications for the following families of systems: a 1 Mbps baseband system, 10 Mbps baseband and broadband systems, a 100 Mbps baseband system, a 1000 Mbps baseband system, and a 10000 Mbps baseband system. In addition, it specifies a method for linearly incrementing a system's data rate by aggregating multiple physical links of the same speed into one logical link.

The IEEE Project 802 develops LAN and Metropolitan Area Network (MAN) standards, mainly for the lowest two layers of the Reference Model for Open Systems Interconnection (OSI). It coordinates with other national and international standards groups, with some standards now published by ISO as international standards. There is strong international participation, and some meetings are held outside the United States.[4] The first meeting of the IEEE Computer Society "Local Network Standards Committee (LMSC)", Project 802, was held in February of 1980. There was going to be one LAN standard. It was divided into media or Physical layer (PHY), Media Access Control (MAC), and Higher Level Interface (HILI). The access method was similar to that for Ethernet, as was the bus topology. By the end of 1980, a token access method was added, and a year later there were three MACs: CSMA/CD, Token Bus, and Token Ring. In the years since, other MAC and PHY groups have been added, and one for LAN security as well. The unifying theme has been a common upper interface to the Logical Link Control (LLC) sublayer, common data-framing elements, and some commonality in media interfaces. The scope of work has grown to include MANs and Wide Area Networks (WANs) and higher data rates have been added. An organizational change gave the team the "LAN/MAN Standards Committee" name and more involvement in the standards sponsorship and approval process. Table 8.4 identifies the key standards that have been produced over the years.

Figure 8.11 depicts the various sublayers that comprise the MAC and PHY apparatus. The basic IEEE 802.3 standard provides for two distinct modes of operation: half duplex and full duplex. A given IEEE 802.3 instantiation operates in either half- or full-duplex mode at any one time. The term "CSMA/CD MAC" is used throughout the standard synonymously with "802.3 MAC," and may represent an instance of either half-duplex or full-duplex mode data terminal equipment (DTE),

---

[4]Overview And Guide To The IEEE 802 LMSC, July, 2002, http://grouper.ieee.org/groups/802/overview_07_12_2002.pdf.

Table 8.4 Key IEEE LAN efforts

IEEE 802.1

The IEEE 802.1 Working Group is chartered to concern itself with and develop standards and recommended practices in the following areas: 802 LAN/MAN architecture, internetworking among 802 LANs, MANs and other wide-area networks, 802 overall network management, and protocol layers above the MAC and LLC layers.

Active Projects
802a—Ethertypes
802.1s—Multiple Spanning Trees
802.1y—802.1D Maintenance
802.1z—802.1Q Maintenance
802.1aa—802.1X Maintenance

Projects under Discussion
802.1AB—Station and Media Access Control Connectivity Discovery
802.1ac—Media Access Control Service revision

Archived Projects
802—Overview and Architecture
802.1D—MAC bridges
802.1G—Remote MAC bridging
802.1Q—Virtual LANs
802.1t—802.1D Maintenance
802.1u—802.1Q Maintenance
802.1v—VLAN Classification by Protocol and Port
802.1w—Rapid Reconfiguration of Spanning Tree
802.1x—Port Based Network Access Control

Withdrawn Projects
802.1r—GARP Proprietary Attribute Registration Protocol (GPRP)

IEEE 802.3   CSMA/CD (Ethernet)

The IEEE 802.3 Working Group develops standards for CSMA/CD- (Ethernet-) based LANs. They have a number of active projects, as listed below:

P802.3ah—Ethernet in the First Mile
P802.3af—DTE Power via MDI
P802.3—Static Discharge in Copper Cables, ad hoc

---

even though full-duplex mode DTEs do not implement the CSMA/CD algorithms traditionally used to arbitrate access to shared-media LANs.

Half-Duplex Operation. In half-duplex mode, the CSMA/CD media access method is the means by which two or more stations share a common transmission medium. To transmit, a station waits (defers) for a quiet period on the medium (that is, no other station is transmitting) and then sends the intended message in bit-serial form. If, after initiating a transmission, the message collides with that of another station, then each transmitting station intentionally transmits for an additional predefined period to ensure propagation of the collision throughout the system. The

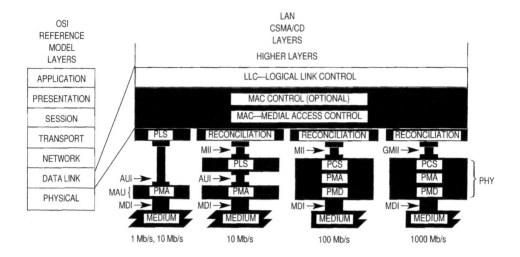

Figure 8.11 LAN standard relationship to the ISO/IEC Opens Systems Interconnection (OSI) reference model.

station remains silent for a random amount of time (backoff) before attempting to transmit again. Each aspect of this access method process is specified in detail in subsequent clauses of the IEEE 802.3 standard. Half-duplex operation can be used with all media and configurations.

Full-Duplex Operation. Full-duplex operation allows simultaneous communication between a pair of stations using point-to-point media (dedicated channel). Full-duplex operation does not require that transmitters defer, nor do they monitor or react to receive activity, as there is no contention for a shared medium in this mode. Full-duplex mode can only be used when all of the following are true:

1. The physical medium is capable of supporting simultaneous transmission and reception without interference.
2. There are exactly two stations connected with a full-duplex point-to-point link. Since there is no contention for use of a shared medium, the multiple access (i.e., CSMA/CD) algorithms are unnecessary.
3. Both stations on the LAN are capable of, and have been configured to use, full-duplex operation.

The most common configuration envisioned for full-duplex operation consists of a central bridge (also known as a switch) with a dedicated LAN connecting each bridge port to a single device. Full-duplex operation constitutes a proper subset of the MAC functionality required for half-duplex operation.

8.3.2 Key Concepts

Compatibility Interfaces. Five important compatibility interfaces are defined within what is architecturally the Physical Layer.

1. *Medium Dependent Interfaces (MDI).* To communicate in a compatible manner, all stations will adhere rigidly to the exact specification of physical-media signals defined in Clause 8 (and beyond) in the IEEE 802.3 standard, and to the procedures that define correct behavior of a station. The medium-independent aspects of the LLC sublayer and the MAC sublayer should not be taken as detracting from this point; communication by way of the ISO/IEC 8802-3 (ANSI/IEEE Std 802.3) Local Area Network requires complete compatibility at the Physical Medium interface (that is, the physical cable interface).

2. *Attachment Unit Interface (AUI).* It is anticipated that most DTEs will be located some distance from their connection to the physical cable. A small amount of circuitry will exist in the Medium Attachment Unit (MAU) directly adjacent to the physical cable, whereas the majority of the hardware and all of the software will be placed within the DTE. The AUI is defined as a second compatibility interface. Although conformance with this interface is not strictly necessary to ensure communication, it is highly recommended, since it allows maximum flexibility in intermixing MAUs and DTEs. The AUI may be optional or not specified for some implementations of the IEEE 802.3 standard that are expected to be connected directly to the medium and so do not use a separate MAU or its interconnecting AUI cable. The Physical Layer Signaling (PLS) and Physical Medium Attachment (PMA) are then part of a single unit, and no explicit AUI implementation is required.

3. *Media Independent Interface (MII).* It is anticipated that some DTEs will be connected to a remote PHY, and/or to different medium-dependent PHYs. The MII is defined as a third compatibility interface. While conformance with implementation of this interface is not strictly necessary to ensure communication, it is highly recommended, since it allows maximum flexibility in intermixing PHYs and DTEs. The MII is optional.

4. *Gigabit Media Independent Interface (GMII).* The GMII is designed to connect a gigabit-capable MAC or repeater unit to a gigabit PHY. Although conformance with implementation of this interface is not strictly necessary to ensure communication, it is highly recommended, since it allows maximum flexibility in intermixing PHYs and DTEs at gigabit speeds. The GMII is intended for use as a chip-to-chip interface. No mechanical connector is specified for use with the GMII. The GMII is optional.

5. *Ten-Bit Interface (TBI).* The TBI is provided by the 1000BASE-X PMA sub-layer as a physical instantiation of the PMA service interface. The TBI is highly recommended for 1000BASE-X systems, since it provides a convenient partition between the high-frequency circuitry associated with the PMA sublayer and the logic functions associated with the PCS and MAC sublayers. The TBI is intended for use as a chip-to-chip interface. No mechanical connector is specified for use with the TBI. The TBI is optional.

Table 8.5 summarizes some key definitions used in the IEEE 802.3 standards.

Cabling. Over the years, a variety of cabling technologies have been used for LANs, as defined below. Almost universally, nearly all enterprise networks today use Category 5 cabling (data center cabling may be fiber-optic-based). Cabling can be shielded or unshielded. Market surveys indicate that Category 5 balanced copper cabling is the predominant installed intrabuilding, horizontal networking medium today.

Unshielded Twisted-Pair Cable (UTP) is an electrically conducting cable, comprising one or more pairs, none of which is shielded. There may be an overall shield, in which case the cable is referred to as unshielded twisted-pair with overall shield.

Shielded Twisted-Pair (STP) Cable is an electrically conducting cable, comprising one or more elements, each of which is individually shielded. There may be an overall shield, in which case the cable is referred to as shielded twisted-pair cable with an overall shield (from ISO/IEC 11801: 1995). Specifically for IEEE 802.3 100BASE-TX, 150 ohm balanced inside cable with performance characteristics specified to 100 MHz (i.e., performance to Class D link standards as per ISO/IEC 11801: 1995). In addition to the requirements specified in ISO/IEC 11801: 1995, IEEE 802.3 Clauses 23 and 25 provide additional performance requirements for 100BASE-T operation over STP.

Category 3 Balanced Cabling. Balanced 100 ohms and 120 ohms cables and associated connecting hardware whose transmission characteristics are specified up to 16 MHz (i.e., performance meets the requirements of a Class C link as per ISO/IEC 11801: 1995). Commonly used by IEEE 802.3 10BASE-T installations. In addition to the requirements outlined in ISO/IEC 11801: 1995, IEEE 802.3 Clauses 14, 23, and 32 specify additional requirements for cabling when used with 10BASE-T, 100BASE-TX, and 1000BASE-T.

Category 4 Balanced Cabling. Balanced 100 ohm and 120 ohm cables and associated connecting hardware whose transmission characteristics are specified up to 20 MHz as per ISO/IEC 11801: 1995. In addition to the requirements outlined in ISO/IEC 11801: 1995, IEEE 802.3 Clauses 14, 23, and 32 specify additional requirements for this cabling when used with 10BASE-T, 100BASE-T4, and 100BASE-T2, respectively.

Table 8.5 Key definitions used in IEEE 802.3 standards

Concept	Definition	Clause
10BASE2	IEEE 802.3 Physical Layer specification for a 10 Mbps CSMA/CD local-area network over RG 58 coaxial cable.	Described in IEEE 802.3, Clause 10.
10BASE5	IEEE 802.3 Physical Layer specification for a 10 Mbps CSMA/CD local-area network over coaxial cable (i.e., thicknet).	Described in IEEE 802.3, Clause 8.
10BASE-F	IEEE 802.3 Physical Layer specification for a 10 Mbps CSMA/CD local-area network over fiber optic cable.	Described in IEEE 802.3, Clause 15.
10BASE-FB port	A port on a repeater that contains an internal 10BASE-FB Medium Attachment Unit (MAU) that can connect to a similar port on another repeater.	Described in IEEE 802.3, Clause 9.
10BASE-FB segment	A fiber-optic link segment providing a point-to-point connection between two 10BASE-FB ports on repeaters.	
10BASE-FL segment	A fiber-optic link segment providing point-to-point connection between two 10BASE-FL Medium Attachment Units (MAUs).	
10BASE-FP segment	A fiber-optic mixing segment, including one 10BASE-FP Star and all of the attached fiber pairs.	
10BASE-FP Star	A passive device that is used to couple fiber pairs together to form a 10BASE-FP segment. Optical signals received at any input port of the 10BASE-FP Star are distributed to all of its output ports (including the output port of the optical interface from which it was received). A 10BASE-FP Star is typically comprised of a passive-star coupler, fiber-optic connectors, and a suitable mechanical housing.	Described in IEEE 802.3, 16.5.
10BASE-T	IEEE 802.3 Physical Layer specification for a 10 Mbps CSMA/CD local-area network over two pairs of twisted-pair telephone wire.	Described in IEEE 802.3, Clause 14.
100BASE-FX	IEEE 802.3 Physical Layer specification for a 100 Mbps CSMA/CD local-area network over two optical fibers.	Described in IEEE 802.3, Clauses 24 and 26.
100BASE-T	IEEE 802.3 Physical Layer specification for a 100 Mbps CSMA/CD local-area network.	Described in IEEE 802.3, Clauses 22 and 28.
100BASE-T2	IEEE 802.3 specification for a 100 Mbps CSMA/CD local-area network over two pairs of Category 3 or better balanced cabling.	Described in IEEE 802.3, Clause 32.
100BASE-T4	IEEE 802.3 Physical Layer specification for a 100 Mbps CSMA/CD local-area network over four pairs of Categories 3, 4, and 5 unshielded twisted-pair (UTP) wire.	Described in IEEE 802.3, Clause 23.

(continued)

Table 8.5 *Continued*

Concept	Definition	Clause
100BASE-TX	IEEE 802.3 Physical Layer specification for a 100 Mbps CSMA/CD local-area network over two pairs of Category 5 unshielded twisted-pair (UTP) or shielded twisted-pair (STP) wire.	Described in IEEE 802.3, Clauses 24 and 25.
100BASE-X	IEEE 802.3 Physical Layer specification for a 100 Mbps CSMA/CD local-area network that uses the Physical Medium Dependent (PMD) sublayer and Medium Dependent Interface (MDI) of the ISO/IEC 9314 group of standards developed by ASC X3T12 (FDDI).	Described in IEEE 802.3, Clause 24.
1000BASE-CX	1000BASE-X over specialty shielded balanced copper jumper cable assemblies.	Described in IEEE 802.3, Clause 39.
1000BASE-LX	1000BASE-X using long-wavelength laser devices over multimode and single-mode fiber.	Described in IEEE 802.3, Clause 38.
1000BASE-SX	1000BASE-X using short-wavelength laser devices over multimode fiber.	Described in IEEE 802.3, Clause 38.
1000BASE-T	IEEE 802.3 Physical Layer specification for a 1000 Mbps CSMA/CD LAN using four pairs of Category 5 balanced copper cabling.	Described in IEEE 802.3, Clause 40.
1000BASE-X	IEEE 802.3 Physical Layer specification for a 1000 Mbps CSMA/CD LAN that uses a Physical Layer derived from ANSI X3.230-1994 (FC-PH).	Described in IEEE 802.3, Clause 36.
10BROAD36	IEEE 802.3 Physical Layer specification for a 10 Mbps CSMA/CD local-area network over single road and cable.	Described in IEEE 802.3, Clause 11.
1BASE5	IEEE 802.3 Physical Layer specification for a 1 Mbps CSMA/CD local-area network over two pairs of twisted-pair telephone wire.	Described in IEEE 802.3, Clause 12.
4D-PAM5	The symbol encoding method used in 1000BASE-T. The four-dimensional quinary symbols (4D) received from the 8B1Q4 data encoding are transmitted using five voltage levels (PAM5). Four symbols are transmitted in parallel each symbol period.	Described in IEEE 802.3, Clause 40.
8B/10B transmission code	A dc-balanced, octet-oriented data encoding.	
8B1Q4	For IEEE 802.3, the data encoding technique used by 1000BASE-T when converting GMII data (8B-8 bits) to four quinary symbols (Q4) that are transmitted during one clock (1Q4) period.	Described in IEEE 802.3, Clause 40.

Table 8.5 *Continued*

Concept	Definition	Clause
Attachment Unit Interface (AUI)	In 10 Mbps CSMA/CD, the interface between the Medium Attachment Unit (MAU) and the data terminal equipment (DTE) within a data station. Note that the AUI carries encoded signals and provides for duplex data transmission.	Described in IEEE 802.3, Clauses 7 and 8.
bit time (BT)	The duration of one bit as transferred to and from the Media Access Control (MAC). The bit time is the reciprocal of the bit rate. For example, for 100BASE-T the bit rate is 10^{-8} s or 10 ns.	
bridge	A Layer 2 interconnection device that does not form part of a CSMA/CD collision domain but conforms to the ISO/IEC 15802-3:1998 (ANSI/IEEE 802.1D, 1998 Edition) International Standard. A bridge does not form part of a CSMA/CD collision domain but, rather, appears as a Media Access Control (MAC) to the collision domain.	Described in IEEE Std 100-1996.
collision	A condition that results from concurrent transmissions from multiple data terminal equipment (DTE) sources within a single collision domain.	
collision domain	A single, half-duplex mode CSMA/CD network. If two or more Media Access Control (MAC) sublayers are within the same collision domain and both transmit at the same time, a collision will occur. MAC sublayers separated by a repeater are in the same collision domain. MAC sublayers separated by a bridge are within different collision domains.	
collision presence	A signal generated within the Physical Layer by an end station or hub to indicate that multiple stations are contending for access to the transmission medium.	Described in IEEE 802.3, Clauses 8 and 12.
Physical Coding Sublayer (PCS)	Within IEEE 802.3, a sublayer used in 100BASE-T, 1000BASE-X, and 1000BASE-T to couple the Media Independent Interface (MII) or Gigabit Media Independent Inter-face (GMII) and the Physical Medium Attachment (PMA). The PCS contains the functions to encode data its into code groups that can be transmitted over the physical medium. Three PCS structures are defined for 100BASE-T-one for 100BASE-X, one for 100BASE-T4, and one for 100BASE-T2 (described in IEEE 802.3, Clauses 23, 24, and 32). One PCS structure is defined for 1000BASE-X and one PCS structure is defined for 1000BASE-T.	Described in IEEE 802.3, Clauses 36 and 40.

(continued)

Table 8.5 *Continued*

Concept	Definition	Clause
Physical Layer entity (PHY)	Within IEEE 802.3, the portion of the Physical Layer between the Medium Dependent Interface (MDI) and the Media Independent Interface (MII), or between the MDI and GMII, consisting of the Physical Coding Sublayer (PCS), the Physical Medium Attachment (PMA), and, if present, the Physical Medium Dependent (PMD) sublayers. The PHY contains the functions that transmit, receive, and manage the encoded signals that are impressed on and recovered from the physical medium.	Described in IEEE 802.3, Clauses 23-26, 32, 36, and 40.
Physical Medium Attachment (PMA) sublayer	Within 802.3, that portion of the Physical Layer that contains the functions for transmission, reception, and (depending on the PHY) collision detection, clock recovery, and skew alignment.	Described in IEEE 802.3, Clauses 7, 12, 14, 16, 17, 18, 23, 24, 32, 36 and 40.
Physical Medium Dependent (PMD) sublayer	In 100BASE-X, that portion of the Physical Layer responsible for interfacing to the transmission medium. The PMD is located just above the Medium Dependent Interface (MDI).	Described in IEEE 802.3, Clause 24.
repeater	Within IEEE 802.3, a device as specified in Clauses 9 and 27 that is used to extend the length, topology, or interconnectivity of the physical medium beyond that imposed by a single segment, up to the maximum allowable end-to-end transmission line length. Repeaters perform the basic actions of restoring signal amplitude, waveform, and timing applied to the normal data and collision signals. For wired star topologies, repeaters provide a data distribution function. In 100BASE-T, a device that allows the interconnection of 100BASE-T Physical Layer (PHY) network segments using similar or dissimilar PHY implementations (e.g., 100BASE-X to 100BASE-X, 100BASE-X to 100BASE-T4, etc.). Repeaters are only for use in half-duplex mode networks.	Described in IEEE 802.3, Clauses 9 and 27.

Category 5 Balanced Cabling. Balanced 100 ohm and 120 ohm cables and associated connecting hardware whose transmission characteristics are specified up to 100 MHz (i.e., cabling components meet the performance specified in ISO/IEC 11801:1995). In addition to the requirements outlined in ISO/IEC 11801:1995, IEEE 802.3 Clauses 14, 23, 25, and 40 specify additional requirements for this cabling when used with 10BASE-T and 100BASE-T.

CHAPTER 9

Communication Systems
for National Grids

In this chapter, we examine a small set of recently developed WAN technologies that may be of importance for grid computing in the next few years. There is a plethora of technologies that range from Time Division Multiplexing dedicated lines based on SONET/SDH/OTN/ASON technology to Asynchronous Transfer Mode (ATM). These services range in bandwidth from 1 to 10 Mbps at the "medium end," to 155 to 2400 Mbps at the "high end." In this chapter, we look at two baseline technologies: Frame Relay MultiLink Frame Service (MLF) and Multi-Protocol Label Switching (MPLS).[1] These technologies have applicability to the intragrid environment.

9.1 MULTILINK FRAME SERVICE

This section covers what we perceive to be a possibly important service for future intranet development with regard to communication support, particularly for interconnection of secondary locations to a main location, like a data center. These satellite locations may well have available and underutilized servers, storage, and so on, that can be put to use by a grid system. This service is standardized *inverse multiplexing (imux) service.* Inverse multiplexing allows several low-speed channels to be combined transparently into one higher-speed channel. Typically, one combines a few T1s into a 3.2 or 6.4 Mbps stream. Backbone nodes, say the data center, a backup disaster recovery center, and, perhaps, some major gatewaying node, may be typically interconnected by a high-capacity backbone network comprised, possibly, of ATM or high-speed point-to-point SONET links. But secondary locations are usually connected with T1 or frame relay links. This is even more the case for international locations. This section looks at one promising inverse multiplexing technology that can support graceful growth to higher speeds while keeping cost-effectiveness in mind.

Although there has been a lot of hype about gigabit (Ethernet) communications in the metro environment, users continue to have plain and simple needs that hover around the equivalent of a few T1s. The author has documented in other publica-

---

[1]This section is based on [77].

A Networking Approach to Grid Computing. By Daniel Minoli
ISBN 0-471-68756-1 © 2005 John Wiley & Sons, Inc.

tions a quantitatively demonstrable end-user need for a *typical office environment* of 6 Mbps at this time and around 66 Mbps in 5 years, based on traffic scenarios [74]. This implies that while a T1 worth of speed is potentially limiting at this time, user needs are not at the gigabit range but at the single-digit megabit range. There is a large bandwidth, price, and availability gap between T1 and the next step up in network access technology: fiber-based DS-3 (45 Mbps) facilities. These considerations drive to "affordable" bandwidth. We already alluded to in Chapter 1 that the IT budget of a company can range from 2 to 10% of the company's top revenue line, with 6% being the average. The networking portion of the IT budget is generally around 10% (with 30% of that 10% being for bandwidth, 30% for amortized equipment expenses, and 40% for network operations). Hence, a $10B/yr company may have a $60M/yr telecom/networking budget, and $18M/yr for transmission costs. Very clearly, there is a strong desire to keep those costs in check.

One way to solve the DS1–DS3 gap is to bundle together multiple T1 lines into one larger channel. Inverse multiplexing can be supported by a number of technologies, including point-to-point TDM equipment; it is more advantageous, however, when it is used with a switched service like Frame Relay (FR) or ATM (see Figure 9.1). Inverse Multiplexed ATM (IMA) saw some deployment in the late 1990s. We focus on Frame Relay because there is a significant embedded base of applications and customers that may benefit from an upgrade. The MPLS/Frame Relay Forum FRF.16.1, Multilink Frame Relay (MFR) UNI/NNI (User-to-Network Interface/Network-to-Network Interface), that was ratified in 2002, provides economical multilink solutions for increasing bandwidth without using a higher bandwidth transmission facility.[2] Customer premise equipment (CPE) devices and central office [CO, also called point of presence (POP)] concentrators can be interoperable based on the FRF.16 MFR specification. Users of MFR can increase their bandwidth without making costly changes to their network and can move forward with confidence that FRF.16-compliant equipment is interoperable. CO concentrators enable the carrier to add Frame Relay services without changing the switch. Carrier tariffs make MFR an attractive alternative to moving up the digital hierarchy to the next-faster access loop; it is less expensive to add a few T1s than to move up to a DS3 facility. This is especially important to medium-sized hub sites and/or branch locations, and interest seems greatest around 6 to 9 Mbps. These locations could have underutilized IT resources that can be made part of a computational or data grid.

Done correctly, an NxT1 solution will enable network service providers to address the emerging demand for multimegabit service while maximizing the return on investment in the existing T1 infrastructure. After all, T1 lines are affordable and readily available, tariffs already exist, and T1 provides guaranteed symmetrical transport and "relatively" secure connectivity (high-quality encryption is the ultimate key to network-oriented security). Existing bundled T1 solutions, however, were developed for niche markets, are not interoperable, and were not intended for the large-scale service deployment required for the Internet. Bit-interleaved inverse multiplexing, for example, is a technology implemented in proprietary products that typically make point-to-point connections between two routers at speeds faster than

---

[2]In 2004 MPLS/Frame Relay Forum and the ATM Forum merged operations, reflecting the convergence of the technologies in the marketplace.

T1. Products that employ this proprietary technology do not support switching and multipoint connectivity as shown in Figure 9.1. Load balancing, used in most routers to intelligently direct packets down parallel paths, can provide higher-capacity WAN access for up to two T1s. However, this solution becomes less efficient as more WAN circuits are used. Neither of these solutions is ideal for providing the type of scalable, standards-based solution that the market needs [5].

9.1.1 Motivations and Scope

The MPLS/Frame Relay Forum created Multilink Frame Relay as a standard for inverse multiplexing over traditional T1 or E1 circuits from the customer's location to

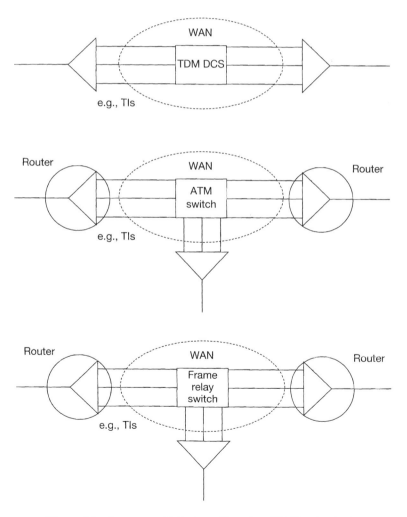

Figure 9.1 Inverse multiplexing with various WAN technologies.

the FR service-provider POP. The Multilink Frame Relay implementation agreements were created by the Forum in response to end users requests for a standards-based solution to bridge the bandwidth gap that exists between T1/E1 and DS3/E3 [28].

The continued success of Frame Relay technology as a proven, low-latency, and secure access service has driven customer demand for more flexibility in implementations. However, T1 local loops at those sites are not (always) fast enough to meet the demands of growing traffic volume and new applications [25]. When the T1/E1 is filled, it a major financial step to a T3/E3, even if these facilities are available where the user requires them. Changing the local loop to the next line rate up the traditional digital hierarchy (56 Kbps to 1.5 Mbps to 45 Mbps), represents a large cost increase, by a factor of almost 30. Many corporate planners cannot justify the jump in the monthly recurring charge (MRC) for network access, particularly at smaller branch locations. The three- to eight-time increase for the local loop is typical; higher port charges on the switch and other fees can raise the network access portion of the monthly bill by as much as ten times. Monthly access charges for T1 have dropped below $300 or less in many locations around the United States. DS3 local access circuits, on the other hand, typically cost $4,000 or more per month, and most business customers do not require the speed of a DS3. Moreover, local-access DS3 circuits are unavailable to a nontrivial fraction of businesses (or, business locations; there are approximately 750,000 commercial office buildings in the United States) and, where available, may take months to be provisioned [142]. Many of today's high-speed access solutions either do not address the T1–DS3 gap or are not appropriate for business subscribers. ATM imposes a "cell tax" that can reduce circuit payload. Asymmetrical DSL (ADSL), with its low up-link speeds is aimed at consumers.

Until recently, a second frame relay access line represented not just more bandwidth, but added a separate network path. That approach, therefore, required reengineering the traffic, assigning part to each local loop. An application was confined to its assigned link, and could not burst to the top speed represented by the total of the links. If one link were idle, applications assigned to another link could not use the bandwidth. If a router were involved, it would need another IP subnet. Addressing that specific part of the problem, carriers offer various proprietary forms of physical layer inverse multiplexing based on multiple loops (56/64 Kbps or T1/E1). These solutions, in practice, are difficult to order and slow to install. The handoff to the user is a HSSI (high-speed serial interface), one of the more expensive router cards that is typically low density (only one port per slot). Some forms of inverse muxing do not load balance; that is, if one link fails, the entire aggregate goes down. Conversely, adding an additional link to increase bandwidth involves taking down and reprovisioning the aggregate on the inverse mux, and this causes a disruption in the service [25, 142].

Frame Relay connections at access speeds of 1.544 Mbps (T1) or 2.048 Mbps (E1) are widely available from service providers today. Nevertheless, with applications such as storage backup, audio/video streaming, collaboration and conferencing, distributed software development, or possibly grid computing (particularly, data grids), bandwidth requirements have grown, in many cases, beyond the capacity of that single T1or E1 link. The obvious solution is to upgrade the customer's frame relay access link, into the "cloud," to the next higher speed in the digital hierarchy: DS3 (45 Mbps) for North America or E3 (34 Mbps) offered in Europe, South America, and Asia. As noted, in many cases, these transmission facilities are too expensive or may not be available in many small-to-medium metropolitan areas. Assuming that these lines may be locally tariffed, the required bandwidth, from the enterprise, may be far below DS3/E3 speeds and the link would, therefore, be underutilized. Unless the telecommunications carrier is offering specific private line services at speeds between T1/E1 and DS3/E3, which is rare, upgrading the link to the higher speed is often not feasible, economically and logistically. Hence, there is a gap in the continuity of speed solutions for FR access connections greater than T1 or E1 speeds but less than DS3/E3 [13]. Again, it might appear to government-funded recipients that bandwidth ought not to be a "big deal"; but in commercial settings, indeed it is.[3]

Figure 9.2 depicts an example of the use of MLF services to allow remote offices to become part of a grid environment.

---

[3]Occasionally, bandwidth costs may take a second-row seat in favor of overreaching functionality. This could well be indicated for the DoD's "Global Information Grid" (GIG), described next, as an illustrative example; but it never takes a second-row seat in commercial situations. The "Global Information Grid" [9] is a globally interconnected end-to-end set of information capabilities, associated processes, and personnel for collecting, processing, storing, disseminating, and managing information on demand to war fighters, policy makers, and support personnel. (GIG is *not exactly a grid* in the sense that we have discussed in this book; it is more a global ubiquitous communication apparatus, like a private Internet, a "private label IP network," although, in some sense it could also be seen as a grid.) Currently, the GIG concept is supported by the Department of Defense Chief Information Office (DoD CIO) 1999 memorandum "Global Information Grid" [145]. The memorandum describes GIG as the globally interconnected, end-to-end set of information capabilities, associated processes, and personnel for collecting, processing, storing, disseminating, and managing information on demand to war fighters, policy makers, and support personnel. While the GIG is composed of various network technologies, major elements are ATM, MPLS, and optical networks, which are the critical networking technologies for the GIG in the near- to mid-term [9]. The GIG includes all owned and leased communications and computing systems and services, software (including applications), data, security services, and other associated services necessary to achieve information superiority. It also includes national security systems as defined in Section 5142 of the Clinger-Cohen Act of 1996. The GIG supports all DoD, national security, and related intelligence community missions and functions (strategic, operational, tactical, and business), in war and in peace. The GIG provides capabilities from all operating locations (bases, posts, camps, stations, facilities, mobile platforms, and deployed sites). The GIG provides interfaces to coalition, allied, and non-DoD users and systems [76]. As part of the GIG, the DoD operates many systems that transmit information over commercial network infrastructures between enclaves. The network infrastructure contains components, such as routers and switches, which direct the flow of information through the infrastructures. Today, commercial carriers provide over 95% of all the transmission service for all GIG communications. Additionally, many networks used by Government agencies within the GIG have outsourced their network management services. On the other hand, commercial carriers view network security as a business issue. They will not simply add security features without financial or market incentive. For them, a business case must be made, and this begins with customers' demand for these services. The DoD is designing and deploying an enterprise-wide Information Assurance (IA) architectural overlay to the GIG [146] that is consistent with the overall GIG Architecture and implements a defense-in-depth strategy [9].

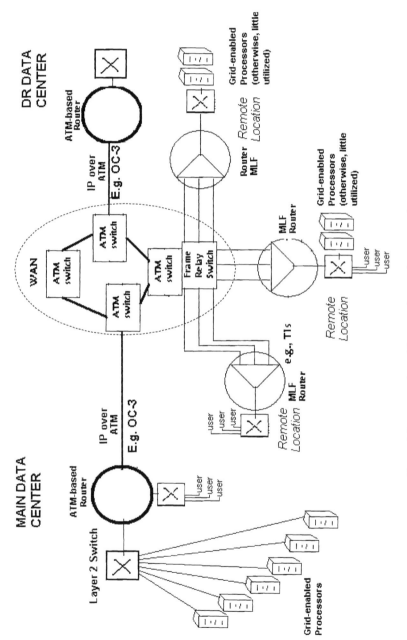

Figure 9.2 Enterprise grid computing environment using MLF.

318

9.1.2 Multilink Frame Relay Basics

The solution to achieving greater than T1/E1 bandwidth is to "bond" or "bundle" up to eight parallel T1/E1 circuits and create a "logical" link" with a maximum wire-line speed of 12 Mbps (T1) or 16 Mbps (E1). What is critical, though, is that the technology does this in a vendor-independent manner. The MPLS Frame Relay Forum has developed an implementation agreement for inverse multiplexing over standard T1 or E1 circuits from the customer's location to the FR service-provider POP. This approach will allow CPE equipment and FR switch vendors to all interoperate with the same N × T1/E1 inverse muxing protocol and provide customers with an economical and more widely available high-bandwidth solution alternative [13].

The MPLS/Frame Relay Forum standardized two forms of inverse multiplexing, defined in implementation agreements (IAs) FRF.15 and FRF.16. Both IAs show how to add increments of bandwidth to network access at a site while preserving the ability of any application to burst to the aggregate speed of all the physical links (it is still one logical channel):

- FRF.15 is a point-to-point method between CPE devices; the carrier does not participate. That is, the end devices know that they are doing MFR, but the carrier knows only that they are moving frames across the customer's links.
- FRF.16 is single-ended solution, between one site and the frame relay network or at the NNI; MFR terminates on the serving switch.

Both of these IAs define how an enterprise user, alone or in cooperation with its frame relay carrier, can increase the effective speed of the access link into a site— exactly what is wanted. To accomplish this goal, MFR aggregates multiple physical links into a single logical path.

FRF.15, the End-to-End Multilink Frame Relay Implementation Agreement, and the original FRF.16, the Multilink Frame Relay UNI/NNI Implementation Agreement, were published in 1999 and created the basis for the Multilink Frame Relay (MFR) protocol. MFR specifies the procedures and frame format to be used by CPE to offer an aggregated virtual circuit (AVC) service. AVC service allows frame relay CPE to use multiple virtual circuits (VCs) for transport of a single stream of sequenced frames. The UNI/NNI Implementation Agreement provides physical interface emulation for frame relay devices that consists of one or more physical links aggregated into a single bundle of bandwidth. The MFR standard ensures that the multimegabit access solution for frame relay integrates seamlessly with existing network hardware, creating an aggregated path that looks and feels like T1 to end users, but is faster. The bundle will be compatible with the frame relay infrastructure, and will result in additional bandwidth and increased transport resiliency by supporting distribution of data traffic over multiple underlying VCs. These circuits will not require load balancing or future maintenance to get the most efficient use of each T1 circuit. Service Providers and business users will be able deploy the MFR solution with minimal changes to their existing network [142].

MFR connections add and drop links gracefully. During a change in the number of constituents in an aggregate, frames flow without interruption. The ability to migrate smoothly between links also creates at least a potential to shop for less expensive or more reliable capacity. The real benefit (besides bandwidth) is resiliency: the aggregate channel can change bandwidth "on the fly" without interrupting traffic flow. Load balancing shifts traffic automatically from a failed link to active links. MFR's inherent load balancing across multiple paths provides resiliency that recovers quickly from a lost link. For optimum protection, access links at both ends will have diverse routing and, for FRF.15 end-to-end MFR, the backbone will have diverse routing as well [25].

Individual frames may take any of the physical links; therefore, successive frames could take different paths with different delays. The MFR function applies a fragmentation header to ensure proper order of the frames on both sides of the aggregate path. Due to the effect of fragmentation, each frame sees reduced latency as well as increased bandwidth. The multilink solution adds some complexity that is similar to adding an additional line without MFR. However, MFR preserves the logical simplicity of a single virtual path, allowing traffic to burst to a speed that is almost the total of all the constituent links. (The fragmentation headers in MFR require a modest amount of additional overhead.)

Adding bandwidth at a site using FRF.16 MFR at the UNI, means placing one or more additional local loops between the serving frame relay switch and the CPE. MFR software, in both the switch and CPE, combines a bundle of physical links into one virtual or logical link. Any PVC the application sees in the CPE appears in the backbone as the same single Permanent Virtual Connection (PVC). One can use the MFR virtual link in any way one would employ a real link. Specifically, one can add a PVC to the network without changing the MFR configuration. The arrangement with the carrier—bundling the physical links into a logical link—is transparent to applications, which see normal but faster connections. This UNI access arrangement (FRF.16) depends on carrier support in the serving switch that has started to roll out but is not yet widespread. There will be a small premium for the MFR termination service in the switch.

Opting for end-to-end MFR, and not involving the carrier, may have drawbacks for some applications precisely because FRF.15 is a point-to-point connection. This means that CPE at the remote site must also support MFR (FRF.15) to terminate those PVCs as an aggregate. In contrast to FRF.15, FRF.16 at the UNI carries a connection across the backbone on a single virtual circuit. When the carrier terminates MFR, and deals with a single fast PVC, it more easily can:

- Route to any location, (including a site whose access link is fast enough not to need MFR)
- Apply services like rerouting to an alternate data center
- Provision a connection to another frame relay carrier at an NNI

The logical bundling aspect of MFR can provide sites with more bandwidth, and the resiliency aspect adds reliability. These two factors make MFR a very powerful access option.

To ensure low latency within a virtual multimegabit access circuit, packets are segmented into individual fragments, and each fragment is transported over a separate member of the T1 bundle. The size of the packet fragments can be adjusted to optimize bandwidth efficiency and latency. The fragments are reassembled at the other end of the T1 link using MFR sequence numbers, thus ensuring packet order. If a T1 circuit within a MFR connection fails, the bandwidth is downshifted but service is not interrupted. When the T1 line comes back up, it is automatically added back to the bundle. In addition, the T1 lines in a bundle can be connected to the POP and CPE in any order, eliminating the risk of intrabundle wiring problems throughout the network. Service can be made more robust by using T1 circuits from different carriers. This enables continuous operation of the bundle if a carrier experiences problems with its T1 circuits. Bundling multiple copper lines to create a virtual multimegabit circuit may introduce differential delay between the different T1 links, particularly if different carriers are used within the bundle. The ability to handle these delays and still deliver complete packets, in sequence, is built into MFR protocols [142].

9.2 MPLS TECHNOLOGY

This section provides a short overview of MPLS-based technology and standards. During the past 25 years, corporations have sought improved packet technologies to support intranets, extranets, and public switched data networks such as the Internet. The progression went from X.25 packet-switched technology to Frame Relay technology, and also, on a parallel track, cell relay/ATM technology. In the meantime, throughout the 1980s and 1990s, IP-based connectionless packet services (a Layer 3 service) continued to make major inroads. IP, however, has limited QoS capabilities by itself. Therefore, the late 1990s and early 2000s saw the development of MPLS as a way to, perhaps, provide a better QoS framework based on improved packet handling. MPLS is a hybrid Layer 2/Layer 3 service that attempts to bring together the best of both worlds: Layer 2 and Layer 3, ATM and IP [79].

MPLS is a late-1990s set of specifications that provides a link-layer-independent transport mechanism for IP. The specification-development work is carried out by the IETF. MPLS protocols allow high-performance label switching of IP packets: network traffic is forwarded using a simple label apparatus as described in RFC3031 [136]. By combining the attributes of Layer 2 switching and Layer 3 routing into a single entity, MPLS provides [88] (i) enhanced scalability by way of switching technology; (ii) support of Class of Service (CoS) and QoS-based services (Differentiated Services/diffserv, as well as Integrated Services/intserv); (iii) elimination of the need for an IP-over-ATM overlay model and its associated management overhead; and, (iv) enhanced traffic shaping and engineering capabilities. In addition, MPLS provides a gamut of features in support of VPNs.

The basic idea of MPLS involves assigning short fixed-length labels to packets at the ingress to an MPLS cloud (based on the concept of forwarding equivalence classes). Throughout the interior of the MPLS domain, the labels attached to packets are used to make forwarding decisions (usually without recourse to the original packet headers). A set of powerful constructs to address many critical issues in the (eventually) emerging diffserv Internet can be devised from this relatively simple paradigm. One of the most significant initial applications of MPLS is in traffic engineering (TE). (It should be noted that even though the focus is on Internet backbones, the capabilities described in MPLS TE are equally applicable to Traffic Engineering in enterprise networks [4].)

The key MPLS RFCs are named next. IETF RFC 2702, Requirements for Traffic Engineering over MPLS, identifies the functional capabilities required to implement policies that facilitate efficient and reliable network operations in an MPLS domain; these capabilities can be used to optimize the utilization of network resources and to enhance traffic-oriented performance characteristics [4]. IETF RFC 3031, Multiprotocol Label Switching Architecture, specifies the architecture of MPLS [136]. IETF RFC 3032, MPLS Label Stack Encoding, specifies the encoding to be used by a Label Switch Router (LSR) in order to transmit labeled packets on Point-to-Point Protocol (PPP) data links, on LAN data links, and, possibly, on other data links [137]. Also, this RFC specifies rules and procedures for processing the various fields of the label stack encoding. An array of supplementary Internet Drafts support the various aspects of MPLS.

9.2.1 Approaches[4]

As a packet of a traditional connectionless network-layer protocol travels from one router to the next, each router makes an independent forwarding decision for that packet. That is, each router analyzes the packet's header, and each router runs a network-layer routing algorithm. Each router independently chooses a next hop for the packet, based on its analysis of the packet's header and the results of running the routing algorithm.

Packet headers contain considerably more information than is needed simply to choose the next hop. Choosing the next hop can, therefore, be thought of as the composition of two functions. The first function partitions the entire set of possible packets into a set of "Forwarding Equivalence Classes (FECs)." The second maps each FEC to a next hop. Insofar as the forwarding decision is concerned, different packets that get mapped into the same FEC are indistinguishable. All packets that belong to a particular FEC and that travel from a particular node will follow the same path (or, if certain kinds of multipath routing are in use, they will all follow one of a set of paths associated with the FEC).

In conventional IP forwarding, a particular router will typically consider two packets to be in the same FEC if there is some address prefix X in that router's rout-

---

[4]This subsection is based on [136].

ing tables such that X is the "longest match" for each packet's destination address. As the packet traverses the network, each hop in turn reexamines the packet and assigns it to a FEC.

In MPLS, the assignment of a particular packet to a particular FEC is done just once, as the packet enters the network. The FEC to which the packet is assigned is encoded as a short fixed-length value known as a "label." When a packet is forwarded to its next hop, the label is sent along with it; that is, the packets are "labeled" before they are forwarded. At subsequent hops, there is no further analysis of the packet's network-layer header. Rather, the label is used as an index into a table that specifies the next hop, and a new label. The old label is replaced with the new label, and the packet is forwarded to its next hop.

In the MPLS forwarding paradigm, once a packet is assigned to a FEC, no further header analysis is done by subsequent routers; all forwarding is driven by the labels. This has a number of advantages over conventional network layer forwarding:

- MPLS forwarding can be done by switches that are capable of doing label lookup and replacement, but are either not capable of analyzing the network layer headers or are not capable of analyzing the network layer headers at adequate speed.

- Since a packet is assigned to a FEC when it enters the network, the ingress router may use, in determining the assignment, any information it has about the packet, even if that information cannot be gleaned from the network layer header. For example, packets arriving on different ports may be assigned to different FECs. Conventional forwarding, on the other hand, can only consider information that travels with the packet in the packet header.

- A packet that enters the network at a particular router can be labeled differently than the same packet entering the network at a different router, and, as a result, forwarding decisions that depend on the ingress router can be easily made. This cannot be done with conventional forwarding, since the identity of a packet's ingress router does not travel with the packet.

- The considerations that determine how a packet is assigned to a FEC can become ever more and more complicated, without any impact at all on the routers that merely forward labeled packets.

- Sometimes, it is desirable to force a packet to follow a particular route which is explicitly chosen at or before the time the packet enters the network, rather than being chosen by the normal dynamic routing algorithm as the packet travels through the network.[5] This may be done as a matter of policy, or to support traffic engineering. In conventional forwarding, this requires the packet to carry an encoding of its route along with it ("source routing"). In MPLS, a label can be used to represent the route, so that the identity of the explicit route need not be carried with the packet.

---

[5]For example, there may be policy reasons for doing so, such as law enforcement access for wiretaps at a convenient location.

Some routers analyze a packet's network-layer header not merely to choose the packet's next hop, but also to determine a packet's "precedence" or "class of service." They may then apply different discard thresholds or scheduling disciplines to different packets. MPLS allows (but does not require) the precedence or class of service to be fully or partially inferred from the label. In this case, one may say that the label represents the combination of a FEC and a precedence or class of service.

MPLS's techniques are applicable to any network-layer protocol. In this discussion; however, we focus on the use of IP as the network-layer protocol. A router that supports MPLS is known as a "label switching router" or LSR. Table 9.1 identifies key terms used in MPLS.

9.2.2 MPLS Operation

MPLS runs over ATM, Frame Relay, Ethernet, and point-to-point packet-mode links. MPLS-based networks use existing IP mechanisms for addressing of elements and for routing of traffic. MPLS adds a sort of connection-oriented capabilities to the connectionless IP architecture. It is the industry-accepted manifestation of the ""Network Layer/Layer 3/Tag/IP Switching" technology that was developed by various constituencies in the mid-to-late 1990s. MPLS integrates the label-swapping forwarding paradigm with network-layer routing. In an MPLS environment, when a stream of data traverses a common path, a label-switched path (LSP) can be established using MPLS signaling protocols. At the ingress LSR, each packet is assigned a label and is transmitted downstream. At each LSR along the LSP, the label is used to forward the packet to the next hop. To deliver reliable service, MPLS requires a set of procedures to provide protection of the traffic carried on different paths. This requires that the LSRs support fault detection, fault notification, and fault recovery mechanisms, and that MPLS signaling support the configuration of recovery [133]. QoS support is where MPLS can find its sweet technical spot in supporting multimedia and voice applications. The improved traffic management, the QoS capabilities, and the expedited packet forwarding via the label mechanism can be a significant technical advantage in delay-sensitive applications.

The LSRs know what label values to use because this information will have been propagated by a label distribution protocol. (The MPLS architecture allows several different methods and protocols for label distribution—we discuss this later.) In essence, the label-switched path is established once all the LSRs have valid label entries in their forwarding tables.

The packet flow across a network with an illustrative Ingress LSR1 and an Egress LSR4 is as follows (see Figure 9.3):

1. An IP packet arrives at LSR1.
2. LSR1 examines the IP header and the IP destination address.
3. The IP packet is then labeled; the label value given to the packet is associated with a label-switched path across the network to the egress point (LSR4).

Table 9.1 Key terms used in MPLS

DLCI	A label used in Frame Relay networks to identify frame relay circuits
Forwarding equivalence class	A group of IP packets that are forwarded in the same manner (e.g., over the same path, with the same forwarding treatment).
Frame merge	Label merging, when it is applied to operation over frame-based media, so that the potential problem of cell interleave is not an issue.
Label	A short fixed-length, physically contiguous identifier that is used to identify a FEC, usually of local significance.
Label merging	The replacement of multiple incoming labels for a particular FEC with a single outgoing label.
Label stack	An ordered set of labels.
Label swap	The basic forwarding operation consisting of looking up an incoming label to determine the outgoing label, encapsulation, port, and other data handling information.
Label swapping	A forwarding paradigm allowing streamlined forwarding of data by using labels to identify classes of data packets that are treated indistinguishably when forwarding.
Label switched hop	The hop between two MPLS nodes, on which forwarding is done using labels.
Label switched path	The path through one or more LSRs at one level of the hierarchy followed by a packets in a particular FEC.
Label switching router	An MPLS node that is capable of forwarding native L3 packets
Layer 2	The protocol layer under Layer 3 (which, therefore, offers the services used by Layer 3). Forwarding, when done by the swapping of short fixed-length labels, occurs at Layer 2, regardless of whether the label being examined is an ATM VPI/VCI, a frame relay DLCI, or an MPLS label.
Layer 3	The protocol layer at which IP and its associated routing protocols operate; link layer synonymous with Layer 2.
Loop detection	A method of dealing with loops in which loops are allowed to be set up, and data may be transmitted over the loop, but the loop is later detected.
Loop prevention	A method of dealing with loops in which data is never transmitted over a loop.
Merge	Point on a node at which label merging is done.
MPLS domain	A contiguous set of nodes that operate MPLS routing and forwarding and that are also in one routing or administrative domain.
MPLS edge node	An MPLS node that connects an MPLS domain with a node that is outside of the domain, either because it does not run MPLS, and/or because it is in a different domain. Note that if an LSR has a neighboring host that is not running MPLS, that LSR is an MPLS edge node.

(continued)

Table 9.1 *Continued*

MPLS egress node	An MPLS edge node in its role of handling traffic as it leaves an MPLS domain.
MPLS ingress node	An MPLS edge node in its role of handling traffic as it enters an MPLS domain.
MPLS label	A label that is carried in a packet header, and that represents the packet's FEC.
MPLS node	A node that is running MPLS. An MPLS node will be aware of MPLS control protocols, will operate one or more L3 routing protocols, and will be capable of forwarding packets based on labels. An MPLS node may optionally be also capable of forwarding native L3 packets.
Network layer	Synonymous with Layer 3.
Stack	Synonymous with label stack.
Switched path	Synonymous with label switched path.
VC merge	Label merging where the MPLS label is carried in the ATM VCI field (or combined VPI/VCI field), so as to allow multiple VCs to merge into one single VC.
Virtual circuit	A circuit used by a connection-oriented Layer 2 technology such as ATM or Frame Relay, requiring the maintenance of state information in Layer 2 switches.
VP merge	Label merging in which the MPLS label is carried in the ATM VPI field, so as to allow multiple VPs to be merged into one single VP. In this case, two cells would have the same VCI value only if they originated from the same node. This allows cells from different sources to be distinguished via the VCI.
VPI/VCI	A label used in ATM networks to identify circuits.

4. Now that the packet has been labeled, the label associates the packet with a particular path. There is no longer any need to look up the IP address inside the network as the other LSRs can forward the labeled packet based on the label value.

5. A point to note is that the labels are rewritten at each switch.

6. When the packet reaches the last or egress LSR (LSR4), the label is stripped (popped), thus exposing the IP packet.

7. LSR 4 can then deliver the IP packet to the destination using normal IP forwarding.

Originally, the main benefit of label switching was facilitating high-speed switching in Layer 3 devices. However, this is no longer perceived as the main benefit of MPLS, since ASIC-based routers, can now perform line-speed routing on most interfaces. Now, the major benefits of MPLS are perceived to be [26]:

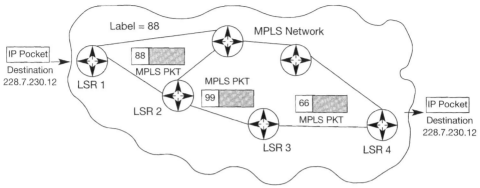

1. IP Packet enter MPLS Network
2. IP Address examined
3. Packet Associated with path
4. Packet is labelled (88)
5. Packet forwarded to LSR 2
6. LSR 2 examines label
7. LSR 2 forwards packet based on
   label value—writes new label value (99)
8. Packet reaches egress (LSR 4)
9. Strip label and deliver IP Packet

Time sequence

Figure 9.3 An MPLS network.

- Simplifying packet forwarding. Since the routing decision is made only once at the edge of the network, the core could retain only minimal routing information, thus reducing the overall complexity of the network (e.g., BGP could be run at the edge only, but there would be no need for it in the core).
- Aggregation of Protocol Data Unit (PDU) Streams. In MPLS, the label stacking mechanism can be used to perform the aggregation within Layer 2 itself. Typically, when multiple streams have to be aggregated for forwarding into a switched path, processing is required at both Layer 2 and Layer 3. The top label of the MPLS label stack is used to switch PDUs along the label-switched path, whereas the rest of the label stack is "application specific."
- Explicit/Improved Routes. MPLS supports explicit routes (a route that has not been set up by normal IP hop-by-hop routing, but rather an ingress/egress node has specified all or some of the downstream nodes of that route).
- Improved Performance. MPLS enables higher data transmission performance due to simplified packet forwarding and switching mechanisms.
- Link Layer Independence. MPLS works with any type of link layer medium such as ATM, Frame Relay, Packet-over-SONET, Ethernet, and so on.
- QoS Support. Using MPLS' inherent mechanisms for traffic prioritization and traffic path control, a service provider could create a network that deliv-

ers QoS, facilitates offering differentiated services to customers, and fulfills the offered service-level agreements.

- Scalability of Network Layer Routing. A key MPLS desideratum was to achieve an improved and more efficient transfer of PDUs in the current IP networks. Combining the routing knowledge at Layer 3 with the ATM Switching capability in ATM devices results in a better solution. In the MPLS scenario, it is sufficient to have adjacencies with the immediate peers. The edge LSRs interacts with the adjacent LSR and this is sufficient for the creation of LSPs for the transfer of data.
- Traffic Engineering. MPLS supports traffic engineering (a process of selecting the paths chosen by data traffic in order to balance the traffic load on the various links, routers, and switches in the network). MPLS offers capabilities to control the paths taken by different flows. Using these capabilities, traffic could be rerouted to avoid congestion points in a network. Key performance objectives of TE are (a) traffic-oriented: includes those aspects that enhance the QoS of traffic streams; and (b) resource-oriented: includes those aspects that pertain to the optimization of resource utilization.
- Virtual Private Network Support. VPN is an application that uses the label stacking mechanisms. At the VPN ingress node, the VPN label is mapped onto the MPLS label stack and packets are label-switched along the LSP within the VPN until they emerge at the egress. At the egress node, the label stack is used to determine further forwarding of the PDUs.

In particular, traffic engineering is concerned with performance optimization of operational networks. In general, it encompasses the application of technology and scientific principles to the measurement, modeling, characterization, and control of Internet traffic, and the application of such knowledge and techniques to achieve specific performance objectives. The aspects of traffic engineering that are of interest concerning MPLS are measurement and control. A major goal of Internet traffic engineering is to facilitate efficient and reliable network operations while simultaneously optimizing network resource utilization and traffic performance. Traffic engineering has become an indispensable function in many large autonomous systems because of the high cost of network assets and the commercial and competitive nature of the Internet. These factors emphasize the need for maximal operational efficiency [4]. Note that MPLS is not a routing protocol. In fact, MPLS needs the reachability information provided by the current routing protocols in order to calculate the paths that it uses. MPLS augments the functionality of the routing protocols, but does not replace them.

9.2.3 Key Mechanisms Supporting MPLS

MPLS requires a set of procedures for augmenting network layer packets with "label stacks," thereby turning them into "labeled packets." Routers that support MPLS are known as "label-switching routers." In order to transmit a labeled packet on a particular data link, an LSR must support an encoding technique that, given a label stack and a network layer packet, produces a labeled packet [137]. MPLS can be logically and functionally partitioned into two elements to provide the label-switching functionality:

1. MPLS forwarding/label-switching mechanism
2. MPLS label-distribution mechanism

MPLS Forwarding/Label-Switching Mechanism. The key mechanism of MPLS is the forwarding/label-switching function. This is an advanced form of packet forwarding that replaces the conventional longest-address-match-forwarding with a more efficient label-swapping forwarding algorithm. The IP header analysis is performed once at the Ingress of the LSP for the classification of PDUs. PDUs that are forwarded via the same next hop are grouped into a FEC based on one or more of the following parameters: Address Prefix; Host Address; and/or Host Address and QoS.

The FEC to which the PDU belongs is encoded at the edge LSRs as a short fixed-length value known as a "label." When the PDU is forwarded to its next hop, the label is sent along with it. At downstream hops, there is no further analysis of the PDU's network-layer header. Instead, the label is used as an index into a table; the entry in the table specifies the next hop and a new label. The incoming label is replaced with this outgoing label, and the PDU is forwarded to its next hop. Labels usually have a local significance and are used to identify FECs based on the type of the underlying network. For example, in ATM networks, the VPI and VCI are used in generating the MPLS label; in Frame Relay networks, the Data Link Control Identifier (DLCI) is used. In ATM environments, the labels assigned to the FECs (PDUs) are the VPI/VCI of the virtual connections established as a part of the LSP. In Frame Relay environments, the labels assigned to the FECs (PDUs) are the DCLIs.

So an FEC is a class of packets that should be forwarded in the same manner (i.e., over the same path) [26]. A FEC is not a packet, nor is it a label. A FEC is a logical entity created by the router to represent a class (category) of packets. When a packet arrives at the ingress router of an MPLS domain, the router parses the packet's headers, and checks to see if the packet matches a known FEC (class). Once the matching FEC is determined, the path and outgoing label assigned to that FEC are used to forward the packet. FECs are typically created based on the IP destinations known to the router, so for each different destination a router might create a different FEC, or if a router is doing aggregation, it might represent multiple destinations with a single FEC (for example, if those destinations are reachable through the same immediate next hop anyway). The MPLS framework, however, allows for the creation of FECs using advanced criteria like source and destination address pairs, destination address and TOS, and so on.

Label Switching has been designed to leverage the Layer 2 switching function done in the current data link layers such as ATM and FR. It follows that the MPLS

Forwarding mechanism should be able to update the switching fabric(s) in ATM and FR hardware in the LSR for the relevant sets of LSPs, that can be switched at the hardware level [29]. In the Ethernet-based networks, the labels are short headers placed between the data link headers and the data link layer PDUs.

MPLS Label-Distribution Mechanism. In an MPLS environment, the distribution of labels in MPLS is accomplished in two ways:

1. Utilizing the Resource ReSerVation Protocol (RSVP) signaling mechanism to distribute labels mapped to the RSVP flows [138]
2. Utilizing the Label Distribution Protocol (LDP) [80]

Label Distribution Using RSVP. RSVP defines a '"session" to be a data flow with a particular destination and transport-layer protocol [138]. From the early 1990s to the late 1990s, RSVP was being considered for QoS support in IP networks. When RSVP and MPLS are combined, a flow or session can be defined with greater generality. The ingress node of an LSP can use a variety of means to determine which PDUs are assigned a particular label. Once a label is assigned to a set of PDUs, the label effectively defines the "flow" through the LSP. Such an LSP is referred to as an "LSP tunnel" because the traffic flowing though it is "opaque" to intermediate nodes along the label-switched path. The label-request information for the labels associated with RSVP flows will be carried as part of the RSVP Path messages and the label-mapping information for the labels associated with RSVP flows will be carried as part of the RSVP Resv messages [29]. The initial implementors of MPLS chose to extend RSVP into a signaling protocol to support the creation of LSPs that could be automatically routed away from network failures and congestion. An Internet Draft defines the extension to RSVP for establishing LSPs in MPLS networks [139].

The use of RSVP as a signaling protocol for traffic engineering is different from that envisioned by its original developers in the mid-1990s [11]:

- A number of extensions were added to the base RSVP specification (RFC 2205 and RFC 2209) to support the establishment and maintenance of explicitly routed LSPs.
- RSVP signaling takes place between pairs of routers (rather than pairs of hosts) that act as the ingress and egress points of a traffic trunk. Extended RSVP installs state that applies to a collection of flows that share a common path and a common pool of shared network resources, rather than a single host-to-host flow. By aggregating numerous host-to-host flows into each LSP tunnel, extended RSVP significantly reduces the amount of RSVP state that needs to be maintained in the core of a service provider's network.
- RSVP signaling installs distributed state related to packet forwarding, including the distribution of MPLS labels.

- The scalability, latency, and traffic overhead concerns regarding RSVP's soft-state model are addressed by a set of extensions that reduce the number of refresh messages and the associated message processing requirements.
- The path established by RSVP signaling is not constrained by conventional destination-based routing, so it is a good tool to establish traffic engineering trunks [81].

The initial implementors of MPLS had a number of reasons to choose to extend RSVP rather than design an entirely new signaling protocol to support traffic engineering requirements [11]:

- By implementing the proposed extensions, RSVP provides a unified signaling system that delivers everything that network operators need to dynamically establish LSPs.
- Extended RSVP creates an LSP along an explicit route to support the traffic engineering requirements of large service providers.
- Extended RSVP establishes LSP state by distributing label-binding information to the LSRs in the LSP.
- Extended RSVP can reserve network resources in the LSRs along the LSP (the traditional role of RSVP). Extended RSVP also permits an LSP to carry best-effort traffic without making a specific resource reservation.

Hence, RSVP can serve a dual role in MPLS: for label distribution and for QoS support.

Label Distribution Protocol. LDP is a set of procedures and messages by which LSRs establish LSPs through a network by mapping network-layer routing information directly to data-link layer-switched paths. These LSPs may have an endpoint at a directly attached neighbor (this being comparable to IP hop-by-hop forwarding), or may have an endpoint at a network egress node, enabling switching via all intermediary nodes. LDP associates an FEC with each LSP it creates. The FEC associated with an LSP specifies which PDUs are "mapped" to that LSP. LSPs are extended through a networks as each LSR "splices" incoming labels for an FEC to the outgoing label assigned to the next hop for the given FEC.

The messages exchanged between the LSRs are classified into the four categories:

1. Discovery messages. Used to announce and maintain the presence of an LSR in a network.
2. Session messages. Used to establish, maintain, and terminate sessions between LSP peers.
3. Advertisement messages. Used to create, change, and delete label mappings for FECs.
4. Notification messages. Used to provide advisory information and to signal error information.

The LDP uses the TCP for session, advertisement, and notification messages. TCP is utilized to provide reliable and sequenced messages. Discovery messages are transmitted by using the UDP. These messages are sent to the LSP port at the "all routers on this subnet" group multicast address.

Discovery messages provide a mechanism for the LSRs to indicate their presence in a network. LSRs send the Hello message periodically. When an LSR chooses to establish a session with another LSR discovered via the Hello message, it uses the LDP initialization procedure (this is done using TCP). Upon successful completion of the initialization procedure, the two LSRs are LSP peers, and may exchange advertisement messages. The LSR requests a label mapping from a neighboring LSR when it needs one, and advertises a label mapping to a neighboring LSR when it wishes the neighbor to use a label.

9.2.4 Service Availability

Almost every major Tier 1 communication provider in the United States has announced that they have or will be bringing out MPLS-based services. In the midterm, it is expected that MPLS services will either replace or at least cap the new deployment of ATM services. Clearly, MPLS has the potential of being superior to ATM for data applications because it eliminates the "cell tax" and the need for laborious IP frame segmentation and reassembly. We think the service will play a role in future grid computing applications.

CHAPTER 10

Communication Systems for Global Grids

This chapter takes a rather limited look at the kinds of lower-layer networking technologies that can be deployed to support intergrids. It should be noted that any traditional network of sufficient capacity and security could be employed for this application. However, the global nature of the GAN required makes it less likely that users can employ a cost-effective dedicated, high-speed network to interconnect all of the resources dispersed on a global scale. Therefore, the use of Internet-based services, and in particular VPN, is likely to be the pragmatic solution of choice for most (nongovernment funded) grids.

Given these observations, this final chapter looks at VPN technologies, also providing a comparison with emerging Layer 2 VPNs that are facilitated by MPLS technology discussed in the previous chapter.

Traditionally, VPNs have been achieved by encapsulating encrypted Layer 3 packets (e.g., using IPSec) inside the payload of a regular IP packet that is then transmitted through an IP-based network, such as the Internet. While VPN "tunnels" suffer from issues such as being point-to-point in nature, requiring encryption key administration, and providing rather poor QoS through the Internet (in terms of bandwidth, latency, and jitter, the service is strictly "best effort"), they are relatively inexpensive. Cost and reach are the two key charactecteristics that make VPNs popular.

Now, VPNs based on MPLS are also becoming practical. As we discussed briefly in the previous chapter, MPLS has certain mechanisms that in theory can assist in the administration of QoS, such as traffic engineering (TE). The question is, when are MPLS VPNs better implemented at Layer 3, using BGP (Border Gateway Protocol)-based VPNs, and when at Layer 2, using MPLS tunneling technologies?

In the sections that follow,[1] we explain what underlies the choice between Layer 2 and Layer 3 MPLS VPNs. Neither will always be the "right" choice for every network. The nature of existing network architectures and desired applications and services are what ultimately decide the matter. And, of course, some network operators may deploy both types of VPN, or salutary combinations of the two technologies.

---

[1] The rest of this chapter is based on a White Paper by Tim Wu and Andy Walden, and included with their permission [154].

A Networking Approach to Grid Computing. By Daniel Minoli
ISBN 0-471-68756-1 © 2005 John Wiley & Sons, Inc.

Layer 3 MPLS VPNs will likely remain most appealing to Internet service providers that already use BGP extensively and have already deployed high-end IP/MPLS routing equipment at the edge. However, for carriers with existing Layer 2 VPN deployments or those accustomed to delivering transport services, Layer 2's MPLS "overlay" model should prove much more attractive. This follows because such carriers are unlikely to be interested in the degree of IP routing and (more to the point) high-end IP-equipment expenditures that Layer 3 VPNs call for. In addition, it is clear that where direct interoperability with existing Layer 2 VPN deployments is important, Layer 2 VPNs have the advantage.

MPLS interfaces on new equipment brought to market by routing vendors (particularly those developing edge devices) offer complete Layer 2 or Layer 3 VPN solutions based on Martini-draft tunneling and RFC 2547-bis, and these companies play a leading role in developing joint Layer 2/Layer 3 VPN solutions.

10.1   THE BASICS OF LAYER 2 AND LAYER 3 VPNs

It is easy to lose sight of the purpose of MPLS VPN technology in the first place. The goal is simple: to build a network that, as much as possible, acts like an extension of the private corporate network on a service provider's shared network infrastructure. The result, ideally, is a fast and efficient means of making scattered places seem just like local sites, from workers' homes to branch offices.

MPLS, designed to scale IP networks, is a natural choice for virtual private networks. Supporting multiple private networks on a shared infrastructure suggests immediate scaling problems for both Layer 3 and Layer 2 networks. On a Layer 3 network, asking each router on the network to potentially support thousands of different routing tables (one for each virtual private network, in addition to those of the public network) is an interminable option. Layer 2 networks, on the other hand, have a different scaling problem: they lack the scope of routed networks, limiting a Layer 2 implementation to the confines of the transport medium. Certain link-layer protocols, like Ethernet, also have scaling limits that reflect their LAN origins (i.e., the 4095 VLAN limit).

10.2   THE LAYER 3 APPROACH

The Layer 3 VPN MPLS implementation is an early leader. The BGP model is based on an IETF Request for Comments (RFC) 2547, and these "2547 VPNs" have already been implemented in several major carrier networks, including parts of the IP/MPLS backbones of AT&T and Bell Canada, among others.

How does a RFC 2547 VPN work? As the RFC explains, "MPLS is used for forwarding packets over the backbone, and BGP is used to distribute routes over the backbone" (see Figure 10.1). Each RFC 2547 VPN is really a private IP network, with modified private IP addresses for each of the provider edge (PE) routers imme-

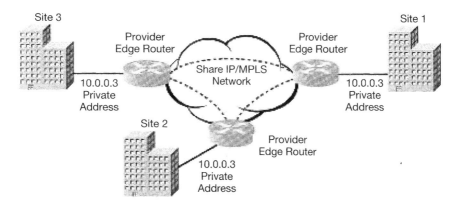

Figure 10.1 A private BGP network with private IP addresses.

diately connected to the customer site. The route to each of the sites on the private network is distributed using the familiar BGP routing protocol.

The relationship between the PE router and the customer edge (CE) router is the truly distinctive aspect of RFC 2547 VPNs. The CE router becomes a peer of the PE router (and not a peer to the other CE routers). The CE router provides the PE router with route information for the private network. The PE router, in turn, must be capable of storing multiple private routing tables—one for each customer connection—along with the usual public Internet forwarding information (see Figure 10.2).

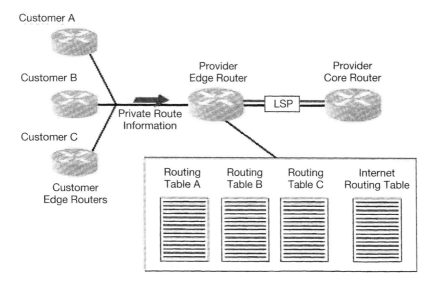

Figure 10.2 The provider edge/customer edge router relationship.

MPLS handles the forwarding between the nodes on a RFC 2547 network (in this respect, Layer 2 and 3 VPN approaches are identical). This MPLS forwarding role is crucial because it means the routers in the core of the network ("P" routers) need not know about the routes connecting the RFC 2547 private network. A RFC 2547 network uses a two-level label stack—the ingress PE router pushes both a Next-Hop BGP header (for the private network) and a Next-Hop Interior Gateway Protocol (IGP) header (for the shared infrastructure) onto the packet. After reaching the egress PE router via one or more MPLS label-switched paths (LSPs), the PE pops the MPLS headers and delivers a normal IP packet to the customer.

What is to be made of the RFC 2547 approach? It has great potential. It takes advantage of the ubiquity of IP networks and, like IP, runs over multiple transport networks. It also has strong automatic route discovery; this is important for dynamic VPNs.

On the other hand, several comparative limitations are also clear. The RFC 2547 approach can be very demanding of provider edge routers. Although not all RFC 2547 deployments will necessarily require anything but a number of static routes, the potential for overburdening the network exists. For 2547-bis to truly scale, the next generation of routers will need to be ubiquitous.

10.3 LAYER 2 MPLS VPNs—A DIFFERENT PHILOSOPHY

A different philosophy underlies Layer 2 MPLS virtual private networks [also known as transparent LAN services or virtual private LAN services (VPLS)]. The goal is the extension, rather than replacement, of existing Layer 2 VPN services. Instead of building a separate, private IP network and running traffic across it, Layer 2 VPNs take existing Layer 2 traffic and sends it through point-to-point tunnels on the MPLS network backbone.

Both Layer 2 and Layer 3 MPLS VPNs rely on MPLS transport through the core. The principal difference lies in how PE-CE router relations are handled. In a Layer 2 MPLS VPN, the PE router is not a peer to the CE router and does not maintain separate routing tables. Rather, it simply maps incoming Layer 2 traffic onto the appropriate point-to-point tunnel. The result is best described as an "overlay" model as opposed to the Layer 3 "peer" model.

Crucial to the Layer 2 VPN model is a method for establishing simple point-to-point tunnels on an MPLS network that can handle various forms of Layer 2 traffic. Today, the industry is standardizing on the Martini drafts (named after Luca Martini of Level 3 Communications) that define point-to-point encapsulation mechanisms for Ethernet, frame relay, ATM, TDM, and PPP/HDLC traffic. Indeed, Martini interoperability between many MPLS vendors was conclusively demonstrated starting in 2002. Still other Internet drafts have built on the Martini draft encapsulations to define frame relay and ATM operations and to define Ethernet transparent LAN services (VPLS).

10.4 WHICH WORKS BETTER WHERE?

The Layer 3 approach, as stated above, is ideally suited to "classic" ISP networks with existing core router deployments. It is a good fit for carriers serving large VPNs with changing locations, making automatic route discovery useful. The Layer 2 approach, on the other hand, is the preferred approach for networks that want to extend and scale legacy Layer 2 VPN deployments, transport-oriented carriers in general, or any situation with few VPN sites and static routes. Many carriers are already providing Layer 2 VPN services (over, say, frame relay or metro Ethernet) and are interested in scaling such services. In that case, the SP does not want a whole new VPN infrastructure, just a way to overlay Layer 2 traffic on MPLS/IP networks. For this task, Layer 2 MPLS VPNs are ideal. See Figure 10.3.

Transport-oriented carriers also should prefer the Layer 2 approach. Again, the main difference with Layer 2 VPNs is at the PE router. Among other things, the Layer 2 approach eliminates the need to peer with CE routers and maintain multiple routing tables. This approach suits carriers that traditionally offer transport services and leave routing to the customer. VPN traffic is carried over an IP/MPLS network, without upgrading to expensive and specialized core routers at the edge. In addition, in a Layer 2 MPLS VPN, reachability is achieved in the data plane through address learning, rather than in the control plane through BGP route exchange.

Finally, where routes are likely to be static and private networks simple, the relative simplicity of the Layer 2 approach is appealing. In a metropolitan TLS scenario, for example, a network usually needs only to interconnect a few sites; a RFC 2547 MPLS VPN may be overkill, from both cost and complexity standpoints.

In the end, as MPLS VPNs are deployed, it is likely that carriers will choose Layer 2 and Layer 3 VPNs for many of the same reasons they decided to deploy

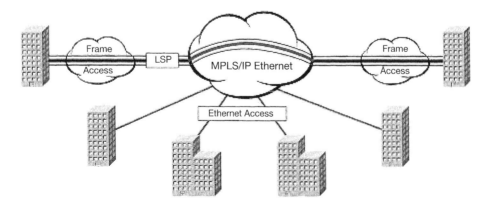

Figure 10.3 Using Layer 2 MPLS VPNs to scale existing Layer 2 VPNs.

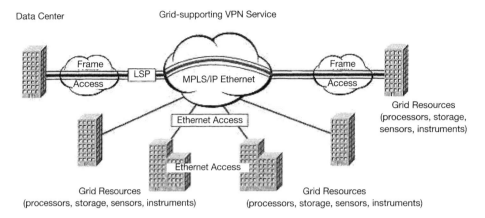

Figure 10.4 Example of a computing grid using MPLS VPNs.

Layer 2 or Layer 3 networks. The question of Layer 2 or Layer 3 deployment is likely to stay with networking for quite some time.

10.5 A GRID COMPUTING APPLICATION

Figure 10.4 depicts an application of the VPN concept to a grid environment. The reader should be able to generalize this to his or her environment.

References

1. 451 Group, *Grids 2004: From Rocket Science To Business Service,* Special Report, 2003, New York.

2. W. Allcock (Ed.), *GWD-R (Recommendation): GridFTP Protocol Spec, GridFTP: Protocol Extensions to FTP for the Grid,* April 2002, Revised April 2003. Copyright © Global Grid Forum (2003). The Global Grid Forum, 9700 South Cass Avenue, Bldg. 221/A142, Lemont, IL, 60439, USA. All Rights Reserved. This document and translations of it may be copied and furnished to others, and derivative works that comment on or otherwise explain it or assist in its implementation may be prepared, copied, published and distributed, in whole or in part, without restriction of any kind, provided that the above copyright notice and this paragraph are included on all such copies and derivative works.

3. I. Foster, C. Kesselman, and S. Tuecke, "The Anatomy of the Grid: Enabling Scalable Virtual Organizations," *International Journal of High Performance Computing Applications, 15* (3), 200–222, 2001.

4. D. Awduche, J. Malcolm, J. Agogbua, M. O'Dell, and J. McManus, "RFC 2702, MPLS Traffic Engineering," IETF, September 1999. Copyright © The Internet Society. All Rights Reserved. This document and translations of it may be copied and furnished to others, and derivative works that comment on or otherwise explain it or assist in its implementation may be prepared, copied, published and distributed, in whole or in part, without restriction of any kind, provided that the above copyright notice and this paragraph are included on all such copies and derivative works.

5. E. Hudson, "Multilink Frame Relay: Expanding the Limits of T1," Tiara Networks, *FRF News,* 4th Quarter 1999.

6. D. Minoli, *A Collection of Potential Network-Based Data Services,* Bellcore/Telcordia Special Report, SR-NPL-000790, 1987, Piscataway, NJ.

7. B. Campbell, "Setting the Standards," *Oracle Magazine,* November/December 2003.

8. Staff, "Briefs," *Oracle Magazine,* November/December 2003.

9. G. Buda, D. Choi, R. F. Graveman, and C. Kubic, "Security Standards for the Global Information Grid," in *Military Communications Conference, 2001. MILCOM 2001. Communications for Network-Centric Operations: Creating Information Force. IEEE.* Booz Allen & Hamilton, Linthicum, MD.

10. R. Buyya, *Economic-Based Distributed Resource Management and Scheduling for Grid Computing,* Ph.D Thesis, Monash University, Melbourne, Australia, April 12, 2002.

11. C. Semeria, "RSVP Signaling Extensions for MPLS Traffic Engineering," White Paper, Juniper Networks, Inc., 2000.

A Networking Approach to Grid Computing. By Daniel Minoli
ISBN 0-471-68756-1 © 2005 John Wiley & Sons, Inc.

12. The Storage Networking Industry Association's (SNIA) IP Storage Forum, "Clearing the Confusion: A Primer on Internet Protocol Storage," SNIA Archives and Promotional Materials. SNIA was incorporated in December 1997 and is a registered 501-C6 nonprofit trade association. 500 Sansome Street, Suite #504, San Francisco, CA 94111.

13. R. J. Ruby, FRF MFR (FRF.16) Interop Event Chairperson, Frame Relay Forum MFR Event, "Giving Frame Relay Users Higher Speed On and Off Ramps to 'The Cloud'," MPLS/Frame Relay Forum. The MPLS and Frame Relay Alliance is an industry-wide association of networking and telecommunication companies focused on advancing the deployment of multivendor, multiservice label-switching networks and associated applications. MPLS & Frame Relay Alliance, 39355 California Street #307, Fremont, CA 94538.

14. K. Czajkowski, S. Fitzgeraldz, I. Foster, and C. Kesselman, "Grid Information Services for Distributed Resource Sharing," in *Proceedings 10th IEEE International Symposium on High- Performance Distributed Computing (HPDC-10)*, IEEE Press, 2001.

15. http://searchwebservices.techtarget.com.

16. S. Hege, and J. E. Refsnes, "Glossary and Tutorials," W3Schools, Web Developers Site On The Net, http://www.w3schools.com.

17. developerWorks staff, "Start Here to Learn about Grid Computing," IBM Developer-Works, August 2003. developerWorks is IBM's resource for developers offering a range of tools, code, and education that enable developers to take full advantage of the IBM tools and technical library. www.ibm.com/developmentworks

18. E. Balusamy, "Web Services Development Made Easy, Learn How to Easily Access Web services Using Oracle JDeveloper 10g," *Oracle Magazine*, November/December 2003.

19. ANSI INCITS, "Fibre Channel Arbitrated Loop (FC-AL-2)," revision 7.0, INCITS Project 1133D, April 1999.

20. TR-20-199X, "Fibre Channel Fabric Loop Attachment (FC-FLA)," revision 2.7, IN-CITS Project 1235-D, August 199.7

21. ANSI INCITS, "Fibre Channel Framing and Signaling (FC-FS)," Rev 1.70, INCITS Project 1331D, Draft Standard, Rev. 1.9, April, 2003.

22. ANSI INCITS, "Fibre Channel Generic Services -3 (FC-GS3)," revision 7.01, INCITS Project 1356-D, November 2000.

23. ANSI INCITS, "Fibre Channel Protocol for SCSI, Second Version," revision 8, IN-CITS Project 1144D, September 2002.

24. ANSI INCITS, "Fibre Channel Switch Fabric -2 (FC-SW2)," revision 5.2, INCITS Project 1305-D, May 2001.

25. W. A. Flanagan, "The Case for Multilink Frame Relay Access, Outgrown Your T1, but a T3 is Too Much? MFR Fills the Gap in Access Speeds, Adds Resiliency to Improve Uptime," *MPLS/Frame Relay Forum Newsletter*, Winter 2000. The MPLS and Frame Relay Alliance is an industry-wide association of networking and telecommunication companies focused on advancing the deployment of multivendor, multiservice label-switching networks and associated applications. MPLS & Frame Relay Alliance, 39355 California Street #307, Fremont, CA 94538.

26. Promotional material from Foundry Networks, 2100 Gold Street, P.O. Box 649100, San Jose, CA 95164-9100.

27. Frank Gens, "IDC Predictions 2004: Top 10 Trends for the IT Industry," IDC Executive Telebriefing, IDC, Boston, Mass. December 4, 2003. (IDC is a subsidiary of IDG.)

28. MPLS and Frame Relay Forum, "FRF.16 Standard." The MPLS and Frame Relay Alliance is an industry-wide association of networking and telecommunication companies focused on advancing the deployment of multivendor, multiservice label-switching networks and associated applications. MPLS & Frame Relay Alliance, 39355 California Street #307, Fremont, CA 94538.

29. Future Software Limited, "MultiProtocol Label Switching White Paper," Chennai, India, 2001.

30. S. Newhouse and J. MacLaren, *Grid Economic Services Architecture (GESA)*, 2003. Copyright © Global Grid Forum (2003). The Global Grid Forum, 9700 South Cass Avenue, Bldg. 221/A142, Lemont, IL, 60439, USA. All Rights Reserved. This document and translations of it may be copied and furnished to others, and derivative works that comment on or otherwise explain it or assist in its implementation may be prepared, copied, published and distributed, in whole or in part, without restriction of any kind, provided that the above copyright notice and this paragraph are included on all such copies and derivative works.

31. G. Fox, M. Pierce, D. Gannon, and M. Thomas, *Overview of Grid Computing Environments*, GFD-I.9, Feb 2003. Copyright © Global Grid Forum (2002). The Global Grid Forum, 9700 South Cass Avenue, Bldg. 221/A142, Lemont, IL, 60439, USA. This document and translations of it may be copied and furnished to others, and derivative works that comment on or otherwise explain it or assist in its implementation may be prepared, copied, published and distributed, in whole or in part, without restriction of any kind, provided that the above copyright notice and this paragraph are included on all such copies and derivative works.

32. OGSI Technology Preview Overview, The Globus Project™, Argonne National Laboratory, USC Information Sciences Institute, 2002. This presentation is licensed for use under the terms of the Globus Toolkit Public License. (See http://www.globus.org/toolkit/download/license.html for the full text of this license.)

33. The Global Grid Forum, 9700 South Cass Avenue, Bldg. 221/A142, Lemont, IL, 60439, USA, http://www.ggf.org.

34. G. Tramacere, "Utility Model," Gartner Research, Gartner, Inc., 56 Top Gallant Road, Stamford, CT 06904, www.gartner.com.

35. Comments by Dr. Ian Foster, Globus Project leader, Argonne National Laboratory and the University of Chicago. Mathematics and Computer Science, Argonne National Laboratory, 9700 Cass Ave, Argonne, IL, 60439, Tel: 630 252-4619, Fax: 630 252-5986, foster@mcs.anl.gov.

36. The Globus Alliance, "The Globus Alliance is a research and development project focused on enabling the application of Grid concepts to scientific and engineering computing," Press Releases, c/o Carl Kesselman, USC/Information Sciences Institute, 4676 Admiralty Way, Suite 1001, Marina del Rey, CA 90292-6695, Tel: 310 822-1511 x338, Fax: 310 823-6714, carl@isi.edu, http://www.globus.org, info@globus.org.

37. GWD-R (draft-ggf-ogsi-gridservice-33) "Globus: A Toolkit-Based Grid Architecture," June 27, 2003.

38. P. Gralla, "What Is Service-Oriented Architecture?" *The Web Services Advisor*, 06 May 2003, http://searchwebservices.techtarget.com.

39. The Global Alliance, "The Globus Toolkit," The Globus Alliance Press Release, c/o Carl Kesselman, USC/Information Sciences Institute, 4676 Admiralty Way, Suite 1001, Marina del Rey, CA 90292-6695, Tel: 310 822-1511 x338, Fax: 310 823-6714, carl@isi.edu, http://www.globus.org, info@globus.org.

40. Foster, I., and Kesselman, C., "Globus: A Toolkit-Based Grid Architecture," in Foster, I. and Kesselman, C. (eds.), *The Grid: Blueprint for a New Computing Infrastructure,* Morgan Kaufmann, 1999, pp. 259–278.

41. The Storage Networking Industry Association's (SNIA), Press Releases, Promotional Material, and Archives. SNIA was incorporated in December 1997 and is a registered 501-C6 non-profit trade association. 500 Sansome Street, Suite #504, San Francisco, CA 94111.

42. R. Housley, W. Ford, W. Polk, and D. Solo, *Internet X.509 Public Key Infrastructure Certificate and CRL Profile,* Request for Comments: 2459, January 1999. Copyright © The Internet Society (1999). This document and translations of it may be copied and furnished to others, and derivative works that comment on or otherwise explain it or assist in its implementation may be prepared, copied, published and distributed, in whole or in part, without restriction of any kind, provided that the above copyright notice and this paragraph are included on all such copies and derivative works.

43. IBM Press Releases. IBM Corporation, 1133 Westchester Avenue, White Plains, New York 10604, www.ibm.com.

44. The Globus Project™, *Introduction to Grid Computing,* Argonne National Laboratory USC Information Sciences Institute, 2002. This material is licensed for use under the terms of the Globus Toolkit Public License. (See http://www.globus.org/toolkit/download/license.html for the full text of this license.)

45. The Globus Project™, *Introduction to Grid Computing,* Argonne National Laboratory, USC Information Sciences Institute. Copyright (c) 2002 University of Chicago and The University of Southern California. All Rights Reserved. This presentation is licensed for use under the terms of the Globus Toolkit Public License. (See http://www.globus.org/toolkit/download/license.html for the full text of this license.)

46. J. Shurtleff, "IP storage: A review of iSCSI, FCIP, and iFCP," *iSCSI Storage/IP network Storage Trend and News,* iSCSI Storage Publications, P.O. Box 7317, Golden, CO, 80304-0100, info@iscsistorage.com, http://www.iscsistorage.com.

47. Sun Microsystems, JavaTM API for XML-Based RPC (JAX-RPC). A Sun Developer Network Site (code in all technical manuals herein, including articles, FAQs, samples) is provided under this License, http://java.sun.com/xml/jaxrpc/docs.html. Sun Microsystems, Inc.; 4150 Network Circle; Santa Clara, CA 95054; Phone: US 1-800-555-9SUN; International 1-650-960-1300.

48. J. Spicer, "In Search Of," *Oracle Magazine,* November/December 2003.

49. J. Joseph, "A Developer's Overview of OGSI and OGSI-Based Grid Computing Get an In-Depth Look at the Open Grid Service Infrastructure," IBM Archives, April 7, 2003.

50. J. Unger, and Matt Haynos, "A Visual Tour Of Open Grid Services Architecture: Examine The Component Structure of OGSA," IBM Achives, August 2003, Updated October 2003.

51. D. Booth, H. Haas, F. McCabe, E. Newcomer, M. Champion, C. Ferris, and D. Orchard, *Web Services Architecture,* W3C, Working Draft, 2003.

52. S. Tuecke, K. Czajkowski, I. Foster, J. Frey, S. Graham, and C. Kesselman, *Grid Service Specification, 2002.* Copyright © Global Grid Forum (2002). The Global Grid Fo-

rum, 9700 South Cass Avenue, Bldg. 221/A142, Lemont, IL, 60439, USA. This document and translations of it may be copied and furnished to others, and derivative works that comment on or otherwise explain it or assist in its implementation may be prepared, copied, published and distributed, in whole or in part, without restriction of any kind, provided that the above copyright notice and this paragraph are included on all such copies and derivative works.

53. F. Cabrera, G. Copeland, B. Cox, T. Freund, J. Klein, T. Storey, and S. Thatte, *Web Services Transaction (WS-Transaction)*, Microsoft Archives, 2002.

54. K. Czajkowski, A. Dan, J. Rofrano, S. Tuecke, and M. Xu, *Agreement-Based Grid Service Management (WS-Agreement)*, Global Grid Forum, Draft, 2003. The Global Grid Forum, 9700 South Cass Avenue, Bldg. 221/A142, Lemont, IL, 60439, USA. Copyright © Global Grid Forum. This document and translations of it may be copied and furnished to others, and derivative works that comment on or otherwise explain it or assist in its implementation may be prepared, copied, published and distributed, in whole or in part, without restriction of any kind, provided that the above copyright notice and this paragraph are included on all such copies and derivative works.

55. I. Foster, D. Gannon, H. Kishimoto, and J. Von Reich, *Open Grid Services Architecture Use Cases*, Global Grid Forum OGSA-WG, Draft draft-ggf-ogsa-usecase-2, 2003. Copyright © Global Grid Forum. The Global Grid Forum, 9700 South Cass Avenue, Bldg. 221/A142, Lemont, IL, 60439, USA. This document and translations of it may be copied and furnished to others, and derivative works that comment on or otherwise explain it or assist in its implementation may be prepared, copied, published and distributed, in whole or in part, without restriction of any kind, provided that the above copyright notice and this paragraph are included on all such copies and derivative works.

56. V. A. Vyssotsky, F. J. Corbató, and R. M. Graham, in "Structure of the Multics Supervisor," *Fall Joint Computer Conference, AFIPS Conference Proceedings, vol. 27,* 203 (1965).

57. I. Foster, C. Kesselman, and S. Tuecke, "The Anatomy of the Grid: Enabling Scalable Virtual Organizations," *International Journal of Supercomputer Applications, 15* (3), 200–222. 2001.

58. J. C. R. Licklider and R. W. Taylor, "The Computer as a Communication Device," *Sci. Technol.,* 21–31, April (1968).

59. IBM, Microsoft, RSA Security, and VeriSign, "Web Services Secure Conversation Language (WS-SecureConversation) Version 1.0," 2002.

60. IBM, Microsoft, RSA Security, and VeriSign, "Web Services Trust Language (WS-Trust)," 2002.

61. P. V. Mockapetris and K. Dunlap, "Development of the Domain Name System," in *SIGCOMM,* 1988, ACM, 123-133.

62. "Kansai EPC Chooses IBM for Grid Computing Development, RBC Insurance, Royal Dutch Shell and Kansai Electric Power Newest IBM Grid Customers," IBM Press Release, 28 Apr 2003. IBM Corporation, 1133 Westchester Avenue, White Plains, New York 10604, www.ibm.com.

63. A. Bednarz and D. Dubie, "How to: How to Get to Utility Computing," *Network World,* December 1, 2000.

64. "IBM and Partners Bring Grid Computing to Linux on the Mainframe," IBM, Press Release, March 31, 2003, N.Y., IBM Corporation, 1133 Westchester Avenue, White Plains, New York 10604, www.ibm.com.

65. The Globus Alliance (a partnership of Argonne National Laboratory's Mathematics and Computer Science Division, the University of Southern California's Information Sciences Institute, the University of Chicago's Distributed Systems Laboratory, the University of Edinburgh in Scotland, and the Swedish Center for Parallel Computers.) This material is licensed for use under the terms of the Globus Toolkit Public License. (See http://www.globus.org/toolkit/download/license.html for the full text of this license.)

66. D. Minoli, *Internet and Intranet Engineering,* McGraw-Hill, 1997.

67. D. Minoli and E. Minoli, *Web Commerce Handbook,* McGraw-Hill, 1998.

68. D. Minoli and A. Schmidt, *Internet Architectures,* Wiley, 1999.

69. I. Foster, D. Gannon, and H. Kishimoto (Eds.), *The Open Grid Services Architecture, GWD-R* (draft-ggf-ogsa-ogsa-011), September 23, 2003. Copyright © Global Grid Forum (2002, 2003). All Rights Reserved. The Global Grid Forum, 9700 South Cass Avenue, Bldg. 221/A142, Lemont, IL, 60439, USA. This document and translations of it may be copied and furnished to others, and derivative works that comment on or otherwise explain it or assist in its implementation may be prepared, copied, published and distributed, in whole or in part, without restriction of any kind, provided that the above copyright notice and this paragraph are included on all such copies and derivative works.

70. J. Linn, "Generic Security Service Application Program Interface (GSS-API)," Version 2, Update 1, (RFC2743), January 2000. Copyright © The Internet Society (2000). This document and translations of it may be copied and furnished to others, and derivative works that comment on or otherwise explain it or assist in its implementation may be prepared, copied, published and distributed, in whole or in part, without restriction of any kind, provided that the above copyright notice and this paragraph are included on all such copies and derivative works.

71. L. J. Zhang, Q. Zhoum, and J.-Y. Chung, "Developing Grid Computing Applications, Part 2, Introduction to a Grid Architecture and Toolkit for Building Grid Solutions," December 3, 2002, http://www-106.ibm.com/developerworks/views/grid/articles.jsp.

72. M. Humphrey and M. Thompson, *Security Implications of Typical Grid Computing Usage Scenarios,* GFD-I.12, October 2000. Copyright © Global Grid Forum (2003). The Global Grid Forum, 9700 South Cass Avenue, Bldg. 221/A142, Lemont, IL, 60439, USA. This document and translations of it may be copied and furnished to others, and derivative works that comment on or otherwise explain it or assist in its implementation may be prepared, copied, published and distributed, in whole or in part, without restriction of any kind, provided that the above copyright notice and this paragraph are included on all such copies and derivative works.

73. M. C. Brown, "Grid Computing—Moving to a Standardized Platform," August 2003, IBM archives, IBM Corporation, 1133 Westchester Avenue, White Plains, New York 10604, www.ibm.com.

74. D. Minoli, P. Johnson, and E. Minoli, *SONET-Based Metro Area Networks,* McGraw-Hill, 2002.

75. M. Lehmann, "Who Needs Web Services Transactions?" *Oracle Magazine,* http://otn.oracle.com/oraclemagazine, November/December 2003.

76. A. Miller, M. Jefferson, and J. Rogers, "Global Information Grid Architecture," Mitre White Papers, MITRE Corporation, 202 Burlington Road, Bedford, MA 01730-1420, (781) 271-2000, http://www.mitre.org.

77. D. Minoli, *Telecommunications Technology Handbook,* 2nd ed., Artech House, 2003.

78. C. Monia, R. Mullendore, F. Travostino, W. Jeong, and M. Edwards, "iFCP—A Protocol for Internet Fibre Channel Networking," IP Storage Working Group December 2002, Internet Draft, draft-ietf-ips-ifcp-14.txt. Copyright © The Internet Society, December 2002. All Rights Reserved. This document and translations of it may be copied and furnished to others, and derivative works that comment on or otherwise explain it or assist in its implementation may be prepared, copied, published and distributed, in whole or in part, without restriction of any kind, provided that the above copyright notice and this paragraph are included on all such copies and derivative works.

79. B. Davie, J. Lawrence, K. McCloghrie, Y. Rekhter, E. Rosen, G. Swallow, and P. Doolan, "MPLS using LDP and ATM VC Switching," RFC 3035, The Internet Society, January 2001.

80. L. Andersson, P. Doolan, N. Feldman, A. Fredette, and B. Thomas, "LDP Specification," RFC 3036, The Internet Society, January 2001.

81. D. Awduche, J. Malcolm, J. Agogbua, M. O'Dell, and J. McManus, "Requirements for Traffic Engineering Over MPLS," RFC 2702, The Internet Society, September 1999.

82. V. Lombardi, "Smarter Content Publishing," *Digital Web Magazine,* August 6 2002.

83. B. Boiko, *Content Management Bible,* Wiley, 2001.

84. S. Tuecke, K. Czajkowski, I. Foster, J. Frey, S. Graham, C. Kesselman, T. Maquire, T. Sandholm, D. Snelling, and P. Vanderbilt, (Eds.), *Open Grid Services Infrastructure (OGSI),* Version 1.0, Global Grid Forum, June 27, 2003. Copyright © Global Grid Forum (2003). All Rights Reserved. The Global Grid Forum, 9700 South Cass Avenue, Bldg. 221/A142, Lemont, IL, 60439, USA. This document and translations of it may be copied and furnished to others, and derivative works that comment on or otherwise explain it or assist in its implementation may be prepared, copied, published and distributed, in whole or in part, without restriction of any kind, provided that the above copyright notice and this paragraph are included on all such copies and derivative works.

85. Oracle Corporation, "Web Services Overview," Programmatic Access to Web Sites and Applications, June 2002. Oracle Corporation, World Headquarters, 500 Oracle Parkway, Redwood Shores, CA 94065.

86. P. J. Gill, "Getting Down to Business with Enterprise Grid Computing," *Oracle Magazine,* November/December 2003.

87. C. Del Prete, "On Demand by IBM," *IDC Viewpoint,* November 2003, IDC Corporate Headquarters, 5 Speen Street, Framingham, MA 01701.

88. R. Pulley and P. Christensen, "A Comparison Of MPLS Traffic Engineering Initiatives," A White Paper by NetPlane Systems, Inc., Southboro Office Park,120 Turnpike Road, Southborough, MA 01772.

89. RFC 2246, The TLS Protocol Version 1.0 January 1999. Copyright © The Internet Society, 1999. All Rights Reserved. This document and translations of it may be copied and furnished to others, and derivative works that comment on or otherwise explain it or assist in its implementation may be prepared, copied, published and distributed, in whole or in part, without restriction of any kind, provided that the above copyright notice and this paragraph are included on all such copies and derivative works.

90. developerWorks staff, "Start Here to learn about Grid Computing," IBM Corporation, 1133 Westchester Avenue, White Plains, New York 10604, August 2003.

91. T. Myer, "Grid Computing: Conceptual Flyover for Developers," IBM Corporation, 1133 Westchester Avenue, White Plains, New York 10604, May 2003.

92. L.-J. Zhang, J.-Y. Chung, and Q. Zhou, "Developing Grid Computing Applications, Part 1: Introduction of a Grid Architecture and Toolkit for Building Grid Solutions," Updated November 20, 2002, IBM Corporation, 1133 Westchester Avenue, White Plains, New York 10604, October 1, 2002.

93. F. Berman, G. Fox, and A. J. Hey (Eds.), *Grid Computing: Making the Global Infrastructure a Reality,* Wiley, 2003.

94. M. Chetty and R. Buyya, "Weaving Computational Grids: How Analogous Are They With Electrical Grids?" *IEEE Computing in Science and Engineering,* July/August 2002.

95. D. Minoli and A. Schmidt, *Client/Server Applications on ATM Networks,* Prentice-Hall/Manning, 1997.

96. I. Foster and C. Kesselman (Eds.), *The Grid: Blueprint for a Future Computing Infrastructure,* Morgan Kaufmann Publishers, 1999.

97. M. Waldrop, "Hook Enough Computers Together and What Do You Get? A New Kind of Utility that Offers Supercomputer Processing on Tap," *MIT Enterprise Technology Review,* May 2002. Technology Review, One Main Street, 7th Floor, Cambridge, MA, 02142, Tel: 617-475-8000, Fax: 617-475-8042.

98. Sun Networks Press Release, "Network Computing Made More Secure, Less Complex With New Reference Architectures, Sun Infrastructure Solution," September 17, 2003. Sun Microsystems, Inc., 4150 Network Circle, Santa Clara, CA 95054, Phone: 1-800-555-9SUN or 1-650-960-1300.

99. Sun Microsystems, Inc., 4150 Network Circle, Santa Clara, CA 95054, Phone: 1-800-555-9SUN or 1-650-960-1300.

100. A. Bednarz and D. Dubie, "How To: How to Get to Utility Computing," *Network World,* December 1, 2003.

101. Grid Computing Info Centre (GRID Infoware), "Grid Computing, Answers to the Enterprise Architect Magazine Query," *Enterprise Architect Magazine,* http://www.cs.mu.oz.au/~raj/GridInfoware/gridfaq.html.

102. M. McCommon, "Letter from the Grid Computing Editor: Welcome to the New developerWorks Grid Computing Resource!" *IBM Grid Computing Resource,* April 7, 2003.

103. I. Foster, "What Is the Grid? A Three Point Checklist," Argonne National Laboratory and University of Chicago, July 20, 2002, Argonne National Laboratory, 9700 Cass Ave, Argonne, IL, 60439, Tel: 630 252-4619, Fax: 630 252-5986, foster@mcs.anl.gov.

104. I. Foster, C. Kesselman, and S. Tuecke, "The Anatomy of the Grid: Enabling Scalable Virtual Organizations," *International Journal of High Performance Computer Applications, 15*(3), 200 (2001).

105. R. Buyya, "Delivering Grid Services as ICT Commodities: Challenges and Opportunities," Grid and Distributed Systems (GRIDS) Laboratory, Dept. of Computer Science and Software Engineering, The University of Melbourne, Melbourne, Australia; also presented at IST (Information Society Technologies) 2003, Milan, Italy.

106. P. Padala (Ed.), *A Survey of Grid File Systems,* GFS-WG (Grid File Systems Working Group), Global Grid Forum, September 19, 2003. The Global Grid Forum, 9700 South Cass Avenue, Bldg. 221/A142, Lemont, IL, 60439.

107. I. Foster and C. Kesselman. "Globus: A Metacomputing Infrastructure Toolkit." *The International Journal of Supercomputer Applications and High Performance Computing, 11*(2):115–128, 1997.

108. A. S. Grimshaw, W. A. Wulf, and the Legion Team, "The Legion Vision of a World-wide Virtual Computer," *Communications of the ACM, 40*(1):39–45, January 1997.

109. V. Huber, "UNICORE: A Grid Computing Environment for Distributed and Parallel Computing," *Lecture Notes in Computer Science, 2127,* 258–266, 2001.

110. http://www.gridcomputing.com/.

111. R. B. Cohen and E. Feser, *Grid Computing, Projected Impact in North Carolina's Economy and Broadband Use Through 2010,* Rural Internet Access Authority, September 2003.

112. P. McDougall, "Offshore 'Hiccups In An Irreversible Trend," *InformationWeek,* CMP Media Publishers, Manhasset, NY, Dec. 1, 2003.

113. I. Foster, "The Grid: A New Infrastructure for 21st Century Science," *Physics Today, 55* (2), 42–47, 2002.

114. I. Foster, C. Kesselman, J. Nick, and S. Tuecke, *The Physiology of the Grid: An Open Grid Services Architecture for Distributed Systems Integration,* Globus Project, 2002.

115. A. Chervenak, I. Foster, C. Kesselman, C. Salisbury, and S. Tuecke, "The Data Grid: Towards an Architecture for the Distributed Management and Analysis of Large Scientific Data Sets," *Journal of Network and Computer Applications: Special Issue on Network-Based Storage Services, 23,* 3, 187–200, 2000.

116. C. Baru, R. Moore, A. Rajasekar, and M. Wan,. "The SDSC Storage Resource broker," in *Proceedings of IBM Centers for Advanced Studies Conference,* IBM, 1998.

117. O. Tatebe, Y. Morita, S. Matsuoka, N. Soda, and S. Sekiguchi, "Grid datafarm Architecture for Petascale Data Intensive Computing," H. E. Bal, K.-P. Lohr, and A. Reinefeld (Eds.), *Proceedings of the Second IEEE/ACM International Symposium on Cluster Computing and the Grid (CCGrid2002),* pp. 102–110, Berlin, Germany, 2002, IEEE Computer Society.

118. T. Hawk, Remarks at Grid Computing Planet Conference and Expo, San Jose, 17 June 2002. Also quoted in Globus Alliance, Press Release, July 1, 2003.

119. L.-J. Zhang, J.-Y. Chung, and Q. Zhou, "Developing Grid Computing Applications, Part 1: Introduction of a Grid Architecture and Toolkit for Building Grid Solutions," October 1, 2002, Updated November 20, 2002. IBM Corporation, 1133 Westchester Avenue, White Plains, New York 10604, www.ibm.com (IBM's developerworks/ views/ grid/articles archives).

120. D. Minoli, *Analyzing Outsourcing, Reengineering Information and Communication Systems,* McGraw-Hill, 1994.

121. I. Foster and A. Iamnitchi, "On Death, Taxes, and the Convergence of Peer-to-Peer and Grid Computing," presented at 2nd International Workshop on Peer-to-Peer Systems (IPTPS'03), February 2003, Berkeley, CA.

122. C. Catlett, "The Rise of Third-Generation Grids," *Grid Connections,* Volume *1,* 3, Fall 2003, The Global Grid Forum, 9700 South Cass Avenue, Bldg. 221/A142, Lemont, IL, 60439.

123. M. C. Brown, "Grid Computing—Moving to a Standardized Platform," August 2003, IBM's Developerworks Grid Library, IBM Corporation, 1133 Westchester Avenue, White Plains, New York 10604, www.ibm.com.

124. Globus Alliance, Press Release, "Globus Project Releases First Major Software Implementation Of Grid Services Architecture, Globus Toolkit 3.0 Reflects Community's Convergence on New, Open Grid Services Infrastructure Specification," July 1, 2003.

C/o Carl Kesselman, USC/Information Sciences Institute, 4676 Admiralty Way, Suite 1001, Marina del Rey, CA 90292-6695, Tel: 310 822-1511 x338, Fax: 310 823-6714, carl@isi.edu, http://www.globus.org, info@globus.org.

125. B. Jacob, "Grid Computing: What are the Key Components? Taking Advantage of Grid Computing for Application Enablement," June 2003, IBM's Developerworks Grid Library, IBM Corporation, 1133 Westchester Avenue, White Plains, New York 10604, www.ibm.com.

126. M. Lurie, "The Federation—Database Interoperability, the Adventure Continues (Part 1)," International Business Machines Corporation. July 2003, IBM's Developerworks Grid Library, IBM Corporation, 1133 Westchester Avenue, White Plains, New York 10604, www.ibm.com.

127. D. Minoli, *Broadband Network Design and Analysis,* Artech House, 1993.

128. L.-J. Zhang, J.-Y. Chung, and Q. Zhou, "Developing Grid Computing Applications, Part 1: Introduction of a Grid Architecture and Toolkit for Building Grid Solutions," October 1, 2002, Updated November 20, 2002, IBM's Developerworks Grid Library, IBM Corporation, 1133 Westchester Avenue, White Plains, New York 10604, www.ibm.com.

129. Globus Alliance, Press Releases, c/o Carl Kesselman, USC/Information Sciences Institute, 4676 Admiralty Way, Suite 1001, Marina del Rey, CA 90292-6695, Tel: 310 822-1511 x338, Fax: 310 823-6714, carl@isi.edu, http://www.globus.org, info@globus.org.

130. M. Haney, "Grid Computing: Making Inroads Into Financial Services," IBM's Developerworks Grid Library, 24 April 2003, Issue No: Volume 4, Number 5, IBM Corporation, 1133 Westchester Avenue, White Plains, New York 10604, www.ibm.com.

131. R. Buyya, "Frequently Asked Questions, Grid Computing Info Centre," *GridComputing Magazine.* www.gridcomputing.com

132. G. von Laszewski, I. Foster, and J. Gawor, "CoG Kits: A Bridge Between Commodity Distributed Computing and High Performance Grids," Argonne National Laboratory Archives, Argonne National Laboratory, 9700 South Cass Avenue, Argonne, IL 60439.

133. http://www.ietf.org/internet-drafts/draft-ietf-mpls-recovery-frmwrk-03.txt.

134. S. Bradner, "Key words for use in RFCs to Indicate Requirement Levels," Internet Engineering Task Force, RFC 2119, March 1997. Internet Engineering Task Force, a division of the Internet Society, Internet Society, 1775 Wiehle Ave. Suite 102, Reston, VA 20190.

135. R. Merritt, "DARPA to Overhaul Supercomputing Benchmarks by 2006," *EE Times,* November 14, 2003.

136. E. Rosen, A. Viswanathan, and R. Callon, "RFC 3031: Multiprotocol Label Switching Architecture," IETF January 2001. Copyright © The Internet Society. All Rights Reserved. IETF is a division of the Internet Society, Internet Society, 1775 Wiehle Ave. Suite 102, Reston, VA 20190. This document and translations of it may be copied and furnished to others, and derivative works that comment on or otherwise explain it or assist in its implementation may be prepared, copied, published and distributed, in whole or in part, without restriction of any kind, provided that the above copyright notice and this paragraph are included on all such copies and derivative works.

137. E. Rosen, D. Tappan, G. Fedorkow, Y. Rekhter, D. Farinacci, T. Li, and A. Conta, "MPLS Label Stack Encoding," RFC 3032, January 2001. Copyright © The Internet Society. All Rights Reserved. IETF is a division of the Internet Society, Internet Society, 1775 Wiehle Ave. Suite 102, Reston, VA 20190. This document and translations of

it may be copied and furnished to others, and derivative works that comment on or otherwise explain it or assist in its implementation may be prepared, copied, published and distributed, in whole or in part, without restriction of any kind, provided that the above copyright notice and this paragraph are included on all such copies and derivative works.

138. B. Braden et al., "Resource ReSerVation Protocol (RSVP)—Version 1 Functional Specification," RFC-2205, September 1997. Copyright © The Internet Society. All Rights Reserved. IETF is division of the Internet Society, Internet Society, 1775 Wiehle Ave. Suite 102, Reston, VA 20190. This document and translations of it may be copied and furnished to others, and derivative works that comment on or otherwise explain it or assist in its implementation may be prepared, copied, published and distributed, in whole or in part, without restriction of any kind, provided that the above copyright notice and this paragraph are included on all such copies and derivative works.

139. D. O. Awduche et al., "Extensions to RSVP for LSP Tunnels," work in progress, draft-ietf-mpls-rsvp-lsp-tunnel-08.txt, February 2001. Copyright © The Internet Society. All Rights Reserved. IETF is a division of the Internet Society, Internet Society, 1775 Wiehle Ave. Suite 102, Reston, VA 20190 This document and translations of it may be copied and furnished to others, and derivative works that comment on or otherwise explain it or assist in its implementation may be prepared, copied, published and distributed, in whole or in part, without restriction of any kind, provided that the above copyright notice and this paragraph are included on all such copies and derivative works.

140. The Storage Networking Industry Association (SNIA), "SNIA Shared Storage Model," SNIA Archives and Promotional Materials, 500 Sansome Street, Suite #504, San Francisco, CA, 94111.

141. Staff, "HP Putting Grid In," *Grid Connections, 1,* 3, Fall 2000.

142. E. Hudson, "Multilink Frame Relay: Expanding the Limits of T1," *Tiara Networks, MPLS and Frame Relay Alliance News,* 4th Quarter 1999. MPLS & Frame Relay Alliance, 39355 California Street #307, Fremont, CA 94538.

143. T. Myer, "Grid Computing: Conceptual Flyover for Developers," May 2003, IBM's Developerworks Grid Library, IBM Corporation, 1133 Westchester Avenue, White Plains, New York 10604, www.ibm.com.

144. SNIA IP Storage Forum White Paper, "Internet Fibre Channel Protocol (iFCP)—A Technical Overview," SNIA Archives and Promotional Materials, 500 Sansome Street, Suite #504, San Francisco, CA, 94111.

145. DoD Chief Information Officer (CIO) Guidance and Policy Memorandum No. 8-8001, "Global Information Grid," March 31, 2000.

146. DoD Chief Information Officer (CIO), Guidance and Policy Memorandum No. 6-8510, "Department of Defense Global Information Grid Information Assurance," June 16, 2000.

147. V. Berstis, "Fundamentals of Grid Computing," Redbooks Paper, 2002, IBM Corporation, 1133 Westchester Avenue, White Plains, New York 10604, (ibm.com/redbooks).

148. S. Graham, S. Simeonov, T. Boubez, G. Daniels, D. Davis, Y. Nakamura, and R. Neyama, *Building Web Services with Java: Making Sense of XML, SOAP, WSDL, and UDDI,* Sams, 2001.

149. Robin C., *The Cover Pages,* Web Services Inspection Language (WSIL). *The Cover Pages* is a comprehensive Web-accessible reference collection supporting the SGML/XML family of (meta) markup language standards and their application. The

principal objective in this public access knowledgebase is to promote and enable the use of open, interoperable, standards-based solutions that protect digital information and enhance the integrity of communication. A secondary objective of *The Cover Pages* is to provide reference material on enabling technologies compatible with descriptive markup language standards and applications: object modeling, semantic nets, ontologies, authority lists, document production systems, and conceptual modeling. xml.coverpages.org, OASIS, Tel. 978-665-5115.

150. "Web Services Description Language (WSDL) Version 1.2," Published W3C Working Draft, World Wide Web Consortium. www.w3.org, W3C at MIT/CSAIL, Massachusetts Institute of Technology (MIT), Computer Science and Artifical Intelligence Laborator (CSAIL), 32 Vassar Street, Room 32-G515, Cambridge, MA 02139, USA. Telephone: 1-617-253-2613; fax: 1-617-258-5999. Site manager: Alan Kotok; administration: Susan Westhaver.

151. "Welcome to WSIF: Web Services Invocation Framework," The Apache Software Fundation, The Apache Software Foundation (www.apache.org) provides support for the Apache community of open-source software projects. The Apache projects are characterized by a collaborative, consensus-based development process, an open and pragmatic software license, and a desire to create high-quality software that leads the way in its field.

152. "*Web Services Inspection Language (WS-Inspection) 1.0*," November 2001. IBM Corporation, 1133 Westchester Avenue, White Plains, New York 10604.

153. T. Appnel, "An Introduction to WSIL," O'Reilly OnJava.com, October 16, 2002.

154. T. Wu and A. Walden, "MPLS VPNs: Layer 2 or Layer 3? Understanding the Choice," River Stone Networks White Paper, Riverstone Networks, 5200 Great America Parkway, Santa Clara, CA 95054, Tel. 408-878-6400, www.riverstonenet.com.

155. B. C. Neuman and T. Ts'o, "Kerberos: An Authentication Service for Computer Networks," *IEEE Communications Magazine, 32*(9):33–38, September 1994.

156. D. Abramson, J. Giddy, and L. Kotler, "High-Performance Parametric Modeling with Nimrod-G: Killer Application for the Global Grid?" in *Proceedings of International Parallel and Distributed Processing Symposium,* IEEE Computer Society Press, 2000.

157. R. Buyya, K. Branson, J. Giddy, and D. Abramson, "The Virtual Laboratory: A Toolset to Enable Distributed Molecular Modelling for Drug Design on the World-Wide Grid," *Journal of Concurrency and Computation: Practice and Experience, 25,* 2002; also available as a technical report from Monash University, Melbourne, Australia.

158. A. Oram (Ed.), *Peer-to-Peer: Harnessing the Power of Disruptive Technologies,* O'Reilly Press, Sebastopol, Calif., 2001.

159. I. Foster, C. Kesselman, and S. Tuecke, "The Anatomy of the Grid: Enabling Scalable Virtual Organizations," *International Journal of Supercomputer Applications, 15,* 3, 2001.

160. R. Buyya et al., "Economic Models for Management of Resources in Peer-to-Peer and Grid Computing," in *Proceedings SPIE International Conference on Commercial Applications for High-Performance Computing,* SPIE, Bellingham, WA, 2001.

161. M. Smetanikov, "HP Virtualization a Step Toward Planetary Network," *Web Host Industry Review* (theWHIR.com) April 15, 2003. Web Host Industry Review, Inc. 552 Church Street, Suite 89, Toronto, Ontario, Canada M4Y 2E3, (phone) 416-925-7264, (fax) 416-925-9421.

162. Hewlett-Packard Company, "HP Virtualization: Computing Without Boundaries or Constraints, Enabling an Adaptive Enterprise," HP white paper, 2002, Hewlett-Packard

Company, 3000 Hanover Street, Palo Alto, CA 94304-1185, Phone: (650) 857-1501, Fax: (650) 857-5518, www.hp.com.

163. Hewlett-Packard Company, "Virtual IT: Smarter, Faster, More Flexible with HP's UDC, mpulse Staff," October 2003. Hewlett-Packard Company, 3000 Hanover Street, Palo Alto, CA 94304-1185 USA, Phone: (650) 857-1501, Fax: (650) 857-5518, www.hp.com.

164. P. Thibodeau, "Rivals See Cooperation As Key To Grid Computing Success: IBM, Sun And HP All Want Grid Computing Standards," *ComputerWorld,* July 21, 2003.

165. Nucleus Research Inc., "Grid Computing: More Questions Than Answers," Research Note D131, 2003. Nucleus Research Inc., 36 Washington Street, Wellesley MA 02481, Phone: +1-781-416-2900, Fax: +1-781-416-5252, www.NucleusResearch.com.

166. E. Scannell and T. Sullivan, "Utility on Tap for 2004, Sun, IBM, HP to Build Out Utility Computing Wares," *InfoWorld,* January 05, 2004, InfoWorld Media Group, 501 Second Street, San Francisco, CA, 94107.

167. Grid Forum, now Global Grid Forum, *Grid Security Infrastructure (GSI),* The Global Grid Forum, 9700 South Cass Avenue, Bldg. 221/A142, Lemont, IL, 60439, USA. All Rights Reserved. This document and translations of it may be copied and furnished to others, and derivative works that comment on or otherwise explain it or assist in its implementation may be prepared, copied, published and distributed, in whole or in part, without restriction of any kind, provided that the above copyright notice and this paragraph are included on all such copies and derivative works.

168. Intel, Promotional Materials, Intel, 2200 Mission College Blvd., Santa Clara, CA 95052.

169. W. Fellows, "IBM'S Grid Computing Push Continues," *Gridtoday, Daily News & Information for for the Global Grid Community, 2,* 40, October 6, 2003. Published by Tabor Communications Inc, 8445 Camino Santa Fe, San Diego, California 92121, (858) 625-0070.

170. M. Otey, "Grading Grid Computing," *SQL Server Magazine,* January 2004. Published by Windows & .Net Magazine Network, a Division of Penton Media Inc., 221 E. 29th St., Loveland, CO 80538.

171. Yankee Group, *Enterprise Computing and Networking Report on Performance Management Road Map for Utility Computing,* February 2004. Yankee Group, 31 St. James Avenue Boston MA 02116, (617) 956-5000.

172. Yankee Group, *Enterprise Computing and Networking Report on Utility Computing in Next-Gen IT Architectures,* August 2003, Yankee Group, 31 St. James Avenue, Boston MA 02116, (617) 956-5000.

173. "Grid Computing using .NET and WSRF.NET Tutorial," GGF11, Honolulu, June 6, 2004

174. OASIS, Post Office Box 455, Billerica, MA 01821, USA, +1 978 667 5115 Voice, +1 978 667 5114 Fax, http://www.oasis-open.org.

175. G. E. Moore, "Cramming More Components Onto Integrated Circuits, With Unit Cost Falling as the Number of Components Per Circuit Rises, by 1975 Economics May Dictate Squeezing as Many as 65,000 Components on a Single Silicon Chip," *Electronics, 38,* 8, April 19, 1965.

176. Momentum Software Promotional Material, 4515 Seton Center Parkway, Suite 175, Austin, TX 78759.

177. Morethan IP Promotional Materials, MorethanIP GmbH, An der Steinernen Bruecke 1, D-85757 Karlsfeld, Germany.

Glossary

.NET	Microsoft's Internet and Web strategy. .NET is an Internet- and Web-based infrastructure that delivers software as Web services and is a framework for universal services. It is a server-centric computing model.
1G Grids	First-generation grids that involved local "Metacomputers" with basic services such as distributed file systems and site-wide single sign-on, upon which early adopter developers created distributed applications with proprietary communications protocols.
2G Grids	Second-generation grids that began with projects in which underlying software services and communications protocols are used as a basis for developing distributed applications and services. 2G grids offered basic building blocks, but deployment involved significant customization and filling-in lacunae. Examples include Condor, I-WAY (the origin of Globus), and Legion (the origin of Avaki).
3G Grids	Third-generation grids in which standards define grids in a consistent way; this enables grid systems to become easily built "off-the-shelf" systems. Grid architectures are defined by the Global Grid Forum.
Autonomic management	Intelligent self-regulation of grid resources.
Binding	A concrete protocol and data format specification for a particular port type. In WSDL, the binding identifies a method for relating a service interface to a service inplementation. It accomplishes this by indentifying a transport binding, an invocation style, a SOAP action, and a message encoding scheme. The binding is then used by a collection of operations (port type) [176]
Broker function	A mechanism that identifies appropriate and available resources that can/should be used within the grid, based on the application and application-related parameters provided by the user of the application. The broker functionality provides information about the

A Networking Approach to Grid Computing. By Daniel Minoli
ISBN 0-471-68756-1 © 2005 John Wiley & Sons, Inc.

available resources on the grid and the working status of these resources.

Clusters

Aggregating of processors in parallel-based configurations, typically in local environment (within a data center); all nodes work cooperatively as a single unified resource. Resource allocation is performed by a centralized resource manager and scheduling system. A cluster is comprised of multiple interconnected, independent nodes that cooperatively work together as a single unified resource; unlike grids, cluster resources are typically owned by a single organization.

Collaborative engineering

Applications that entail high-bandwidth access to shared virtual spaces, utilizing interactive manipulation of shared data sets and management of complex simulations, in order to support collaborative design of high-end systems.

Computational grid

This grid is used to allocate resources specifically for computing power. In this situation, most of the processors are high-performance servers. (Processors are sometimes called nodes, resources, members, donors, clients, hosts, engines, or machines.)

Data grid

This grid is used for housing and providing access to data across multiple organizations. Users are not focused on where this data is located as long as they have access to the data.

Data management function

A mechanism for reliable movement of files and data to various nodes within the grid.

Developer tools, grid tools

Tools for developers of grid-enabled applications focused on file transfer, communications, and environment control; they range from utilities to APIs.

Directory services grid tools

Applications and systems on a grid must be capable of discovering what services are available to them; this is done via a directory service. Typically based on the Lightweight Directory Access Protocol (LDAP).

EMC/ VMWare's VMware

Virtualization technology that aims at lowering the cost of "Intel server farms." Converts the workloads of all of a specified set of servers to run as a single hardware pool without inhibiting any application.

Global Grid Forum (GGF)

An industry advocacy group; a forum for exchanging information and defining standards relating to distributed computing and grid technologies. GGF supports community-driven processes for developing and documenting new standards for grid computing. GGF members include Cisco Systems, Hewlett-Packard, IBM,

Microsoft, Qwest Communications, Silicon Graphics, Sun Microsystems, Oracle, Level(3), and BellSouth, among 46 participants at press time.

Globus Toolkit   The de facto standard for several important connectivity, resource, and collective protocols. The tool kit, a "middleware plus" capability, addresses issues of security, information discovery, resource management, data management, communication, fault detection, and portability

Grid computing   (Virtualized) distributed computing environment that enables the dynamic "runtime" selection, sharing, and aggregation of (geographically) distributed autonomous (autonomic) resources based on the availability, capability, performance, and cost of these computing resources, and, simultaneously, also based on an organization's specific baseline and/or burst-processing requirements.

Enables organizations to transparently integrate, streamline, and share dispersed, heterogeneous pools of hosts, servers, storage systems, data, and networks into one synergistic system, in order to deliver agreed-upon service at specified levels of application efficiency and processing performance.

An approach to distributed computing that spans multiple locations and/or multiple organizations, machine architectures, and software boundaries to provide power, collaboration, and information access.

Infrastructure that enables the integrated, collaborative use of computers, supercomputers, networks, data bases, and scientific instruments owned and managed by multiple organizations.

A network of computation, namely, tools and protocols for coordinated resource sharing and problem solving among pooled assets. Allows coordinated resource sharing and problem solving in dynamic, multiinstitutional virtual organizations.

Like P2P, grid computing allows users to share files (many-to-many sharing). With grid computing, the sharing is not only in reference to files, but also other IT resources.

Grid service   A Web service that conforms to a set of conventions (interfaces and behaviors) that define how a client interacts with a grid capability.

Grid synonyms   Some with slightly different connotations: "computational grid," "computing-on-demand," "on-demand computing," "just-in-time computing," "platform computing," "network computing," "computing utility," "utility computing," "cluster computing," and "high-performance distributed computing."

Grid topologies   Local, metropolitan, regional, national, or international footprint.

Systems may be in the same room, or may be distributed across the globe; they may running on homogenous or heterogeneous hardware platforms; they may be running on similar or dissimilar operating systems; and, they may owned by one or more organizations.

Grid types (i) Computational grids: machines with set-aside resources stand by to "number crunch" data or provide coverage for other intensive workloads. (ii) Scavenging grids: commonly used to locate and exploit CPU cycles on idle servers and desktop machines for use in resource-intensive tasks. (iii) Data grids: a unified interface for all data repositories in an organization, through which data can be queried, managed, and secured.

Computational grids can be local enterprise grids (also called intragrids) and Internet-based grids (also called intergrids). Enterprise grids are middleware-based environments used to harvest unused "machine cycles," thereby displacing otherwise-needed growth costs.

Hewlett-Packard's Utility Data Center (UDC) Hewlett-Packard's virtualization product family focusing on resource management across its server.

IBM's On Demand IBM and its partners offer a number of grid solutions in several vertical industries (e.g., automotive, financial markets, government, and others). IBM's strategy is to grid-enable all of its products: the company states that they will continue to incorporate virtualization technologies into its server software products and plans to incorporate autonomic capabilities into DB2 and associated content management products.

Infrastructure grid tools Grid subsystems that include file systems, schedulers and resource managers, messaging systems, security applications, certificate authorities, and file transfer mechanisms.

Intergrid A global Internet-based grid that spans multiple organizations and geographic locations. Generally, an Intergrid may be used to collaborate on "large" projects of common scientific interest. The stringiest levels of security are usually required in this environment.

Intragrid Also known as enterprise grid. A grid environment completely enclosed within one organization.

Job management and resource management function Provides the services to actually launch a job on a particular resource, to check the job's status, and to retrieve the results when the job is complete. Typically, the management component keeps track of the resources available to the grid and which users are members of the grid.

Metadata	A definition or description of data. In IT, metadata is definitional data that provides information about, or documentation of, other data managed within an application or environment. For example, metadata would document data about data elements or attributes, (name, size, data type, etc.), data about records or data structures (length, fields, columns, etc.), and data about data (where it is located, how it is associated, ownership, etc.).
Microsoft's Virtual Server	Virtual Server is a virtual machine solution for application migration and server consolidation. With Virtual Server, a Windows Server 2003-based server can run multiple operating systems concurrently. The goal is to make it easier to migrate legacy applications. Virtual Server aims at reducing capital expenditures through the use of fewer servers. Virtual Server does not require custom drivers and it does not use any proprietary protocols.
Middleware grid tools	Software plug-ins that facilitate the use of grid technology. For example, the open-source Globus Toolkit 3.0, a mature set of tools useful for building a grid, is the first full-scale implementation of the OGSI standard. The toolkit was developed by the Globus Project, a research and development project focused on enabling the application of grid concepts to scientific and engineering computing.
Node security function	A mechanism that supports secure communication between internal elements of a computational grid. A grid is comprised of a collection of hardware and software resources whose origins may not be obvious to a grid user; hence, strong security mechanisms are required.
OGSA architected grid services layer	Services in this layer include discovery, life cycle, state management, service groups, factory, notification, and handle map. These services are based on the Web Services layer.
OGSA grid applications layer	The user-visible layer. It supports user applications.
OGSA physical and logical resources layer	Resources comprise the capabilities of the grid. Physical resources include servers, storage, and network; logical resources provide additional function by virtualizing and aggregating the resources in the physical layer. General-purpose middleware such as file systems, data base managers, directories, and workflow managers provide these abstract services on top of the physical grid.
OGSA Web Services Layer	The specification defines grid services and builds on standard Web services technology. It exploits the mechanisms of Web ser-

vices such as XML and WSDL to specify standard interfaces, behaviors, and interaction for all grid resources.

OGSI Factory
A mechanism (interface) that provides a way to create new grid services. Factories may create temporary instances of limited function, such as a scheduler creating a service to represent the execution of a particular job, or they may create longer-lived services such as a local replica of a frequently used data set.

OGSI Handle Map
Deals with service identity. When factories are used to create a new instance of a grid service, the factory returns the identity of the newly instantiated service. This identity is composed of two parts, a grid service handle (GSH) and a grid service reference (GSR). A GSH provides a reference the grid service indefinitely; GSR can change within the grid services lifetime.

OGSI Life cycle
A mechanism architected to prevent grid services from consuming resources indefinitely without requiring a large scale distributed "garbage collection" scavenger. Every grid service has a termination time set by the service creator or factory. Because grid services may be transient, grid service instances are created with a specified lifetime.

OGSI Notification
Services interact with one another by exchanging messages based on service invocation. The state information that is modeled for grid services changes as the system runs. Many interactions between grid services require dynamic monitoring of changing state. Notification applies a traditional publish/subscribe paradigm to this monitoring.

OGSI Service Groups
Service groups are collections of grid services that are indexed (using service data) for some specific purpose.

OGSI State Management
A framework for representing grid services' "state" along with a mechanism for inspecting or modifying that state, named Find/SetServiceData.

Open Grid Services Architecture (OGSA)
Defines the standard architecture for grid computing. Describes the overall structure and the services to be provided in grid environments. A distributed interaction and computing architecture that is based on the grid service concept, assuring interoperability on heterogeneous systems and allowing different types of systems to communicate and share information. Building on Web services standards, OGSA takes the view that a grid service is simply a Web service that conforms to a particular set of conventions.

OGSA manages resources across distributed heterogeneous platforms. It supports QoS-oriented service-level agreements; provides a common base for autonomic management; defines open, published interfaces and protocols for interoperability of diverse resources; exploits industry standard integration technologies, and leverages existing solutions where appropriate. The foundation of OGSA is rooted in Web services (e.g.. SOAP and WSDL).

Open Grid Services Infrastructure (OGSI)	Defines the standard interfaces and behaviors of a grid service, building on a Web services baseline. Defines mechanisms for creating, managing, and exchanging information among grid services. Approach provides a common and open standards-based mechanism to access various grid services using existing industry standards such as SOAP, XML, and WS-Security.
Operation	The equivalent to a function call. It is the command that is run when invoked. Operations are given names, have input messages, output messages, and potentially faults. Operations are identified in the Port Type section of the WSDL document [176]
Oracle's 10g family	10g family of "grid-aware" products. Focuses on databases. Oracle's own brand of grid computing: a database system that comprises multiple nodes and lets IT planners shift database resources between them.
Peer-to-peer (P2P)	P2P is concerned with same general problem as grid computing, namely, the organization of resource shared within virtual communities. The grid community focuses on aggregating distributed high-end machines such as clusters, whereas the P2P community concentrates on sharing low-end systems such as PCs connected to the Internet.
Platform Computing Inc.	A company that supplies grid products and provides support for Linux on the zSeries mainframe.
Portal/User interface	A possible access mechanism to a grid; it provides the user with an interface to launch grid applications.
Resources	Processors, data storage, scientific equipment, and so on. Physical resources include servers, storage, and network. Above the physical resources are logical resources that provide additional function by virtualizing and aggregating the resources in the physical layer.
Scavenging (computational) grid	This grid is used to "locate processors–cycles." Grid nodes are exploited for available CPU cycles and other resources. Nodes typically equate to desktop computers and a large number of processors are generally involved. Owners of the desktop processors are usually given control over when their resources are available to participate in the grid
Scheduler function	A grid-based system that is responsible for routing a job to a properly selected processor to be executed. The scheduling software identifies a processor on which to run a specific grid job that has been submitted by a user. Schedulers, along with load balancers,

	provide a function to route jobs to available resources based on SLAs, capabilities, and availability.
Security grid tools	Security covers authentication, authorization, message integrity and message confidentiality.
Service-oriented architecture (SOA)	Architecture that defines how two computing entities interact to enable one entity to perform a unit of work on behalf of another entity. The unit of work is referred to as a service, and the service interactions are defined using a description language. Each interaction is self-contained; each interaction is independent of any other interaction environment software components can be exposed as services on the network, and so can be reused for different applications and purposes.
SOAP (Simple Object Access Protocol)	Transport mechanism that is independent of the underlying platform and protocol. For example, two disparate processes can communicate without the intimate knowledge of systems and platforms on which both of them are running. SOAP provides a mechanism of messaging between a service requestor and a service provider. It is a mechanism for formatting a Web service invocation: a simple enveloping process for XML payloads that defines a remote procedure call convention and a messaging convention. While SOAP is considered a protocol, it is viewed by some as a *protocol construction toolkit;* this is because SOAP enables a developer (or other party) to define a protocol using a few basic mechanisms. Mechanisms include (i) a format (as realized by the SOAP envelope) and its associated processing model, (ii) roles such as "sender," "receiver," and "intermediary," and (iii) message exchange patterns. SOAP assembles all of this information into a "protocol binding"—a protocol binding is a way of saying "this is how one uses the SOAP format in a certain way on top of an underlying protocol, such as HTTP." Protocol bindings also have capabilities (known as Properties and Features) that allow the developer to alter their settings; however, the binding does not by itself provide a way to communicate these settings to remote endpoints. To do that, one needs a description format such as WSDL. One can think of WSDL as a *protocol description format,* a way of defining a protocol without formal standardization process.
Sun Microsystems' N1 Data-center Architecture	Approach based on clusters (Sun Grid Engine). N1 is Sun Microsystems' architecture, products and services for supporting network computing. The marketing angle of the company is that N1 allows "managing *n* computers as *1.*"
Transport Binding	A specified method for taking some payload (like SOAP) and attaching (or embedding) it to a transport protocol. This enables an

application to leverage a common transport protocol to move data and then to accurately read or write some information (the payload) in a consistent fashion by all participants.

Transport Protocol	A network protocol (e.g., HTTP, SMTP) that has the capability of transferring some payload from one node to another, thus providing some level of transparency from the underlying network. Transport protocols are used as a mechanism to move payloads around a network. Typically, a payload will have a transport binding associated with it. For example, SOAP (acting as a payload) has a predetermined binding for HTTP (a transport protocol) [176]. Some mechanisms (e.g., JMS, MQ, instant messaging protocols, JXTA, and others) cross layers in the OSI stack, that is, they provide transport functionality but provide additional functionality as well (such as session, reliability, etc.)
UDDI (Universal Description, Discovery and Integration)	Repository that stores the descriptions of Web services. UDDI is used to create a searchable directory of Web services. An XML-based registry for businesses worldwide to list themselves on the Internet.
User security function	A mechanism to support security. Specifically, it provides authentication, authorization, data confidentiality, data integrity, and availability, particularly from a user's point of view.
Virtualization	An approach that allows several operating systems to run simultaneously on one (large) computer. More generally, it is the practice of making resources from diverse devices accessible to a user as if they were a single, larger, homogenous, resource that appears to be locally available. Dynamically shifting resources across platforms to match computing demands with available resources. The computing environment can become dynamic, enabling autonomic shifting applications between servers to match demand. The abstraction of server, storage, and network resources in order to make them available dynamically for sharing by IT services, both internal to and external to an organization. In combination with other server, storage, and networking capabilities, virtualization offers customers the opportunity to build more efficient IT infrastructures. Virtualization is seen by some as a step on the road to utility computing.
Web services (WS)	Web Services define a technique for describing software components to be accessed, methods for accessing these components, and discovery methods that enable the identification of relevant service providers. Web services are programming-language-, pro-

gramming-model-, and system-software-neutral. In other words, they are Internet-based services that provide standard infrastructure for data exchange between two different distributed applications. Web services are small units of code and are independent of operating systems and programming languages. They are designed to handle a limited set of tasks.

Web services are expected to play a key constituent role in the standardized definition of grid computing, since Web services have emerged as a standards-based approach for accessing network applications (keeping in mind that grids provide an infrastructure for aggregation of high-end resources for solving large-scale problems).

Web services standards are being defined within the W3C and other standards bodies and form the basis for major new industry initiatives such as IBM's Dynamic e-Business, Microsoft's .NET, and Sun's Sun ONE.

WSDL (Web services definition language)	A language that provides a way of describing the specific interfaces of Web services and APIs and is used by UDDI. An XML mechanism for describing Web services as a set of endpoints operating on messages. These messages contain either document-oriented (messaging) or remote procedure call payloads. Service interfaces are defined abstractly in terms of message structures and sequences of message exchanges.
WSDL \<binding\>	The communication protocols used by the Web service. Describes how the operation is invoked by specifying concrete protocol and data format specifications for operations and messages.
WSDL \<message\>	The messages used by the Web service. An abstract definition of the data being communicated.
WSDL \<port\>	Specifies a single endpoint as an address for the binding, thus defining a single communication endpoint.
WSDL \<portType\>	The operations performed by the Web service. An abstract set of operations supported by one or more endpoints. A set of related operations.
WSDL \<service\>	Specifies the port address(es) of the binding. The service is a collection of network endpoints or ports.
WSDL \<types\>	The data types used by the Web service. Provides information about any complex data types used in the WSDL document. When simple types are used, the WSDL document does not need this section.
WSIL (Web	An XML-based format utilized to facilitate the discovery and ag-

services inspection language)	gregation of Web service descriptions in a simple and extensible fashion. A simple, lightweight mechanism for Web service discovery that complements UDDI. WSIL is an XML document format designed to facilitate the discovery and aggregation of Web service descriptions in a simple and extensible fashion. Created by IBM and Microsoft.
XML (eXtensible markup language)	A meta-language used to describe grammatical descriptions of objects and describing data structures in an open manner. It is similar in appearance to HTML, is platform-neutral, and can be used to represent both documents and data.

Index

Printed and bound by CPI Group (UK) Ltd, Croydon, CR0 4YY